Seismic Assessment and Retrofit of Reinforced Concrete Structures

Seismic Assessment and Retrofit of Reinforced Concrete Structures

Editors

Pier Paolo Rossi
Melina Bosco

MDPI • Basel • Beijing • Wuhan • Barcelona • Belgrade • Manchester • Tokyo • Cluj • Tianjin

Editors
Pier Paolo Rossi
University of Catania
Italy

Melina Bosco
University of Catania
Italy

Editorial Office
MDPI
St. Alban-Anlage 66
4052 Basel, Switzerland

This is a reprint of articles from the Special Issue published online in the open access journal *Applied Sciences* (ISSN 2076-3417) (available at: https://www.mdpi.com/journal/applsci/special_issues/Seismic_Assessment_Reinforced_Concrete_Structures).

For citation purposes, cite each article independently as indicated on the article page online and as indicated below:

LastName, A.A.; LastName, B.B.; LastName, C.C. Article Title. *Journal Name* **Year**, *Volume Number*, Page Range.

ISBN 978-3-0365-5057-2 (Hbk)
ISBN 978-3-0365-5058-9 (PDF)

© 2022 by the authors. Articles in this book are Open Access and distributed under the Creative Commons Attribution (CC BY) license, which allows users to download, copy and build upon published articles, as long as the author and publisher are properly credited, which ensures maximum dissemination and a wider impact of our publications.

The book as a whole is distributed by MDPI under the terms and conditions of the Creative Commons license CC BY-NC-ND.

Contents

About the Editors vii

Melina Bosco and Pier Paolo Rossi
Seismic Assessment and Retrofitting of Reinforced Concrete Structures
Reprinted from: *Appl. Sci.* 2022, *12*, 7280, doi:10.3390/app12147280 1

Kenji Fujii and Takumi Masuda
Application of Mode-Adaptive Bidirectional Pushover Analysis to an Irregular Reinforced Concrete Building Retrofitted via Base Isolation
Reprinted from: *Appl. Sci.* 2021, *11*, 9829, doi:10.3390/app11219829 5

Taufiq Ilham Maulana, Patricia Angelica de Fatima Fonseca and Taiki Saito
Application of Genetic Algorithm to Optimize Location of BRB for Reinforced Concrete Frame with Curtailed Shear Wall
Reprinted from: *Appl. Sci.* 2022, *12*, 2423, doi:10.3390/app12052423 47

Diego Alejandro Talledo, Irene Rocca, Luca Pozza, Marco Savoia and Anna Saetta
Numerical Assessment of an Innovative RC-Framed Skin for Seismic Retrofit Intervention on Existing Buildings
Reprinted from: *Appl. Sci.* 2021, *11*, 9835, doi:10.3390/app11219835 71

Antonio Formisano and Ylenia Messineo
Seismic Rehabilitation of Abandoned RC Industrial Buildings: The Case Study of a Former Tobacco Factory in the District of Avellino (Italy)
Reprinted from: *Appl. Sci.* 2022, *12*, 5705, doi:10.3390/app12115705 89

Fujian Yang, Guoxin Wang and Mingxin Li
Evaluation of the Seismic Retrofitting of Mainshock-Damaged Reinforced Concrete Frame Structure Using Steel Braces with Soft Steel Dampers
Reprinted from: *Appl. Sci.* 2021, *11*, 841, doi:10.3390/app11020841 105

Filippo Molaioni, Fabio Di Carlo and Zila Rinaldi
Modelling Strategies for the Numerical Simulation of the Behaviour of Corroded RC Columns under Cyclic Loads
Reprinted from: *Appl. Sci.* 2021, *11*, 9761, doi:10.3390/app11209761 129

Min Sook Kim and Young Hak Lee
Flexural Behavior of Reinforced Concrete Beams Retrofitted with Modularized Steel Plates
Reprinted from: *Appl. Sci.* 2021, *11*, 2348, doi:10.3390/app11052348 151

Zhiwei Shan, Lijie Chen, Kun Liang, Ray Kai Leung Su and Zhao-Dong Xu
Strengthening Design of RC Columns with Direct Fastening Steel Jackets
Reprinted from: *Appl. Sci.* 2021, *11*, 3649, doi:10.3390/app11083649 169

Jing Li, Lizhong Jiang, Hong Zheng, Liqiang Jiang and Lingyu Zhou
Cyclic Tests and Numerical Analyses on Bolt-Connected Precast Reinforced Concrete Deep Beams
Reprinted from: *Appl. Sci.* 2021, *11*, 5356, doi:10.3390/app11125356 189

Theodoros Rousakis, Vachan Vanian, Theodora Fanaradelli and Evgenia Anagnostou
3D FEA of Infilled RC Framed Structures Protected by Seismic Joints and FRP Jackets
Reprinted from: *Appl. Sci.* 2021, *11*, 6403, doi:10.3390/app11146403 205

Barham Haidar Ali, Esra Mete Güneyisi and Mohammad Bigonah
Assessment of Different Retrofitting Methods on Structural Performance of RC Buildings against Progressive Collapse
Reprinted from: *Appl. Sci.* **2022**, *12*, 1045, doi:10.3390/app12031045 . . . . **223**

About the Editors

Pier Paolo Rossi

Pier Paolo Rossi (Associate Professor at the University of Catania) received his degree in civil engineering from the University of Naples "Federico II" (1993) and his doctor of philosophy degree (Ph.D.) from the University of Catania (1998). Prior to joining the Faculty of Engineering of the University of Catania in 1995, he worked some years at the Faculty of Engineering of the University of Naples (1993–94). He has authored or coauthored over 160 publications, including journal papers and articles in conference proceedings. He has been a reviewer of papers proposed for publication in many international journals and is a member of Editorial Boards of international journals. His research interests are in the areas of earthquake engineering: in particular, his attention is focused on the seismic analysis, design and retrofitting of steel or reinforced concrete buildings; seismic design of bridges; numerical modelling of steel and reinforced concrete members; seismic behaviour of in plan and in-elevation irregular buildings; strength verification/the design of reinforced concrete members subjected to combined axial force, bending moment and shear force.

Melina Bosco

Melina Bosco (Associate Professor at the University of Catania) received her degree in civil engineering (2002) and her doctor of philosophy degree (2006) from the University of Catania. She has authored or coauthored over 130 publications, including journal papers and articles in conference proceedings. She has been a reviewer of papers proposed for publication in many international journals and is a member of Editorial Boards of international journals. The main research interests are the proposal of design procedures and assessment of the seismic behaviour of steel and reinforced concrete buildings, assessment of the seismic response of regular and in-plan irregular buildings by nonlinear static methods and dynamic analysis, the proposal of design methods for the seismic upgrade of existing buildings, and numerical modelling of steel and reinforced concrete members.

Editorial

Seismic Assessment and Retrofitting of Reinforced Concrete Structures

Melina Bosco and Pier Paolo Rossi *

Department of Civil Engineering and Architecture, University of Catania, 95123 Catania, Italy; mbosco@dica.unict.it
* Correspondence: pierpaolo.rossi@unict.it

1. Introduction

Many constructions are globally built with reinforced or prestressed concrete and a large part of them are designed or expected to resist earthquake actions in addition to gravity loads. In many cases, these actions also work together with other more specific actions, e.g., road or railway traffic loads in bridges or hydraulic actions in dams, and require a more complicated seismic analysis, design and assessment of such structures. In any case, climatic actions cause the corrosion of reinforcements and undermine the initial seismic performance of such structures, exposing them to a greater damage under earthquake actions. Often, regardless of the detrimental actions of aging of materials, the increase in the evaluated seismic hazard requires interventions of seismic retrofit.

In the attempt to limit the effects of seismic events on reinforced or prestressed concrete structures, many attempts have been made by researchers in order to (i) improve the knowledge of the response of materials (steel bars and concrete) and members to persistent and transient loading conditions by means of laboratory tests, (ii) improve the numerical models for such structures, (iii) improve the capacity models, (iv) improve the procedures for the dynamic analysis, (v) improve the procedures for the correct assessment of the seismic performance, and (vi) suggest innovative interventions for the seismic retrofit of old and damaged reinforced or prestressed concrete structures.

In the last decades, researchers have provided valuable contributions to all the above issues and seismic codes have evolved due to the findings and proposals of the scientific research. The large number of reinforced or prestressed concrete structures and the continuous efforts made by researchers for the comprehension of the seismic behavior of such structures have been among the main reasons of the present Special Issue titled "Seismic assessment and retrofit of Reinforced Concrete Structures".

2. Contributions

Many researchers have contributed to this Special Issue. As a result, this thematic issue is composed by eleven contributions covering important topics for Seismic Assessment and Retrofit of Reinforced Concrete Structures.

One paper [1] is devoted to the evaluation of the nonlinear seismic response of irregular r.c. buildings retrofitted by base-isolation. The effectiveness of advanced nonlinear static methods of analysis in predicting the torsional response due to plan irregularities has been widely studied for non-isolated structures, whereas the response of base-isolated irregular structures has been generally determined by nonlinear time-history analyses. The paper under examination [1] points out that the mode-adaptive bidirectional pushover analysis (MABPA) predicts the peak responses of irregular base-isolated buildings with accuracy even if its application is limited to buildings that can be classified as torsionally stiff.

One study [2] intended to propose a structural solution for buildings with curtailed shear walls. To reduce the drifts of the upper part of the frame under earthquakes, buckling-restrained braces (BRBs) are added. The genetic algorithm is applied to determine the

optimum locations of BRBs by considering three main parameters in the fitness function: the interstory drifts, the damage index of the beams and the total number of BRBs. Numerical analyses, calibrated against the results of experimental tests, prove the effectiveness of the proposal.

Three studies [3–5] are focused on the retrofitting of existing r.c. buildings. The retrofitting technology proposed in [3] consists of a RC frame rigidly connected to the external façade of the existing building by means of anchor rods placed at every floor level. The anchor rods are designed based on capacity design principles. The RC-framed skin is casted on site by using prefabricated expanded polystyrene modules as formwork, which also ensures the thermal enhancement of the building. Pushover analyses show that the proposed retrofitting technology provides the existing structure with significant increases in stiffness, strength and often is accompanied a more ductile failure model. In another paper [4], a steel exoskeleton is, instead, used to perform the combined seismic–energy retrofit of a former reinforced concrete industrial building. Most of the existing retrofitting techniques requires time to be realized. In the case of buildings damaged during mainshock events, quick and effective strengthening interventions are important to reduce further damage during aftershocks. To investigate this issue, in [5], a three-story RC frame building is first subjected to a mainshock excitation. Then, the damaged structure is retrofitted by three retrofitting schemes with combined strip-shaped shear-and-flexural steel dampers installed between chevron braces and upper beams. The response of the three considered retrofitting schemes under aftershocks is finally compared.

An accurate evaluation of the seismic response of existing buildings requires a proper simulation of the degradation of materials. One of the major degradation causes of RC structures consists in the corrosion of steel reinforcements. To evaluate the influence of localized corrosion on the cyclic behavior and failure modes of RC columns, one paper of this Special Issue [6] considers different modelling strategies (based on the reduction in constitutive law by uniform or pitting corrosion; bar discretization and section reduction and bar discretization and morphology-based constitutive law reduction). The accuracy of the modelling strategies is validated against results of experimental tests.

Three papers are devoted to the analysis of structural members [7–9]. Two of these studies are related to retrofitting techniques for beams [7] or columns [8], whereas the third study proposes a bolt-connected precast reinforced concrete deep beam as a lateral resisting component to be used in framed structures to resist seismic loads [9]. Specifically, in [7], the structural capacity of reinforced concrete beams is improved by the use of modularized steel plates. This technique leads to rapid and economical construction and, as proved by experimental tests, to an increase in the ductility capacity of the reinforced concrete beams, which is about 2.5 times that of the non-retrofitted beams. A strengthening method based on fastening steel jackets is introduced for columns in [8]. The steel plate thickness, fastener number and connection spacing are designed to prevent brittle shear failure, ensure a lateral load capacity larger than the lateral load demand, limit the axial load ratio and provide adequate flexural stiffness. In regard to bolt-connected precast reinforced-concrete deep beams [9], the seismic performance of two scaled specimens is determined under cyclic loads by experimental tests. A parametric analysis, carried out on the numerical model calibrated against the results of the experimental tests, shows that shear resistance, elastic lateral stiffness and displacement-based ductility are significantly dependent on the height-to-length ratio of the deep beam.

In one paper [10], recently tested real-scale RC-framed-wall infilled structures with innovative seismic protection through polyurethane joints (PUFJ) or polyurethane impregnated fiber grids (FRPU) are investigated. The frames reveal a highly ductile response while preventing infill collapse. Suitable FE models are developed in order to reproduce the experimental results.

Finally, one paper of the Special Issue is devoted to Progressive Collapse [11]. Three different column removal scenarios are considered with reference to a case study that is upgraded by different retrofitting strategies, e.g., X-braces, diagonal braces, inverted

V-braces, viscous dampers and carbon-fiber-reinforced polymer. The damage levels of members under the above-mentioned scenarios are compared.

3. Conclusions

We were very pleased to guest edit this Special Issue and we thank all the authors of the above papers for their contributions to the advancement of the research in the field of the seismic assessment and retrofit of reinforced concrete structures. We hope this Special Issue can reach the widest audience in the scientific community and contribute to boosting further scientific and technological advances into the attractive world of reinforced concrete structures.

Author Contributions: M.B. and P.P.R. have contributed to the conceptualization, writing, review and editing of this manuscript. All authors have read and agreed to the published version of the manuscript.

Funding: This research received no external funding.

Acknowledgments: The Guest Editors would like to express their thanks to all authors for their valuable contributions and to all peer reviewers for their constructive comments and suggestions.

Conflicts of Interest: The authors declare no conflict of interest.

References

1. Fujii, K.; Masuda, T. Application of Mode-Adaptive Bidirectional Pushover Analysis to an Irregular Reinforced Concrete Building Retrofitted via Base Isolation. *Appl. Sci.* **2021**, *11*, 9829. [CrossRef]
2. Maulana, T.I.; Fonseca, P.A.d.F.; Saito, T. Application of Genetic Algorithm to Optimize Location of BRB for Reinforced Concrete Frame with Curtailed Shear Wall. *Appl. Sci.* **2022**, *12*, 2423. [CrossRef]
3. Talledo, D.A.; Rocca, I.; Pozza, L.; Savoia, M.; Saetta, A. Numerical Assessment of an Innovative RC-Framed Skin for Seismic Retrofit Intervention on Existing Buildings. *Appl. Sci.* **2021**, *11*, 9835. [CrossRef]
4. Formisano, A.; Messineo, Y. Seismic Rehabilitation of Abandoned RC Industrial Buildings: The Case Study of a Former Tobacco Factory in the District of Avellino (Italy). *Appl. Sci.* **2022**, *12*, 5705. [CrossRef]
5. Yang, F.; Wang, G.; Li, M. Evaluation of the Seismic Retrofitting of Mainshock-Damaged Reinforced Concrete Frame Structure Using Steel Braces with Soft Steel Dampers. *Appl. Sci.* **2021**, *11*, 841. [CrossRef]
6. Molaioni, F.; Di Carlo, F.; Rinaldi, Z. Modelling Strategies for the Numerical Simulation of the Behaviour of Corroded RC Columns under Cyclic Loads. *Appl. Sci.* **2021**, *11*, 9761. [CrossRef]
7. Kim, M.S.; Lee, Y.H. Flexural Behavior of Reinforced Concrete Beams Retrofitted with Modularized Steel Plates. *Appl. Sci.* **2021**, *11*, 2348. [CrossRef]
8. Shan, Z.; Chen, L.; Liang, K.; Su, R.K.L.; Xu, Z. Strengthening Design of RC Columns with Direct Fastening Steel Jackets. *Appl. Sci.* **2021**, *11*, 3649. [CrossRef]
9. Li, J.; Jiang, L.; Zheng, H.; Jiang, L.; Zhou, L. Cyclic Tests and Numerical Analyses on Bolt-Connected Precast Reinforced Concrete Deep Beams. *Appl. Sci.* **2021**, *11*, 5356. [CrossRef]
10. Rousakis, T.; Vanian, V.; Fanaradelli, T.; Anagnostou, E. 3D FEA of Infilled RC Framed Structures Protected by Seismic Joints and FRP Jackets. *Appl. Sci.* **2021**, *11*, 640. [CrossRef]
11. Ali, B.H.; Mete Güneyisi, E.; Bigonah, M. Assessment of Different Retrofitting Methods on Structural Performance of RC Buildings against Progressive Collapse. *Appl. Sci.* **2022**, *12*, 1045. [CrossRef]

Article

Application of Mode-Adaptive Bidirectional Pushover Analysis to an Irregular Reinforced Concrete Building Retrofitted via Base Isolation

Kenji Fujii [1,*] and Takumi Masuda [2]

[1] Department of Architecture, Faculty of Creative Engineering, Chiba Institute of Technology, Chiba 275-0016, Japan
[2] Graduate School of Creative Engineering, Chiba Institute of Technology, Chiba 275-0016, Japan; s16b1147bq@s.chibakoudai.jp
* Correspondence: kenji.fujii@p.chibakoudai.jp

Featured Application: Seismic response evaluation of a base-isolated buildings: Seismic rehabilitation design for reinforced concrete buildings using the base-isolation technique.

Abstract: In this article, the applicability of mode-adaptive bidirectional pushover analysis (MABPA) to base-isolated irregular buildings was evaluated. The point of the updated MABPA is that the peaks of the first and second modal responses are predicted considering the energy balance during a half cycle of the structural response. In the numerical examples, the main building of the former Uto City Hall, which was severely damaged in the 2016 Kumamoto earthquake, was investigated as a case study for the retrofitting of an irregular reinforced concrete building using the base-isolation technique. The comparisons between the predicted peak response by MABPA and nonlinear time-history analysis results showed that the peak relative displacement can be properly predicted by MABPA. The results also showed that the performance of the retrofitted building models was satisfactory for the ground motion considered in this study, including the recorded motions in the 2016 Kumamoto earthquake.

Keywords: seismic isolation; asymmetric building; mode-adaptive bidirectional pushover analysis (MABPA); seismic retrofit; momentary energy input

1. Introduction

1.1. Background

Seismic isolation is widely applied to buildings for earthquake protection in earthquake-prone countries [1]. Unlike in the case of traditional (non-isolated) earthquake-resistant structures, seismic isolation ensures that the behaviors of building structures are within the elastic range and reduces acceleration in buildings during large earthquakes [2]. Therefore, this technique is applied to not only new buildings, but also existing reinforced concrete and masonry buildings, including historical structures [3–14]. The isolation layer consists of isolators and dampers. Isolators support the gravity loads of the superstructure. The horizontal flexibility and centering capability of seismically isolated buildings is also achieved by isolators [1]. Dampers provide the energy dissipation capacity for reducing the seismic responses. Although it is very common to use hysteresis dampers (e.g., steel dampers) and/or oil dampers in the isolation layer, there are several studies about the innovative seismic isolation systems in recent years, e.g., [15–22].

In general, there is some degree of irregularity in building structures, whether they are newly designed or preexisting. Therefore, the problem of torsional response due to plan irregularities needs to be studied for base-isolated structures, as well as non-isolated structures. Several researchers have investigated the seismic behavior of base-isolated buildings with plan irregularities [23–36]; these include fundamental parametric studies using idealized models [23–27], as well as studies using more realistic frame building models [28–36]. Most of these studies are based on nonlinear time-history analyses [23–35]. However, there are a few investigations that examine the applicability of nonlinear static procedures to base-isolated buildings with asymmetry. Kilar and Koren [36] examined the applicability of the extended N2 method [37,38], which is one of the variants of the nonlinear static procedures, to the base-isolated asymmetric buildings. They found the nonlinear peak response of base-isolated asymmetric building can be predicted by the extended N2 method. However, they examined only those whose superstructure was regular in elevation. Therefore, further investigations are needed for the applicability of the nonlinear static procedures, especially for base-isolated building with plan and elevation irregularities.

1.2. Motivation

There were two main motivations for this study. The first was to extend nonlinear static analyses to base-isolated buildings with irregularities, and the second was to predict the nonlinear peak responses of base-isolated buildings according to the concept of energy input.

With respect to the first motivation, the authors proposed mode-adaptive bi-directional pushover analysis (MABPA) [39–44]. This is a variant of nonlinear static analysis and was originally proposed for the seismic analysis of non-isolated asymmetric buildings subjected to horizontal bidirectional excitation. The first version of MABPA was proposed for non-isolated asymmetric buildings with regular elevation [39]; it was then updated following the development of displacement-based mode-adaptive pushover (DB-MAP) analysis [40]. This version has been applied to reinforced concrete asymmetric frame buildings with buckling-restrained braces [41] and to building models with bidirectional setbacks [42]. The applicability of MABPA has been discussed and evaluated based on the effective modal mass ratio of the first two modes [43]. The seismic capacity of an existing irregular building severely damaged in the 2016 Kumamoto earthquake (the former Uto City Hall) has been evaluated using MABPA [44]. Looking back on the development of MABPA, the logical next step should be to extend the method to base-isolated buildings with irregularities. Considering the case in which the seismic isolation period is well separated from the natural period of a superstructure, the behavior of the superstructure may be that of a rigid body, as discussed by several researchers, e.g., [27]. In such cases, the effective modal mass ratio of the first and second modes is close to unity, provided that the torsional resistance at the isolation layer is sufficient. It is expected that the seismic response of such a base-isolated building with plan irregularities will be accurately predicted by MABPA.

In the nonlinear static analysis shown in the American Society of Civil Engineers ASCE/SEI 41-17 document [3], the reduced spectrum considering the effective damping is used to predict the target displacement for a nonlinear static analysis. Similarly, as shown in Notification No. 2009 of the Ministry of Construction of Japan [45], the equivalent linearization technique can be used for target displacement evaluations of base-isolated buildings. However, several researchers have examined the responses of long-period building structures subjected to pulse-like ground motions, e.g., Mazza examined a base-isolated building [29,31] and Güneş and Ulucan studied a tall reinforced concrete building [46]. Because the effective damping is calculated based on a steady response having the same displacement amplitude in the positive and negative directions, the use of effective damping for the prediction of the peak responses of base-isolated structures subjected to pulse-like ground motion is questionable. From this viewpoint, an alternative concept for predicting the peak response is required. Accordingly, the second motivation of this study was to investigate the use of the energy concept.

The concept of energy input was introduced by Akiyama in the 1980 s [47] and is implemented in the design recommendation for seismically isolated buildings presented by the Architectural Institute of Japan [2]. In Akiyama's theory, the *total input energy* is a suitable seismic intensity parameter to access the cumulative response of a structure. Instead of the total input energy, Inoue et al. proposed the *maximum momentary input energy* [48–50] as an intensity parameter related to peak displacement; nonlinear peak displacement can be predicted by equating the maximum momentary input energy and the cumulative hysteresis energy during a half cycle of the structural response. Following the work of Inoue et al., the authors investigated the relationship between the maximum momentary input energy and the total input energy of an elastic single-degree-of-freedom (SDOF) model [51]. In addition, the concept of the momentary input energy was extended to consider bidirectional horizontal excitation [52,53]. Specifically, the time-varying function of the momentary energy input has been formulated for unidirectional [51] and bidirectional [52] excitation: This function can be calculated from the transfer function of the model and the complex Fourier coefficients of the ground acceleration. Using the time-varying function, both energy parameters, the total and maximum momentary input energy, can be accurately calculated. In a previous study [53], it was shown that the nonlinear peak displacement and the cumulative energy of the isotropic two-degree-of-freedom model representing a reinforced concrete building can be properly predicted using a time-varying function. Therefore, the use of a time-varying function of the momentary energy input is promising for seismic response predictions for base-isolated buildings.

1.3. Objectives

Based on the above discussion, the following questions were addressed in this paper.

- Is MABPA capable of predicting the peak response of irregular base-isolated buildings?
- The prediction of the peak equivalent displacement of the first two modal responses is an essential step in MABPA. For this, the relationship between the maximum momentary input energy and the peak displacement needs to be properly evaluated. How can this relationship be evaluated from the pushover analysis results?
- In the prediction of the maximum momentary input energy of the first two modal responses, the effect of simultaneous bidirectional excitation needs to be considered. Can the upper bound of the peak equivalent displacement of the first two modal responses be predicted using the bidirectional maximum momentary input energy spectrum [52]?

In this study, the applicability of mode-adaptive bidirectional pushover analysis (MABPA) to base-isolated irregular buildings was evaluated. The point of the updated MABPA is that the peaks of the first and second modal responses are predicted considering the energy balance during a half cycle of the structural response. In the numerical examples, the main building of the former Uto City Hall [44], which was severely damaged in the 2016 Kumamoto earthquake, was investigated as a case study for the retrofitting of an irregular reinforced concrete building using the base-isolation technique. The nonlinear peak responses of two retrofitted building models subjected to bidirectional excitation were investigated via a time-history analysis using artificial and recorded ground motion datasets. Then, their peak responses were predicted by MABPA to evaluate its accuracy. Note that of the many types of dampers used nowadays in base-isolated buildings, only hysteresis dampers were considered in this study; this is because they are easily implemented in nonlinear static analysis. The applicability of MABPA to base-isolated buildings with other types of dampers, such as oil and tuned viscous mass dampers [18], will be the next phase of this study.

The rest of paper is organized as follows. Section 2 presents an outline of MABPA, followed by the prediction procedure of the maximum equivalent displacement conducted using the maximum momentary input energy. Section 3 briefly presents information concerning the original building. Then, two retrofitted building models using the base-isolation technique are presented, as well as the ground motion data used in the nonlinear time-history analysis. The validation of the prediction of the peak response is discussed in Section 4. Discussions focused on (i) the relationship between the maximum equivalent displacement and the maximum momentary input energy, (ii) the predictability of the largest maximum momentary input energy from the bidirectional momentary input energy spectrum, and (iii) the accuracy of the upper bound of the maximum equivalent displacement of the first two modes are presented in Section 5. The conclusions and future directions of the study are discussed in Section 6.

2. Description of MABPA
2.1. Outline of MABPA

Figure 1 shows the flow of the updated MABPA. Here, the U-axis is the principal axis of the first modal response, while the V-axis is the axis perpendicular to the U-axis, following previous studies [39,41]. In this study, the predictions of the peak responses of the first and second modes (steps 2 and 4) were updated as follows:

- The bidirectional momentary input energy proposed in the previous study was applied as the seismic intensity parameter.
- The peak response of each mode was predicted from the energy balance in a half cycle of the structural response.

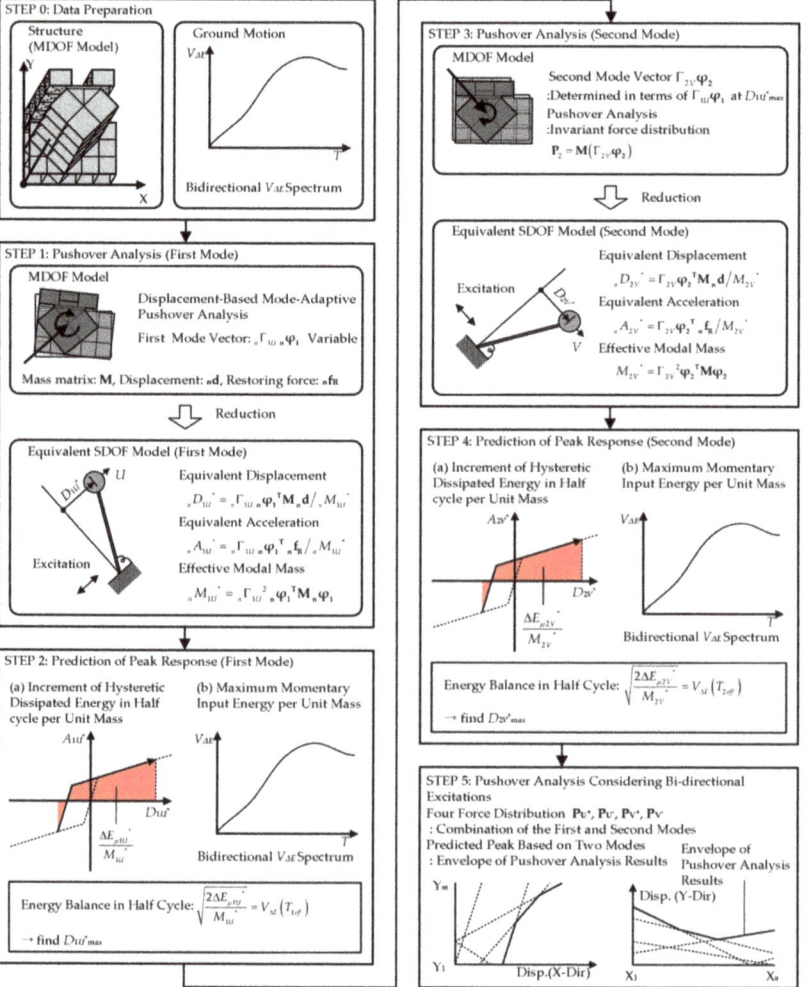

Figure 1. Flow of updated mode-adaptive bi-directional pushover analysis (MABPA).

It was assumed that the upper bound of the unidirectional maximum momentary input energy considering the various directions of the seismic input is approximated by the bidirectional maximum momentary input energy. For the prediction of the peak equivalent displacements of the first and second modal responses, the bidirectional maximum momentary input energy spectrum (bidirectional $V_{\Delta E}$ spectrum) was used. This is because the influences of the simultaneous input of two horizontal components are included in the bidirectional $V_{\Delta E}$ spectrum, as discussed in previous studies [52,53]. Note that the maximum momentary input energy of each modal response varies depending on the angle of incidence of the horizontal ground motion. The assumption that the maximum momentary input energy of each modal response can be predicted from the bidirectional $V_{\Delta E}$ spectrum can lead to conservative predictions. The validation of this assumption is discussed in Section 5.

2.2. Prediction of the Peak Response using the Momentary Energy Input

2.2.1. Calculation of the Bidirectional Momentary Input Energy Spectrum

The bidirectional momentary input energy spectrum, which was proposed in previous studies [52,53], can be calculated as follows:

First, the complex Fourier coefficient of the two orthogonal components of horizontal ground acceleration, $c_{\xi,n}$ and $c_{\zeta,n}$, can be calculated via the discrete Fourier transform of the two horizontal ground acceleration components $a_{g\xi}(t)$ and $a_{g\zeta}(t)$, respectively.

The displacement and velocity transfer function of the linear system with viscous and complex damping can be calculated as follows:

$$H_{CVD}(i\omega_n) = \frac{1}{\omega_0^2 - \omega_n^2 + 2\omega_0\{h\omega_n + \beta\omega_0 \text{sgn}(\omega_n)\}i}, \quad H_{CVV}(i\omega_n) = i\omega_n H_{CVD}(i\omega_n). \quad (1)$$

In Equation (1), $\omega_0 (= 2\pi/T)$ is the natural circular frequency, h and β are the viscous and complex damping of the linear system, respectively; ω_n is the natural frequency of the nth harmonic. From the complex Fourier coefficients ($c_{\xi,n}$ and $c_{\zeta,n}$) and the transfer functions of the linear system ($H_{CVD}(i\omega_n)$ and $H_{CVV}(i\omega_n)$), the duration of a half cycle of response (Δt) can be calculated as:

$$\Delta t = \pi \sqrt{\sum_{n=1}^{N} |H_{CVD}(i\omega_n)|^2 \{|c_{\xi,n}|^2 + |c_{\zeta,n}|^2\} / \sum_{n=1}^{N} |H_{CVV}(i\omega_n)|^2 \{|c_{\xi,n}|^2 + |c_{\zeta,n}|^2\}}. \quad (2)$$

Then, the Fourier coefficient of the time-varying function of the momentary energy input can be calculated as follows:

$$E_{\Delta BI,n}{}^* = \begin{cases} \frac{\sin(\omega_n \Delta t/2)}{\omega_n \Delta t/2} \sum_{n_1=n+1}^{N} \{H_V(i\omega_{n_1}) + H_V(-i\omega_{n_1-n})\} \{c_{\xi,n_1} c_{\xi,-(n_1-n)} + c_{\zeta,n_1} c_{\zeta,-(n_1-n)}\} & : n > 0 \\ 2 \sum_{n_1=1}^{N} \text{Re}\{H_V(i\omega_{n_1})\} \{|c_{\xi,n_1}|^2 + |c_{\zeta,n_1}|^2\} & : n = 0 \\ \overline{E_{\Delta BI,n}{}^*} & : n < 0 \end{cases} \quad (3)$$

The momentary input energy per unit mass at time t can be calculated as follows:

$$\frac{\Delta E_{BI}(t)}{m} = \int_{t-\Delta t/2}^{t+\Delta t/2} \sum_{n=-N+1}^{N-1} E_{\Delta BI,n}{}^* \exp(i\omega_n t) dt. \quad (4)$$

The maximum momentary input energy per unit mass ($\Delta E_{BI,\max}/m$) can be evaluated as the maximum value calculated by Equation (4) over the course of the seismic event.

The bidirectional maximum momentary input energy spectrum (the bidirectional $V_{\Delta E}$ spectrum) can be calculated as:

$$V_{\Delta E} = \sqrt{2\Delta E_{BI,\max}/m}. \qquad (5)$$

The bidirectional $V_{\Delta E}$ spectrum is prepared at the beginning of the MABPA (step 0). The range of the natural period (T) needs to be properly considered. In this study, a range of T from 0.1 to 6.0 s (longer than the seismic isolation period) was considered. The assumption of damping is also important. Because only a hysteresis (displacement-dependent) damper was installed in the isolation layer, complex damping was chosen in this study. The complex damping ratios (β) were set to 0.10, 0.20, and 0.30.

2.2.2. Formulation of the Effective Period and the Hysteretic Dissipated Energy in a Half Cycle

To predict the peak responses of the first two modes, bilinear idealization of the equivalent acceleration-equivalent displacement relationship was made as follows. In the following discussion, the formulations are made for the first modal response. Figure 2 shows the bilinear idealization procedure used in this study.

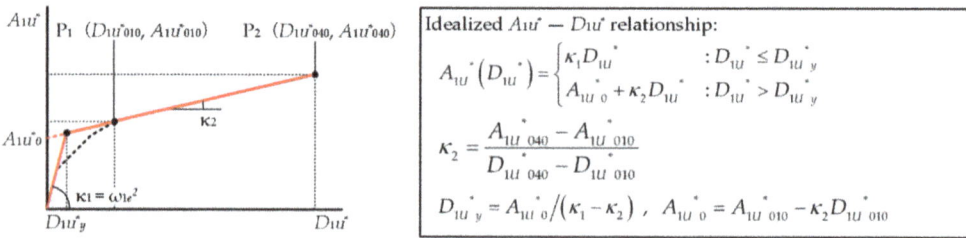

Figure 2. Bilinear idealization of the equivalent acceleration-equivalent displacement relationship.

Then, the effective period (T_{1eff}) and the hysteretic dissipated energy in the half cycle per unit mass ($\Delta E_{\mu1U}^{*}/M_{1U}^{*}$) were formulated. Figure 3 shows the modeling of the structural response in a half cycle used to predict the peak response. In Figure 3, the parameter η is the displacement ratio in the positive and negative directions and was assumed to be in the range from 0 to 1.

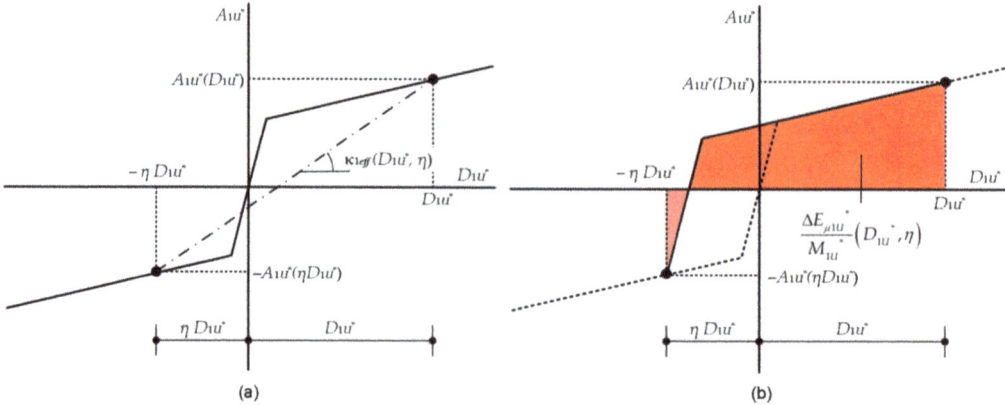

Figure 3. Modeling of the structural response in a half cycle used to predict the peak response: (**a**) The effective slope (κ_{1eff}) and (**b**) the hysteretic dissipated energy in a half cycle per unit mass ($\Delta E_{\mu1U}/M_{1U}^{*}$).

In this study, the effective period of the first mode corresponding to D_{1U}^* was calculated as:

$$T_{1eff}(D_{1U}^*) = 2\pi/\sqrt{K_{1eff}(D_{1U}^*)}. \quad (6)$$

In addition, the equivalent velocity of the hysteretic dissipated energy in a half cycle corresponding to D_{1U}^* was calculated as:

$$V_{\Delta E\mu 1U}^*(D_{1U}^*) = \sqrt{\frac{2\Delta E_{\mu 1U}^*}{M_{1U}^*}(D_{1U}^*)}. \quad (7)$$

The effective slope corresponding to D_{1U}^* was calculated as:

$$K_{1eff}(D_{1U}^*) = \int_0^1 K_{1eff}(D_{1U}^*, \eta) d\eta$$
$$= \begin{cases} \kappa_1 & : D_{1U}^* \leq D_{1U}^*{}_y \\ (2\ln 2)\frac{A_{1U}^*{}_0}{D_{1U}^*} + \kappa_2 - \left\{\ln\left(1 + \frac{D_{1U}^*{}_y}{D_{1U}^*}\right)\right\}\left(\frac{A_{1U}^*{}_0}{D_{1U}^*} + \kappa_1 - \kappa_2\right) + \frac{D_{1U}^*{}_y}{D_{1U}^*}(\kappa_1 - \kappa_2) & : D_{1U}^* > D_{1U}^*{}_y \end{cases} \quad (8)$$

The hysteretic dissipated energy in a half cycle per unit mass corresponding to D_{1U}^* was calculated as:

$$\frac{\Delta E_{\mu 1U}^*}{M_{1U}^*}(D_{1U}^*) = \int_0^1 \frac{\Delta E_{\mu 1U}^*}{M_{1U}^*}(D_{1U}^*, \eta) d\eta$$
$$= \begin{cases} \frac{1}{3}\kappa_1 D_{1U}^{*2} & : D_{1U}^* \leq D_{1U}^*{}_y \\ \frac{1}{3}\kappa_2 D_{1U}^{*2} + \frac{3}{2}A_{1U}^*{}_0 D_{1U}^*\left\{\left(1 - \frac{D_{1U}^*{}_y}{D_{1U}^*}\right)^2 + \frac{2}{3}\frac{D_{1U}^*{}_y}{D_{1U}^*}\left(1 - \frac{2}{3}\frac{D_{1U}^*{}_y}{D_{1U}^*}\right)\right\} & : D_{1U}^* > D_{1U}^*{}_y \end{cases} \quad (9)$$

The predicted peak response $D_{1U}^*{}_{max}$ was obtained from the following equation:

$$V_{\Delta E\mu 1U}^*\left\{T_{1eff}(D_{1U}^*{}_{max})\right\} = V_{\Delta E}\left\{T_{1eff}(D_{1U}^*{}_{max})\right\}. \quad (10)$$

Note that no viscous damping was considered because the isolation layer was assumed to have no viscous damping.

3. Description of the Retrofitted Building Models and Ground Motion Datasets

3.1. Original Building

Figure 4 shows a simplified structural plan and the elevation of the main building of the former Uto City Hall [44]. This five-story reinforced concrete irregular building was constructed in 1965.

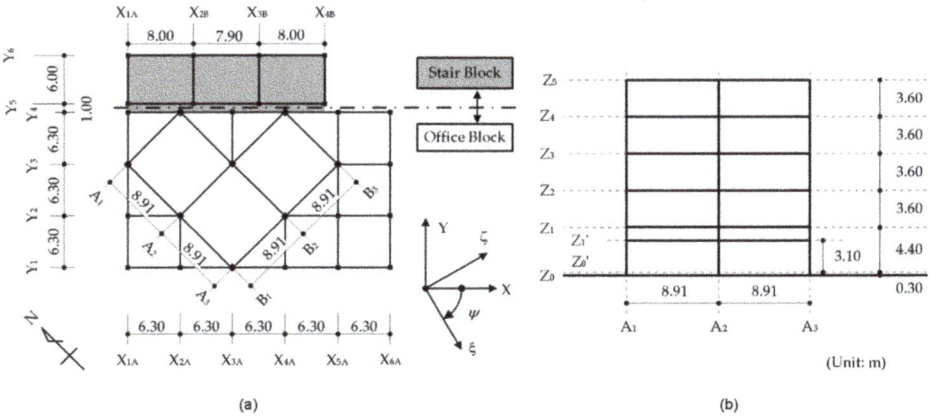

Figure 4. Simplified structural plan and elevation of the main building of the former Uto City Hall [44]: (a) Structural plan (Level Z_0) and (b) simplified plan elevation (frame B_1).

Figure 5 shows the former Uto City Hall after the 2016 Kumamoto earthquake. After the 2016 Kumamoto earthquake, this building was demolished. Details concerning this building can be found in the literature [44].

Figure 5. View of the former Uto City Hall after the 2016 Kumamoto earthquake [44]. Photographs were taken from (**a**) the south, (**b**) the southwest, and (**c**) the north.

Figure 6 shows the soil properties at the former K-NET Uto station, which was within the site of the Uto City Hall at the time of the 2016 Kumamoto earthquake. Note that those soil properties at the former K-NET Uto station are available from the K-NET website [54].

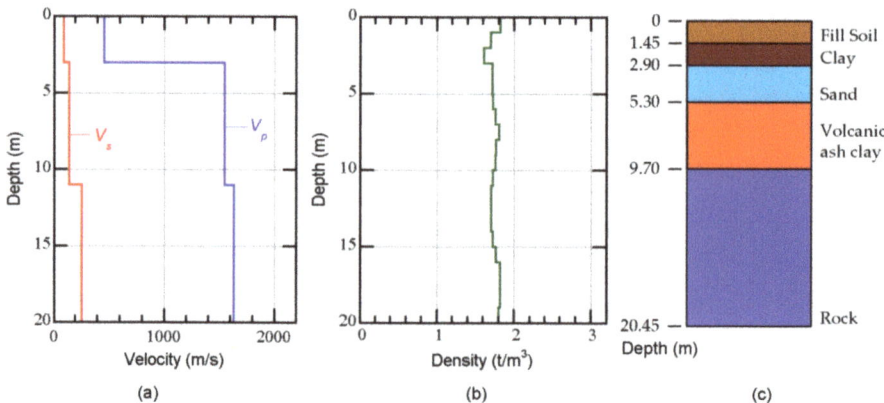

Figure 6. Soil profiles of the former K-NET Uto station. Figures were made from the data provided by the National Research Institute for Earth Science and Disaster Resilience (NIED). (**a**) Primary and shear wave profile (V_p: P-wave velocity, V_s: S-wave velocity); (**b**) density profile; (**c**) soil column.

3.2. Properties of the Isolation Layer

Two retrofitted building models were examined in this study to consider different seismic isolation periods (T_f). The properties of the isolation layer were determined as follows, based on the energy-balanced design method introduced in the design recommendation [2].

First, the mass and the moment of inertia of each floor were calculated. Table 1 shows the floor mass (m_j), moment of inertia (I_j), and the radius of gyration of the floor mass (r_j) for the jth floor. Here, the values of the floor mass and the moment of inertia above level 1 were taken from a previous study [44], while the weight per unit area at level 0 was assumed to be 24 kN/m^2 for the calculation of the floor mass and the moment of inertia of each floor. Therefore, the calculated total mass (M) was 5412 t.

Table 1. Floor mass, moment of inertia, and radius of gyration of the floor mass of each floor level.

Floor Level j	Floor Mass m_j (t)	Moment of Inertia I_j (×10^3 tm^2)	Radius of Gyration of Floor Mass r_j (m)
5	677.8	78.37	10.75
4	548.5	62.85	10.70
3	543.0	62.50	10.73
2	581.0	67.12	10.75
1	1208.1	199.0	12.83
0	1853.7	274.9	12.18

Figure 7 shows the horizontal response spectra of the recorded ground motions of the first (14 April 2016: UTO0414) and second (16 April 2016: UTO0416) earthquakes observed at the K-NET Uto station [54]. Note that the spectral acceleration (S_A) and velocity (S_V) shown in Figure 7a,b were calculated as the absolute (vector) value of the two horizontal components. The total input energy spectrum (the V_I spectrum) was also calculated considering the simultaneous input of the two horizontal components. The viscous damping ratio was set to 0.05 for the calculations of S_A and S_V, while it was set to 0.10 for the calculation of V_I following the work of Akiyama [47].

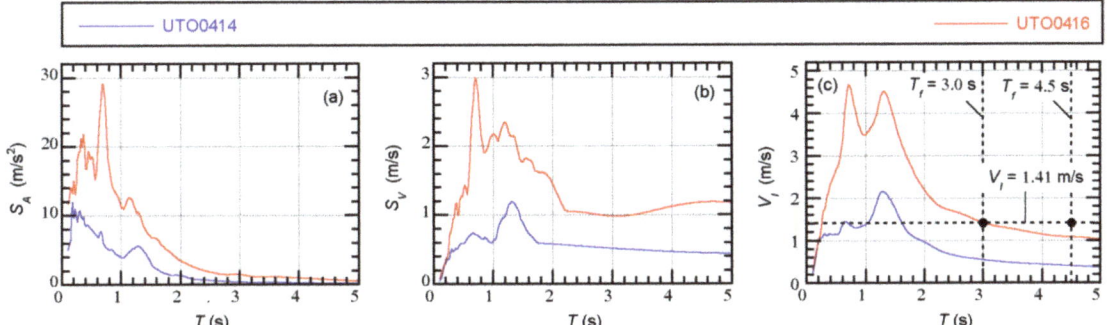

Figure 7. Horizontal response spectra of the recorded ground motions at the K-NET Uto station: (a) Elastic acceleration response spectrum (damping ratio: 0.05); (b) elastic velocity response spectrum (damping ratio: 0.05); (c) elastic total input energy spectrum (damping ratio: 0.10).

In this study, the range of the seismic isolation period (T_f) was considered to be from 3 to 4.5 s when determining the properties of the isolation layer. Based on Figure 7c, the equivalent velocity of the total input energy (V_I) was set to 1.41 m/s to determine the properties of the isolation layer.

Assuming that the superstructure is a rigid body, the equation of the energy balance is expressed as:

$$\frac{1}{2}K_f \delta_{\max}^2 + 4 \, _sn_s Q_y \delta_{\max} = \frac{1}{2}MV_I^2. \tag{11}$$

In Equation (11), $\delta_{\max}$ is the maximum horizontal displacement, $_sn$ is the equivalent number of repetitions, K_f is the total horizontal stiffness of the flexible element, and $_sQ_y$ is the total yield strength of the rigid plastic element. Equation (11) can be rewritten as:

$$\left(\frac{2\pi}{T_f}\right)^2 \delta_{\max}^2 + 8 \, _sn_s \alpha_y g \delta_{\max} = V_I^2, \tag{12}$$

$$\text{where } K_f = M\left(\frac{2\pi}{T_f}\right)^2, \, _s\alpha_y = \frac{_sQ_y}{Mg}. \tag{13}$$

In Equation (13), g is the acceleration due to gravity, assumed to be 9.8 m/s^2, and $_s\alpha_y$ is the yielding shear strength coefficient. In this study, the design-allowable horizontal displacement (δ_a) was set to 0.40 m, while the value of $_s n$ was set to 2 following the design recommendation [2]. Then, the two parameters of the isolation layer, T_f and $_s\alpha_y$, were adjusted, such that the following condition was satisfied:

$$\left(\frac{2\pi}{T_f}\right)^2 \delta_a^2 + 16 \,_s\alpha_y g \delta_a \geq V_I^2. \tag{14}$$

The isolation layer below level 0 comprises natural rubber bearings (NRBs), elastic sliding bearings (ESBs), and steel dampers. Figure 8 shows the layout of the isolators and dampers in the isolation layer of the two retrofitted building models: Model-Tf34, shown in Figure 8a, and Model-Tf44, shown in Figure 8b. As shown in the figures, the steel dampers were placed at the perimeter frames to provide torsional resistance. The point G shown in this figure is the center of mass of the superstructure, point S_0 is the center of stiffness of the isolation layer calculated according to the initial stiffness of the isolators and dampers, while point S_1 is the center of stiffness of the isolation layer calculated according to the secant stiffness of the isolators and dampers considering their displacement (δ) of 0.40 m.

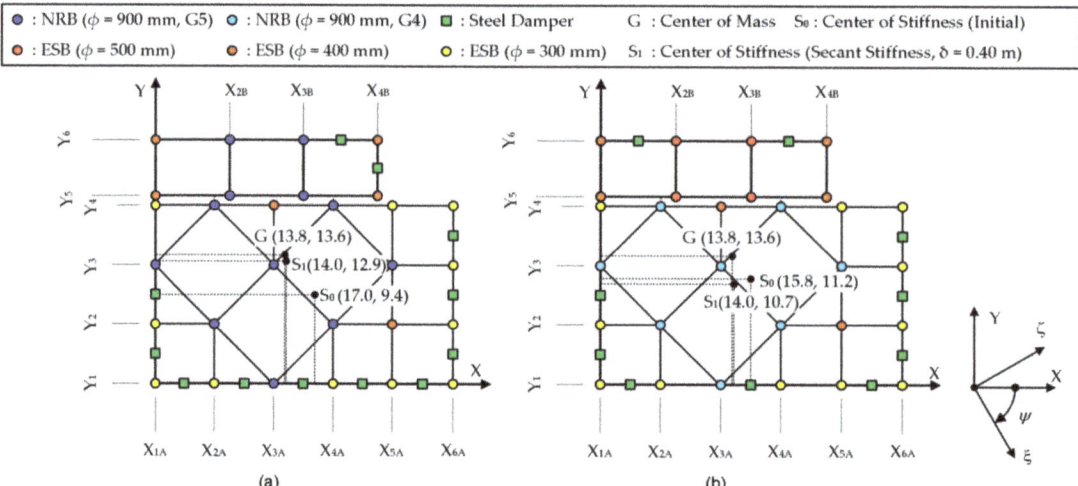

Figure 8. Layout of the isolators and dampers in the isolation layer: (a) Model-Tf34 and (b) Model-Tf44.

For Model-Tf34, $T_f = 3.38$ s and $_s\alpha_y = 0.0448$, while for Model-Tf44, $T_f = 4.40$ s and $_s\alpha_y = 0.0396$. Note that T_f was calculated according to the horizontal stiffness of the NRBs, while $_s\alpha_y$ was calculated according to the yield shear strength of the ESBs and steel dampers.

As shown in Figure 8, the eccentricity at the isolation layer was non-negligible in both models. The eccentricity indices [45] of Model-Tf34, calculated according to the initial stiffnesses in the X and Y directions, were $R_{eX} = 0.285$ and $R_{eY} = 0.216$, respectively, while those calculated according to the secant stiffnesses in the X and Y directions were $R_{eX} = 0.014$ and $R_{eY} = 0.054$, respectively. Similarly, the eccentricity indices of Model-Tf44, calculated according to initial stiffnesses in the X and Y directions, were $R_{eX} = 0.155$ and $R_{eY} = 0.126$, respectively, while those calculated according to the secant stiffnesses in the X and Y directions were $R_{eX} = 0.233$ and $R_{eY} = 0.015$, respectively. In this study, the perimeter frames X_{1A} and Y_6 are referred to as the "flexible-side frames," while the frames X_{6A} and Y_1 are referred to as the "stiff-side frames". Note that no optimization to minimize the torsional response was made to choose the dampers in this study, because such optimization was beyond the scope of this study.

Figure 9 shows envelopes of the force–deformation relationship for the isolators and dampers. The behavior of the NRBs was assumed to be linear elastic, while that of the ESBs and dampers was assumed to be bilinear.

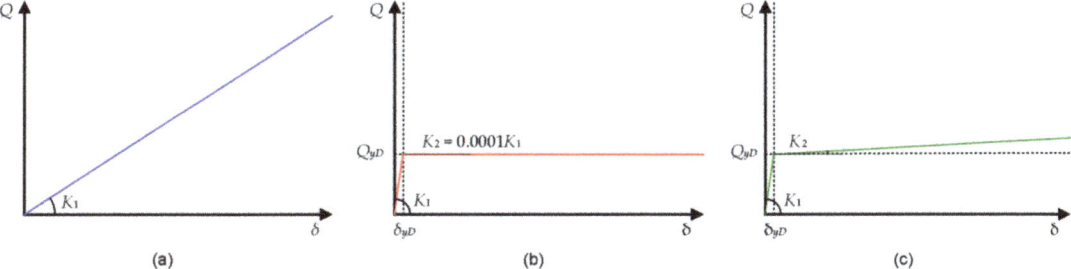

Figure 9. Envelope of the force–deformation relationship for the isolators and dampers: (**a**) Natural rubber bearings (NRBs); (**b**) elastic sliding bearings (ESBs); (**c**) steel dampers.

All isolators were chosen from a catalog provided by the Bridgestone Corporation [55,56], considering that the ultimate horizontal deformation was larger than 150% of δ_a (1.5 × 0.40 m = 0.60 m). Meanwhile, the steel dampers were chosen from a catalog provided by Nippon Steel Corporation Engineering Co. Ltd. (Tokyo, Japan) [57]. Tables 2–4 list the properties of the isolators and dampers used in the two models.

Table 2. Properties of the selected natural rubber bearings (NRBs).

Type	Outer Diameter (mm)	Total Rubber Thickness (mm)	Shear Modulus (MPa)	Horizontal Stiffness K_1 (MN/m)	Vertical Stiffness K_V (MN/m)
NRB (ϕ = 900 mm, G5)	900	180	0.441	1.56	3730
NRB (ϕ = 900 mm, G4)	900	180	0.392	1.38	3420

Table 3. Properties of the selected elastic sliding bearings (ESBs).

Type	Outer Diameter (mm)	Shear Modulus (MPa)	Friction Coefficient μ	Initial Horizontal Stiffness K_1 (MN/m)	Vertical Stiffness K_V (MN/m)
ESB (ϕ = 300 mm)	300	0.392	0.010	0.884	1380
ESB (ϕ = 400 mm)	400	0.392	0.010	1.48	2270
ESB (ϕ = 500 mm)	500	0.392	0.010	2.40	3710

Table 4. Properties of the selected steel dampers.

Initial Stiffness K_1 (MN/m)	Yield Strength Q_{yd} (kN)	Post Yield Stiffness K_2 (MN/m)
7.60	184	0.128

The yield strength of an ESB was calculated as:

$$Q_{yD} = \mu P_V. \tag{15}$$

In Equation (15), P_V is the vertical load of the ESB due to gravity.

3.3. Structural Modeling

The two building models were modeled as three-dimensional spatial frames, wherein the floor diaphragms were assumed to be rigid in their own planes without an out-of-plane stiffness. Figure 10 shows the structural modeling. For the numerical analyses, a nonlinear analysis program for spatial frames developed by the authors in a previous study [58] was used. The structural models were based on Model-RuW4-100 from a previous study [44]. As shown in Figure 10b,c, all of the vertical and rotational springs in the basement of the original model were replaced by isolators, and dampers were installed in the isolation layer. The shear behavior of the isolators was modeled using the multi-shear spring model proposed by Wada and Hirose [59], while their axial behavior was modeled using a linear elastic spring, as shown in Figure 10d. The shear behavior of the steel dampers was also modeled using the multi-shear spring model, as shown in Figure 10e. No bending stiffness was considered in either the isolators or the dampers. The hysteresis rules of the ESBs and steel dampers were modeled following the normal bilinear rule for simplicity of analysis. Details of the structural modeling of the superstructure can be found in the literature [44]. The damping matrix of the superstructure was then assumed to be proportional to the tangent stiffness matrix of the superstructure, with 2% of the critical damping of the first mode. The damping of the isolators and dampers was not considered, assuming that their energy absorption effects were already included in the hysteresis rules. Note that the force–displacement relationships of all members, including those in the superstructures and in the isolation layers, were assumed to be symmetric in the positive and negative loading directions, as in previous studies [44,58].

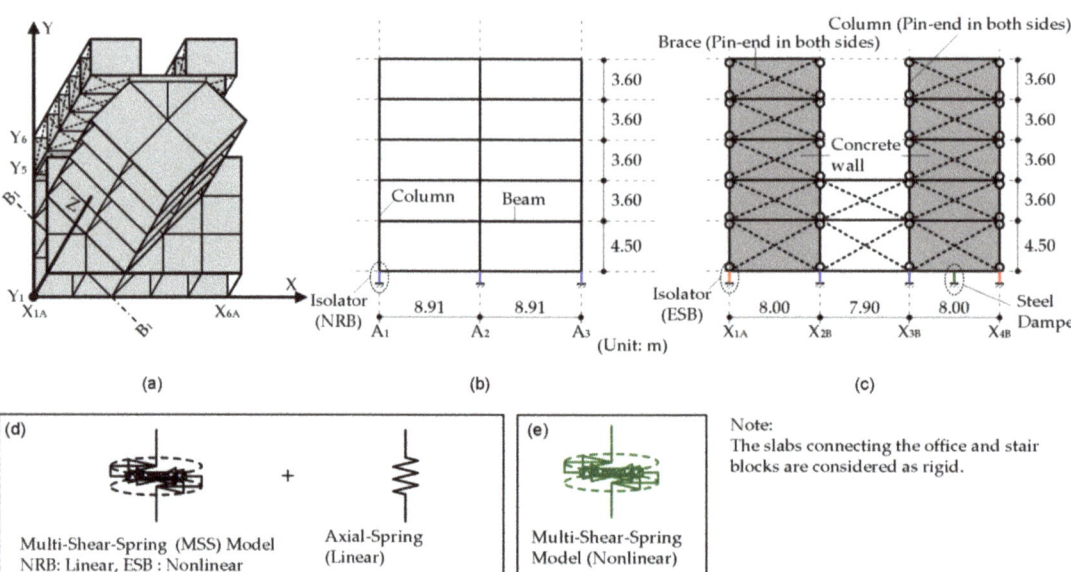

Figure 10. Structural modeling: (**a**) Overview of the structural model; (**b**) frame B_1; (**c**) frame Y_6; (**d**) modeling of the isolators (NRBs and ESBs); (**e**) modeling of the steel dampers.

It should be mentioned that the foundation compliance and kinematic soil–structure interaction were not considered for the simplicity of the analysis. Almansa et al. [60] investigated the performance of a rubber-isolated six-story RC building in soft soil. They concluded that the soil–structure interaction effect is rather negligible in the case of isolated buildings. Based on this, the authors thought the influence of the soil–structure interaction effect to the response of this building would not be significant.

Figures 11 and 12 show the first three natural modes of Model-T34 and Model-T44 in the elastic range. Here, T_{ie} is the natural period of the ith mode in the elastic range (i = 1–3), m_{ie}^* is the effective modal mass ratio of the ith mode with respect to its principal direction in the elastic range, ψ_{ie} is the incidence of the principal direction of the ith modal response in the elastic range, and $R_{\rho i e}$ is the torsional index of the ith mode in the elastic range. The values of m_{ie}^*, ψ_{ie}, and $R_{\rho i e}$ were calculated according to the ith elastic mode vector ($\boldsymbol{\varphi}_{ie}$) as follows:

$$m_{ie}^* = \frac{1}{M} \frac{\left(\sum_{j=0}^{5} m_j \phi_{Xjie}\right)^2 + \left(\sum_{j=0}^{5} m_j \phi_{Yjie}\right)^2}{\sum_{j=0}^{5} m_j \phi_{Xjie}^2 + \sum_{j=0}^{5} m_j \phi_{Yjie}^2 + \sum_{j=0}^{5} I_j \phi_{\Theta jie}^2}, \quad (16)$$

$$\tan \psi_{ie} = -\sum_{j=0}^{5} m_j \phi_{Yjie} / \sum_{j=0}^{5} m_j \phi_{Xjie}, \quad (17)$$

$$R_{\rho i e} = \sqrt{\sum_{j=0}^{5} I_j \phi_{\Theta jie}^2 / \left(\sum_{j=0}^{5} m_j \phi_{Xjie}^2 + \sum_{j=0}^{5} m_j \phi_{Yjie}^2\right)}, \quad (18)$$

$$\boldsymbol{\varphi}_{ie} = \left\{ \phi_{X0ie} \quad \cdots \quad \phi_{X5ie} \quad \phi_{Y0ie} \quad \cdots \quad \phi_{Y5ie} \quad \phi_{\Theta 0ie} \quad \cdots \quad \phi_{\Theta 5ie} \right\}^T. \quad (19)$$

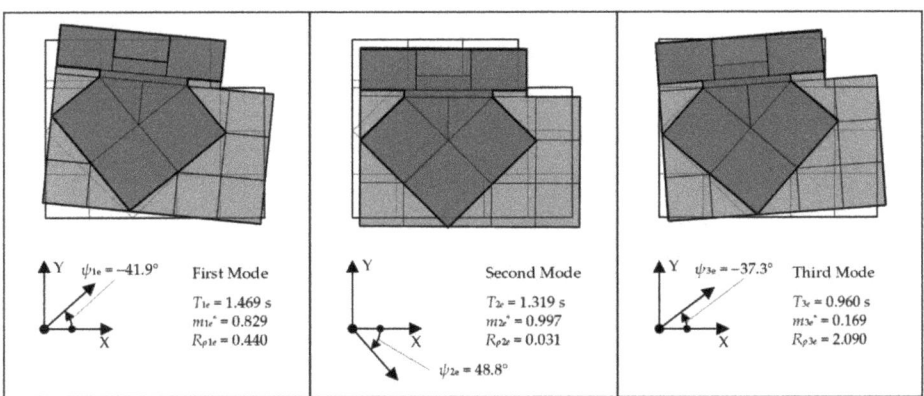

Figure 11. Shape of the first three natural modes of Model-Tf34 in the elastic range.

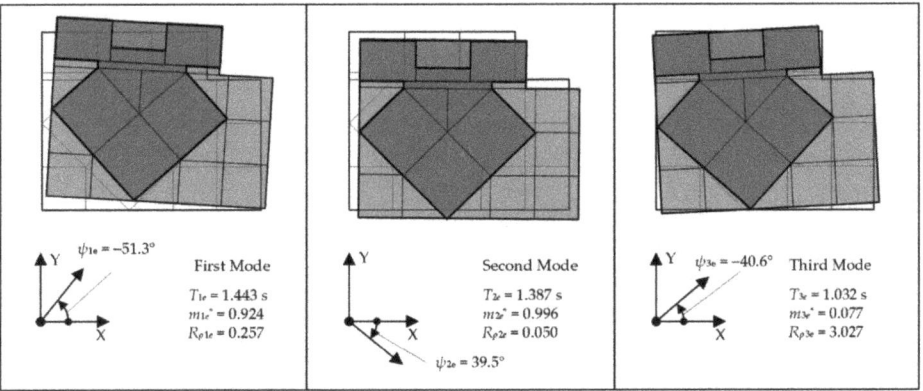

Figure 12. Shape of the first three natural modes of Model-Tf44 in the elastic range.

The behavior of the superstructure can be considered a rigid body in the first three modes in the elastic range, as observed from Figures 11 and 12. In both building models, the effective modal mass ratios of the first two modes were close to 1. In addition, the first mode was predominantly translational ($R_{\rho 1e} < 1$), the second mode was nearly purely translational ($R_{\rho 2e} \ll 1$), and the third mode was predominantly torsional ($R_{\rho 3e} > 1$). In addition, the angles between the principal directions of the first two modes were close to 90° (90.7° for Model-Tf34 and 90.8° for Model-Tf44). Therefore, both building models were classified as torsionally stiff buildings: In a previous study [43], the conditions for classification as a torsionally stiff building were $R_{\rho 1}$ and $R_{\rho 2} < 1$, and $R_{\rho 3} > 1$.

The natural periods of the superstructure in the elastic range, assuming that the entire basement is supported by pins, were 0.373 s (first mode), 0.342 s (second mode), and 0.192 s (third mode). Therefore, the natural period of the superstructure was well separated from the seismic isolation period in both models.

3.4. Ground Motion Data

In this study, the seismic excitation was bidirectional in the X–Y plane, and both artificial and recorded ground motions were considered.

3.4.1. Artificial Ground Motions

In this study, two series of artificial ground motion datasets (the Art-1 and Art-2 series) were generated. As noted in Section 3.2, the equivalent velocity of the total input energy (V_I) was set to 1.41 m/s to determine the properties of the isolation layer. Therefore, the artificial ground motion datasets were generated such that their bidirectional V_I spectrum fit the predetermined target V_I spectrum as follows.

The target bidirectional V_I spectrum with a damping ratio of 0.10 was determined such that:

$$V_I(T) = \begin{cases} 1.41(T/0.75) \text{ m/s} & : T \leq 0.75 \text{ s} \\ 1.41 & : T > 0.75 \text{ s} \end{cases}. \quad (20)$$

To generate each horizontal component, it was assumed that the intensity of V_I in both orthogonal components (ξ and ζ components) was identical. The target equivalent velocities of the total input energy of the ξ and ζ components, $V_{I\xi}(T)$ and $V_{I\zeta}(T)$, respectively, were calculated as:

$$V_{I\xi}(T) = V_{I\zeta}(T) = V_I(T)/\sqrt{2}. \quad (21)$$

The phase angle is given by a uniform random value, and to consider the time-dependent amplitude of the ground motions, a Jenning-type envelope function ($e(t)$) was assumed. In this study, two envelope functions were considered. In Art-1-00, the envelope function was set as in Equation (22), while in Art-2-00, the envelope function was set as in Equation (23).

$$e(t) = \begin{cases} (t/5)^2 & : 0 \text{ s} \leq t \leq 5 \text{ s} \\ 1 & : 5 \text{ s} \leq t \leq 25 \text{ s} \\ \exp\{-0.066(t-25)\} & : 25 \text{ s} \leq t \leq 60 \text{ s} \end{cases} \quad (22)$$

$$e(t) = \begin{cases} (t/2.5)^2 & : 0 \text{ s} \leq t \leq 2.5 \text{ s} \\ 1 & : 2.5 \text{ s} \leq t \leq 12.5 \text{ s} \\ \exp\{-0.132(t-12.5)\} & : 12.5 \text{ s} \leq t \leq 30 \text{ s} \end{cases} \quad (23)$$

Figures 13 and 14 show the time-histories of the two components and the orbits of Art-1-00 and Art-2-00, respectively. Because the same target V_I spectrum was applied while the duration of Art-2-00 is the half that of Art-1-00, the peak ground acceleration of Art-2-00 was larger than that of Art-1-00.

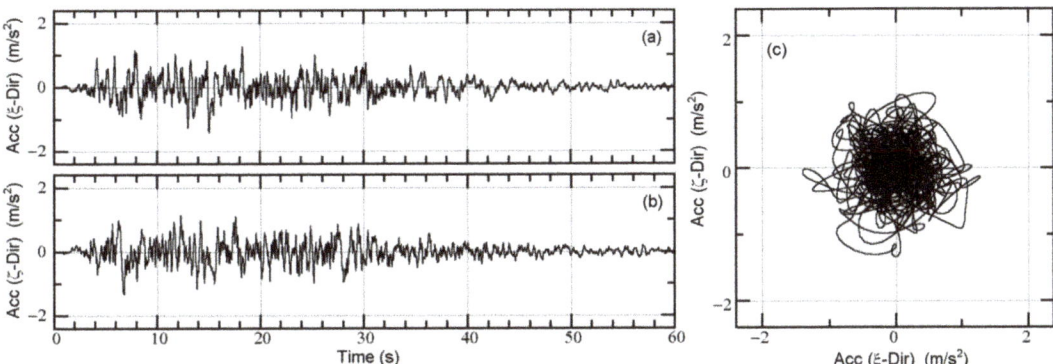

Figure 13. Two components of the generated artificial ground motion (Art-1-00): (**a**) ξ direction; (**b**) ζ direction; (**c**) orbit.

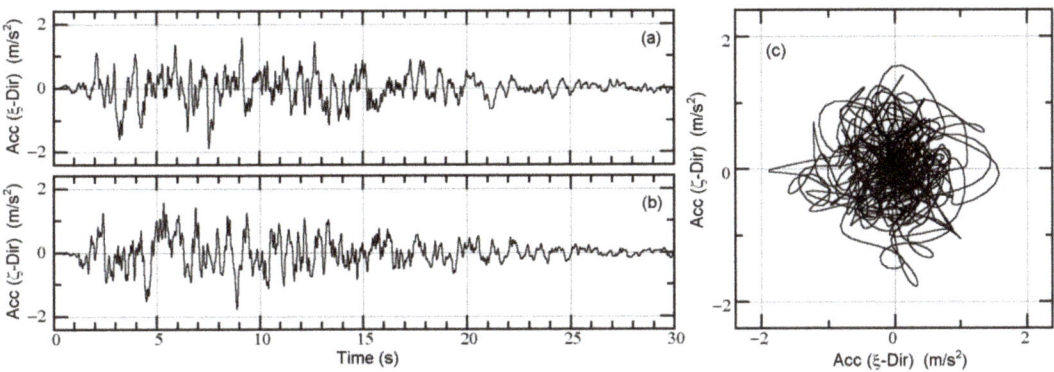

Figure 14. Two components of the generated artificial ground motion (Art-2-00): (**a**) ξ direction; (**b**) ζ direction; (**c**) orbit.

Note that the artificial ground motions used in this study were generated independently. The correlation coefficients of Art-1-00 and Art-2-00 were close to zero, even though the envelope functions of the two components were the same; therefore, the two components can be considered independently of one another.

Next, 11 artificial ground motion datasets were generated from Art-1-00 and Art-2-00 by shifting the phasing angle. The generated artificial ground motion vector ($\mathbf{a_g}(t, \Delta\phi_0)$) is expressed as:

$$\mathbf{a_g}(t, \Delta\phi_0) = \left\{ \begin{array}{c} a_{g\xi}(t, \Delta\phi_0) \\ a_{g\zeta}(t, \Delta\phi_0) \end{array} \right\} = \sum_{n=-N}^{N} \left\{ \begin{array}{c} c_{\xi,n} \\ c_{\zeta,n} \end{array} \right\} \exp[i\{\omega_n t - \mathrm{sgn}(\omega_n)\Delta\phi_0\}]. \qquad (24)$$

In Equation (24), $c_{\xi,n}$ and $c_{\zeta,n}$ are the complex Fourier coefficients of the nth harmonics of $a_{g\xi}(t)$ and $a_{g\zeta}(t)$, respectively, ω_n is the circular frequency of the nth harmonic, and $\Delta\phi_0$ is the constant used to shift the phase angle of all the harmonics. As in previous studies [51,53], the constant $\Delta\phi_0$ was set from $\pi/12$ to $11\pi/12$ with an interval of $\pi/12$. Notably, the phase difference of each ground motion component did not change when shifting the phase angle. In addition, the total input energy spectrum (the V_I spectrum) was independent of the shifting phase angle ($\Delta\phi_0$). This is because the shifting phase angle does not affect the absolute value of the complex Fourier coefficient, and the total input energy per unit mass can be calculated from the Fourier amplitude spectrum without the phase characteristics of the ground motion, as shown by Ordaz [61]. The generated artificial ground motion datasets are numbered from 01 to 11 depending on $\Delta\phi_0$, with a total of $2 \times 12 = 24$ artificial ground motion datasets generated and used in this study.

Figures 15 and 16 show the horizontal response spectra of the generated ground motion datasets. As shown in these figures, the spectral acceleration (S_A) and the spectral velocity (S_V) differed slightly because of the shift in the phase angle. A comparison of Figures 15 and 16 reveals that S_A and S_V of the Art-2 series were larger than those of the Art-1 series, even though the target V_I spectrum was identical.

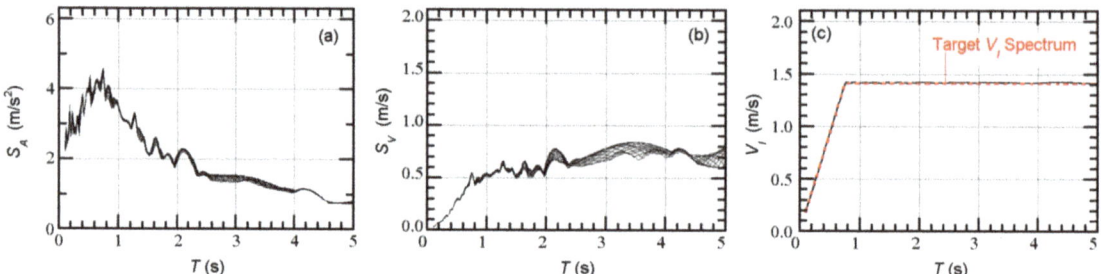

Figure 15. Horizontal response spectra of the artificial ground motion (Art-1 series): (**a**) Elastic acceleration response spectrum (damping ratio: 0.05); (**b**) elastic velocity response spectrum (damping ratio: 0.05); (**c**) elastic total input energy spectrum (damping ratio: 0.10).

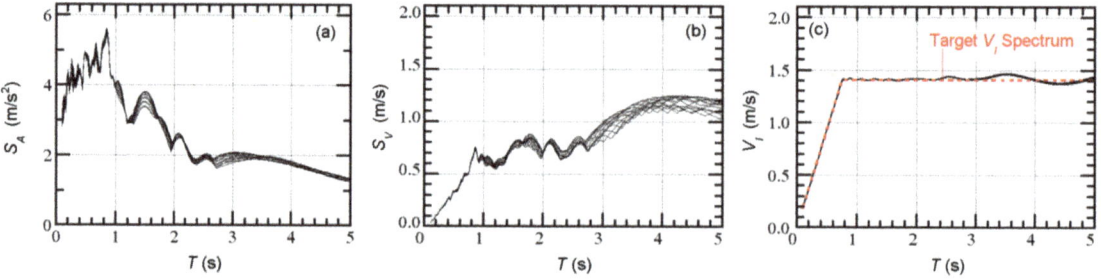

Figure 16. Horizontal response spectra of the artificial ground motion (Art-2 series): (**a**) Elastic acceleration response spectrum (damping ratio: 0.05); (**b**) elastic velocity response spectrum (damping ratio: 0.05); (**c**) elastic total input energy spectrum (damping ratio: 0.10).

3.4.2. Recorded Ground Motions

Table 5 lists the four datasets of recorded ground motions used in this study. Details concerning the original ground motions can be found in the Appendix A. Figure 17 shows the horizontal response spectra of the unscaled recorded ground motion datasets.

Table 5. List of the recorded ground motion datasets used in this study.

Earthquake of the Original Record	Ground Motion ID	Scale Factor	
		Model-Tf34	Model-Tf44
Kumamoto, 14 April 2016	UTO0414	1.000	1.000
Kumamoto, 16 April 2016	UTO0416	1.000	1.000
Chichi, 1999	TCU	0.5540	0.5718
Kocaeli, 1999	YPT	0.4293	0.5057

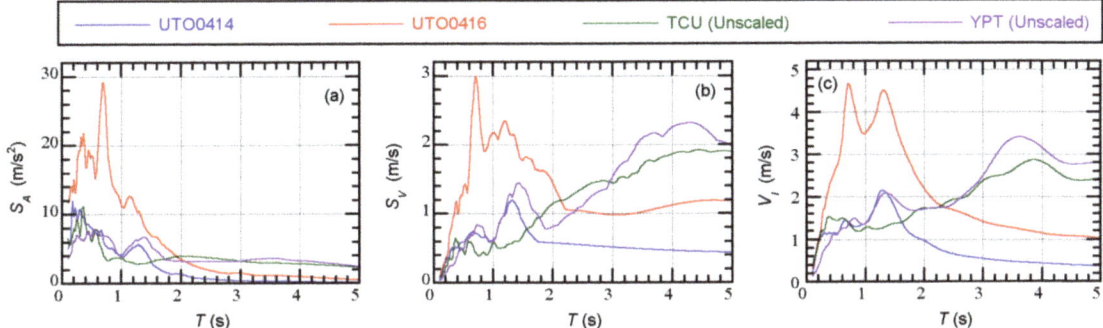

Figure 17. Horizontal response spectra of the recorded ground motions: (**a**) Elastic acceleration response spectrum (damping ratio: 0.05); (**b**) elastic velocity response spectrum (damping ratio: 0.05); (**c**) elastic total input energy spectrum (damping ratio: 0.10).

One of the motivations of this analysis was to evaluate the seismic performance of the two retrofitted models under the ground motions recorded during the 2016 Kumamoto earthquake. Therefore, the two recorded ground motion datasets (UTO0414 and UTO0416) [54] were used without scaling. A further two datasets of ground motions, TCU and YPT, were chosen as examples of long-period pulse-like ground motions, as investigated by Güneş and Ulucan [46]. These two ground motion datasets were scaled such that the equivalent velocity of the total energy input (V_I) at the isolation period (T_f) equaled 1.41 m/s, which is consistent with the intensity considered in the determination of the properties of the isolation layer, as discussed in Section 3.2.

4. Analysis Results

The analysis results are split into two subsections. In the first subsection, an example of the prediction of the peak equivalent displacement using the updated MABPA is shown. In the second subsection, comparisons are made between the nonlinear time-history analysis results and the predicted results.

4.1. Example of a Prediction of the Peak Equivalent Displacement

The structural model shown here as an example is Model-Tf44. Figure 18 shows the nonlinear properties of the equivalent SDOF model calculated according to Section 2.2.2.

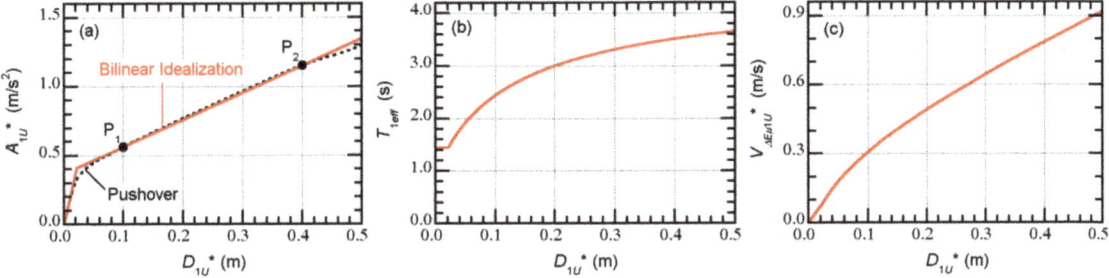

Figure 18. Nonlinear properties of the equivalent single-degree-of-freedom (SDOF) model representing the first modal response (Model-Tf44): (**a**) The A_{1U}^*–D_{1U}^* relationship; (**b**) the T_{1eff}^*–D_{1U}^* relationship; (**c**) the $V_{\Delta E\mu 1U}^*$–D_{1U}^* relationship.

In the results shown in Figure 18b,c, the $V_{\Delta E\mu 1U}^*$–T_{1eff}^* curve was constructed and overlaid with the bidirectional $V_{\Delta E}$ spectrum. Figure 19 shows the prediction of the peak equivalent displacements of the first two modes.

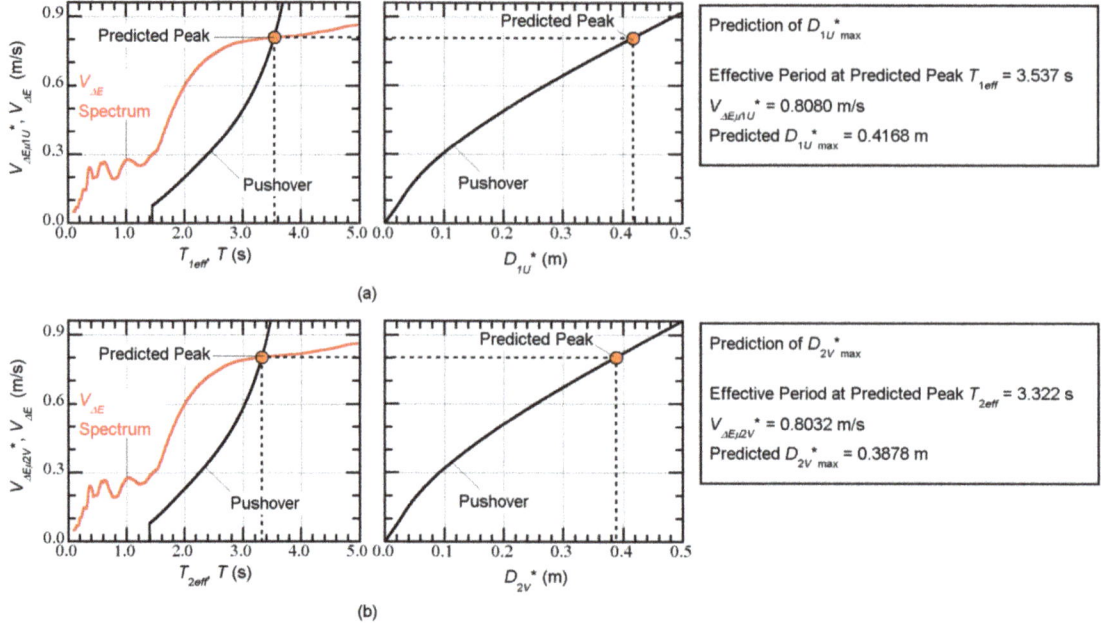

Figure 19. Prediction of the peak equivalent displacements (model: Model-Tf44, ground motion: TCU): (**a**) Prediction of the peak of the first modal response and (**b**) prediction of the peak of the second modal response.

The predicted peak was obtained as the intersection point of the $V_{\Delta E \mu 1U}{}^*$–T_{1eff} curve according to the pushover analysis results and the bidirectional $V_{\Delta E}$ spectrum, as shown on the left side of Figure 19a. Then, the peak equivalent displacement $D_{1U}{}^*{}_{max}$ was predicted from the $V_{\Delta E \mu 1U}{}^*$–$D_{1U}{}^*{}_{max}$ curve. The prediction of the peak equivalent displacement of the second mode, $D_{2V}{}^*{}_{max}$, can be made using the same procedure as that of $D_{1U}{}^*{}_{max}$, as shown in Figure 19b. Note that for the predictions of $D_{1U}{}^*{}_{max}$ and $D_{2V}{}^*{}_{max}$, the same $V_{\Delta E}$ spectrum (bidirectional $V_{\Delta E}$ spectrum) was used.

4.2. Comparisons with the Nonlinear Time-History Analysis

In Section 4.2.1, the nonlinear time-history analysis results using 24 artificial ground motion datasets are compared with the predicted results. In this analysis, 24 × 4 = 96 cases were analyzed for each model: The angle of incidence of the horizontal ground motion (ψ) in Figure 3 was set to −45°, 0°, 45°, and 90° considering the symmetry of the force-deformation relationship assumed in the structural model. Because the characteristics of the two horizontal components were similar, the discussion focuses on comparisons between the envelope of the time-history analysis results and the predicted results.

In Section 4.2.2, the nonlinear time-history analysis results using the four recorded ground motion datasets are compared with the predicted results. In this analysis, 4 × 12 = 48 cases were analyzed for each model: The angle of incidence of the horizontal ground motion (ψ) was set to values from −75° to 90° with an interval of 15°.

4.2.1. Artificial Ground Motion

Figures 20 and 21 show the peak relative displacement at $X_{3A}Y_3$, which is the closest point to the center of the floor mass at each level. In these figures, the nonlinear time-history analysis results are compared with the results predicted by MABPA (for complex damping ratios (β) of 0.10, 0.20, and 0.30). As shown here, the predicted peaks can approximate the envelope of the time-history analysis results, except in the case of Model-Tf34 subjected to the Art-2 series shown in Figure 20b. In addition, the predicted peak with $\beta = 0.10$ is larger than that with $\beta = 0.30$. When the models were subjected to the Art-1 series, the predicted peak closest to the envelope of the time-history analysis results was found with $\beta = 0.10$ for Model-Tf34 (Figure 20a) and with $\beta = 0.30$ for Model-Tf44 (Figure 21b). Differences in the predicted peaks resulting from the value of the complex damping were small in the case of Model-Tf34 but relatively noticeable in the case of Model-Tf44.

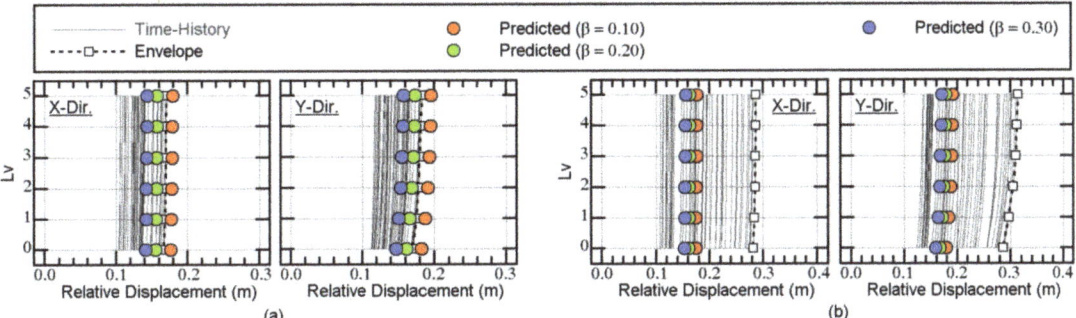

Figure 20. Comparison of the peak relative displacements at $X_{3A}Y_3$ (Model-Tf34) for the (**a**) Art-1 and (**b**) Art-2 series.

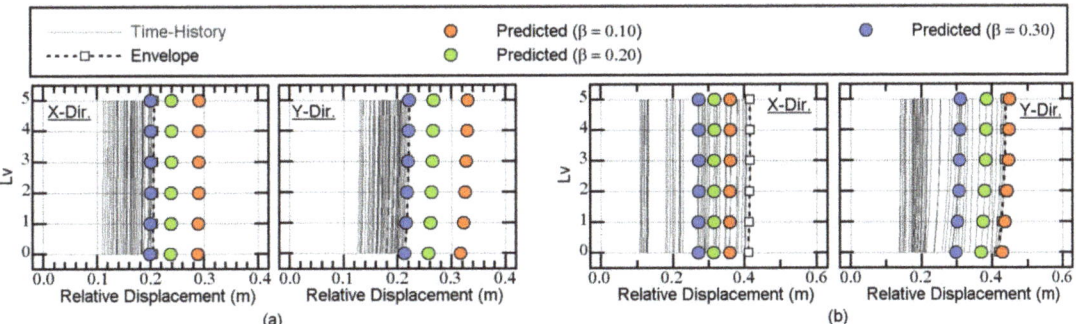

Figure 21. Comparison of the peak relative displacements at $X_{3A}Y_3$ (Model-Tf44) for the (**a**) Art-1 and (**b**) Art-2 series.

Figures 22 and 23 show comparisons of the horizontal distributions of the peak displacement at level 0, which is just above the isolation layer. In Figure 22, the results of Model-Tf34 subjected to the Art-1 series are shown. The predicted peak agrees with the envelope of the time-history analysis results, even though the predicted peak was not conservative at Y_1 and Y_2 (the stiff side in the X direction). A similar observation can be made in the case of Model-Tf44 subjected to the Art-1 series, as shown in Figure 23, where the predicted peak with $\beta = 0.30$ agrees very well with the envelope of the time-history analysis results, except at Y_1 and Y_2.

Based on the above results, the updated MABPA can predict the peak relative displacement at $X_{3A}Y_3$ and the horizontal distribution of the peak displacement at level 0 for both models according to the artificial ground motion datasets.

4.2.2. Recorded Ground Motion

Figures 24 and 25 show the peak relative displacement at $X_{3A}Y_3$. The accuracy of the predicted peak displacement is satisfactory: The predicted peak approximated the envelope of the nonlinear time-history analysis results, even though some cases were overestimated (e.g., model: Model-Tf44, ground motion: UTO0416, Figure 25b). The variation of the predicted peak due to the assumed complex damping ratio (β) depends on the ground motion. In some cases, the largest peak was obtained when β was 0.30 (e.g., model: Model-Tf34, ground motion: UTO0416, Figure 24b), while in other cases, the largest peak was obtained when β was 0.10 (e.g., model: Model-Tf34, ground motion: TCU, Figure 24c).

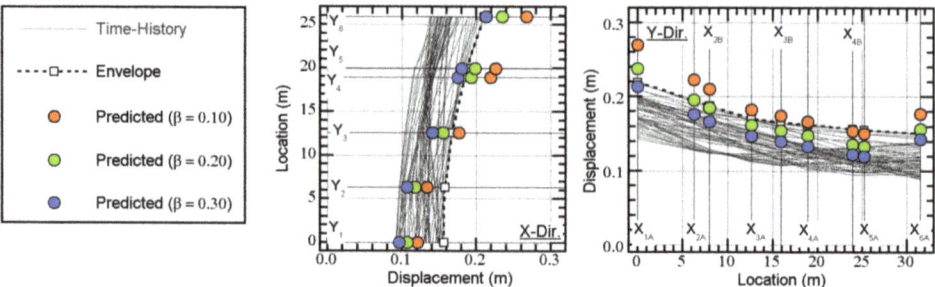

Figure 22. Comparison of the horizontal distributions of the peak displacement at level 0 (model: Model-Tf34, ground motion: Art-1 series).

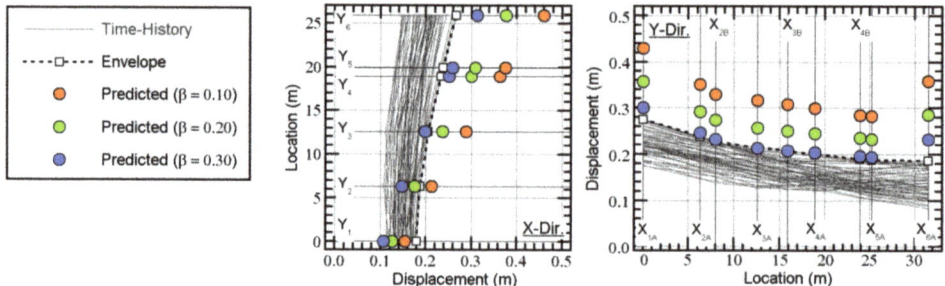

Figure 23. Comparison of the horizontal distributions of the peak displacement at level 0 (model: Model-Tf44, ground motion: Art-1 series).

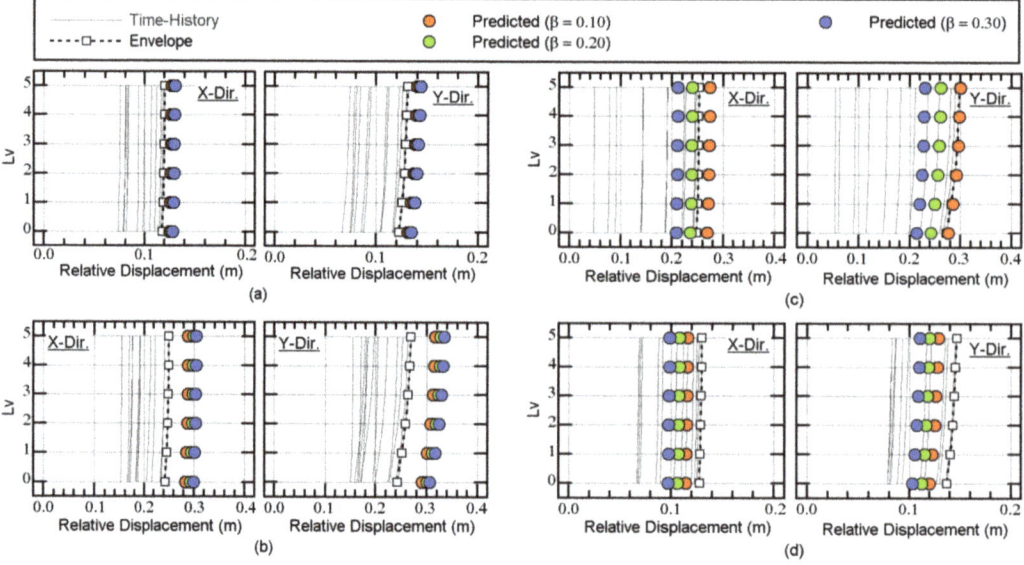

Figure 24. Comparison of the peak relative displacements at $X_{3A}Y_3$ (Model-Tf34): (**a**) UTO0414; (**b**) UTO0416; (**c**) TCU; (**d**) YPT.

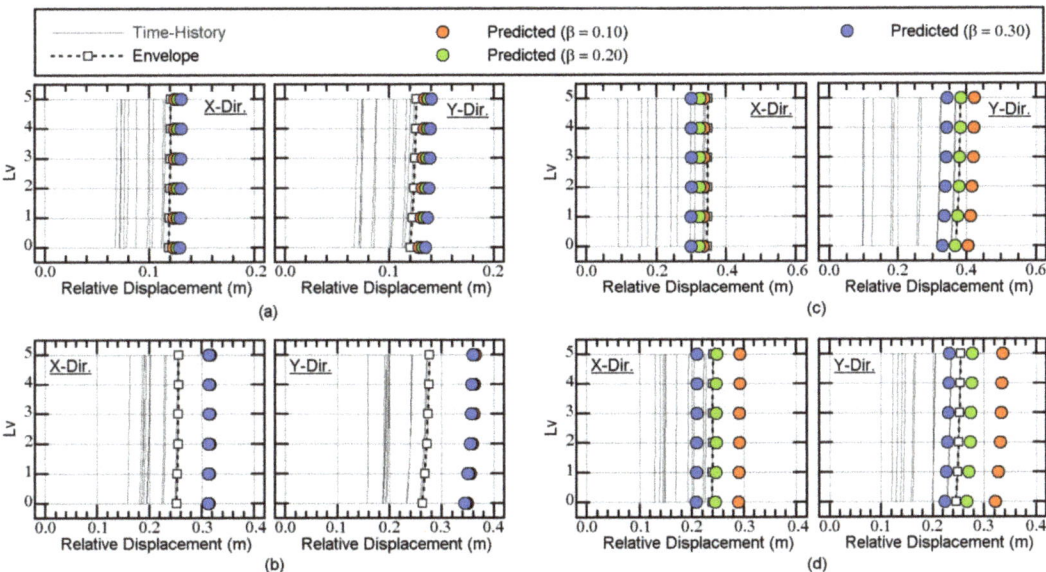

Figure 25. Comparison of the peak relative displacements at $X_{3A}Y_3$ (Model-Tf44): (**a**) UTO0414; (**b**) UTO0416; (**c**) TCU; (**d**) YPT.

Figure 26 shows comparisons of the horizontal distribution of the peak displacement at level 0. In this figure, the Model-Tf34 results are shown, with the results for UTO0414 shown in Figure 26a and those for TCU shown in Figure 26b.

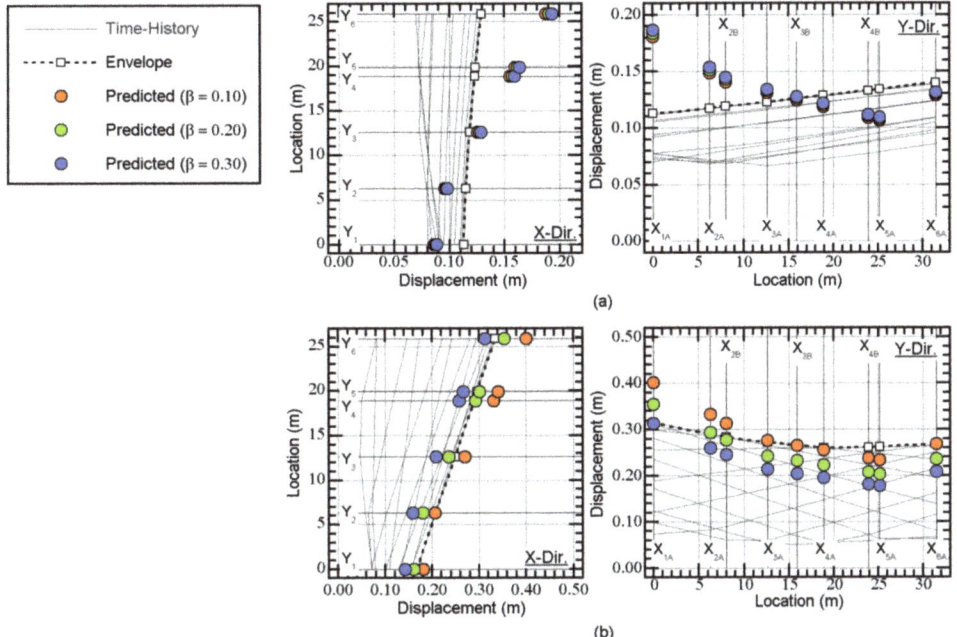

Figure 26. Comparison of the horizontal distributions of the peak displacement at level 0 (model: Model-Tf34) for (**a**) UTO0414 and (**b**) TCU.

The predicted horizontal distribution of the peak displacement in the Y direction was notably different from the envelope of the time-history analysis in the case of UTO0414, as shown in Figure 26a. In the predicted distribution, the largest displacement occurred at X_{1A} (the flexible side in the Y direction); meanwhile, in the envelope of the time-history analysis results, the displacement at X_{1A} was the smallest. Conversely, the predicted horizontal distribution of the peak displacement in the Y direction fits the envelope of the time-history analysis in the case of TCU very well, as shown in Figure 26b. In the predicted distribution, the largest displacement occurred at X_{1A}, which is consistent with the envelope of the time-history analysis results. The modal displacement responses at X_{1A} and X_{6A} (the stiff side in the Y direction) are compared and discussed further in Section 5.5.

One of the motivations of this analysis was to evaluate the seismic performance of the retrofitted building models under the recorded motions from the 2016 Kumamoto earthquake. It has already been reported in a previous study [44] that the seismic capacity evaluation of the main building of the former Uto City Hall indicate that the structure was insufficient to withstand the earthquake that occurred on 14 April 2016. Therefore, the drift of the superstructure was examined as follows.

Figure 27 shows comparisons of the peak drift for the three columns in Model-Tf34 in cases with UTO0414 and UTO0416. The peak drifts for the three columns obtained from the nonlinear time-history analysis results were smaller than 0.4%. The largest peak drift occurred for column A_3B_3 on the second story, with a value of 0.12% in the case of UTO0414 (Figure 27a) and a value of 0.30% in the case of UTO0416 (Figure 27b). Therefore, the seismic performance of Model-Tf34 was excellent for the motions recorded during the 2016 Kumamoto earthquake. Figure 28 shows comparisons of the peak drift of Model-Tf44. Similar observations can be made for Model-Tf44. Comparisons of Figures 27 and 28 indicate that the peak drift of Model-Tf44 was smaller than that of Model-Tf34. These figures also illustrate that the accuracy of the predicted peak depends on the column, with the accuracy in column A_1B_1 being satisfactory and that in column A_3B_3 being less satisfactory.

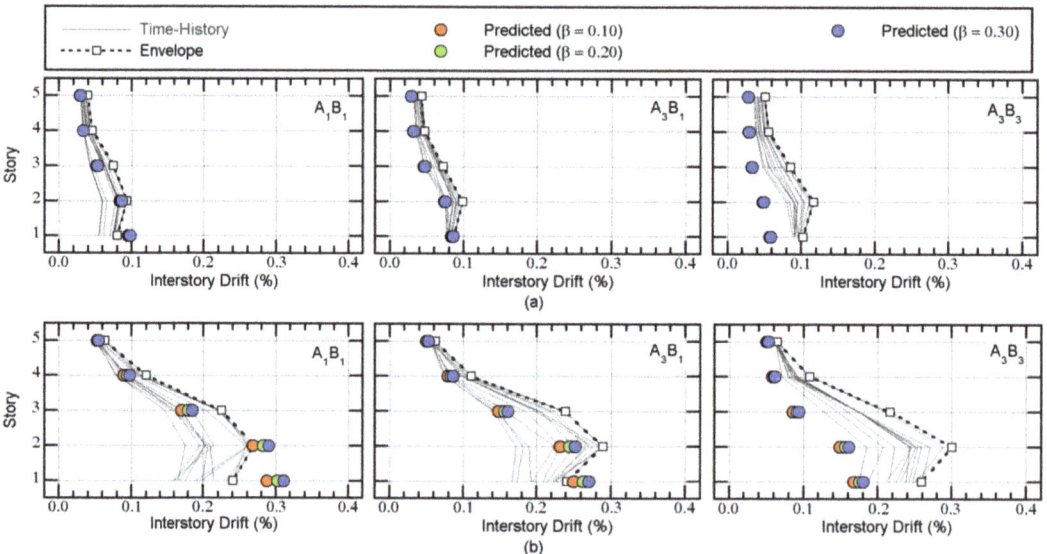

Figure 27. Comparison of the peak drift for the columns (Model-Tf34) with (**a**) UTO0414 and (**b**) UTO0416.

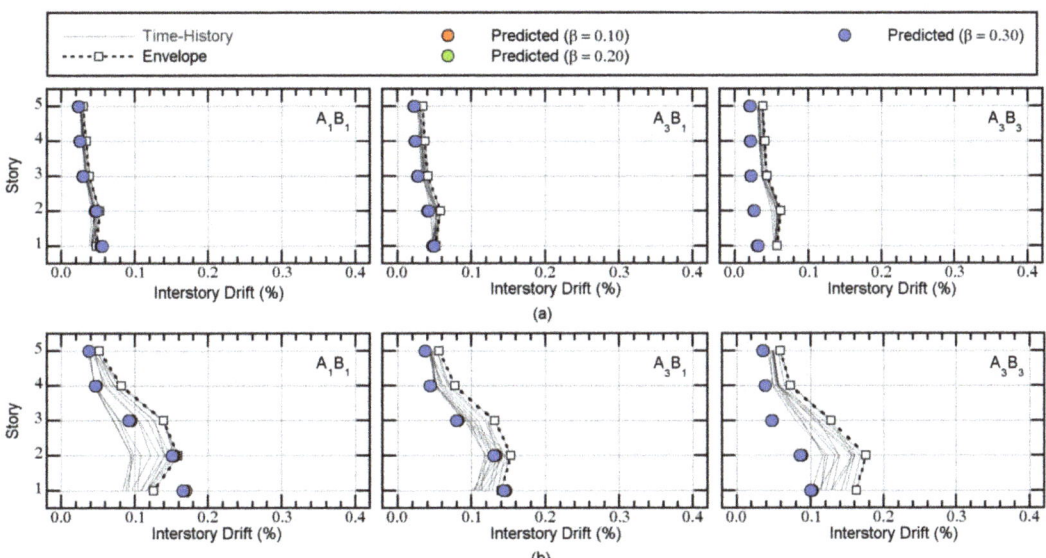

Figure 28. Comparison of the peak drift for the columns (Model-Tf44) with (**a**) UTO0414 and (**b**) UTO0416.

Figures 29 and 30 show the peak displacements at two isolators ($X_{1A}Y_6$ and $X_{6A}Y_1$) for various angles of incidence of seismic input (ψ). In these figures, the displacement of each isolator was calculated as the absolute (vector) value of the two horizontal directions, and the predicted peaks are shown by the colored lines. As shown in these figures, the upper bounds of the peak displacement of the isolators can be satisfactorily evaluated using the updated MABPA presented in this study.

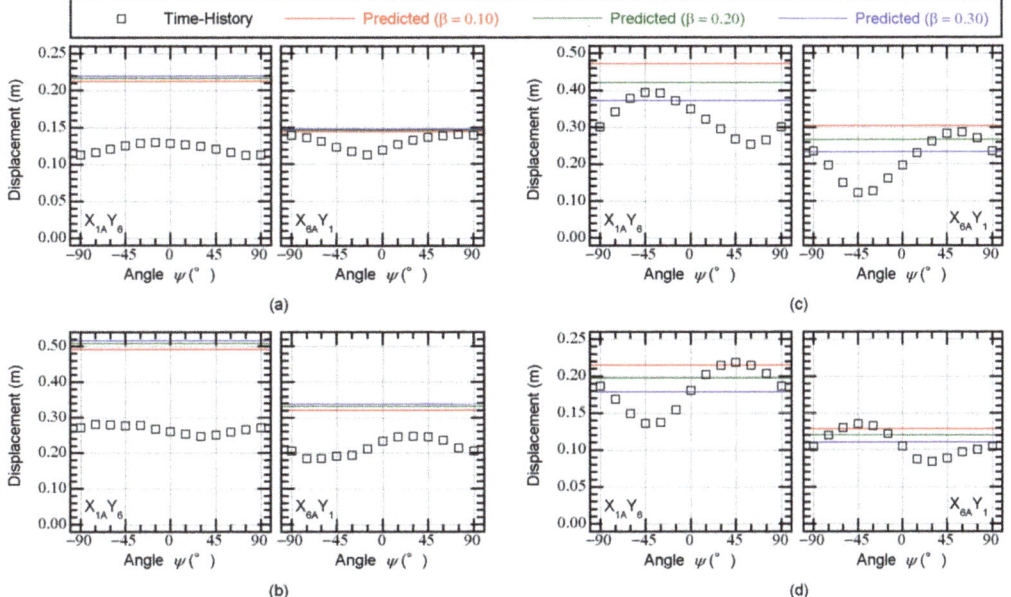

Figure 29. Peak displacement at isolators for various angles of incidence of seismic input (Model-Tf34): (**a**) UTO0414; (**b**) UTO0416; (**c**) TCU; (**d**) YPT.

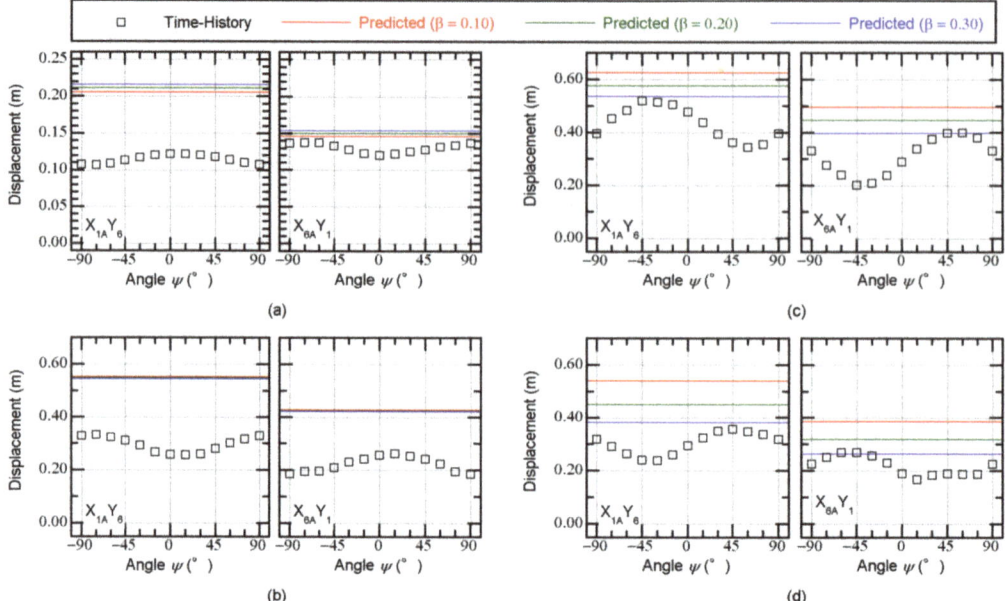

Figure 30. Peak displacement at isolators for various angles of incidence of seismic input (Model-Tf44): (**a**) UTO0414; (**b**) UTO0416; (**c**) TCU; (**d**) YPT.

These figures also indicate that the variation in the peak displacement of the isolators due to the angle of incidence of the seismic input is noticeable, e.g., in the cases of TCU and YPT, for both models. This point is discussed in Section 5.

5. Discussion

Here, there are four points of focus: (i) The relationship between the maximum momentary input energy and the peak equivalent displacement of the first modal response and its predictability from the pushover analysis results, (ii) the predictability of the maximum momentary input energy of the first modal response from the bidirectional $V_{\Delta E}$ spectrum, (iii) the accuracy of the predicted peak equivalent displacements of the first and second modal responses, and (iv) the contribution of the higher mode to the displacement response at the edge of level 0.

5.1. Calculation of the Modal Responses

Figure 31 shows the calculation flow for the first and second modal responses from the nonlinear time-history analysis results. This procedure is based on the procedure originally proposed by Kuramoto [62] for analyzing the nonlinear modal response of a non-isolated planer frame structure extended to analyze that of three-dimensional base-isolated structures.

Next, the momentary input energy of the first modal response per unit mass was calculated as follows. From the time-history of the equivalent displacement of the first modal response ($D_{1U}^*(t)$), the momentary input energy of the first modal response per unit mass was calculated as:

$$\frac{\Delta E_{1U}^*}{M_{1U}^*} = -\int_{t}^{t+\Delta t} \dot{D}_{1U}^*(t) a_{gU}(t) dt. \tag{25}$$

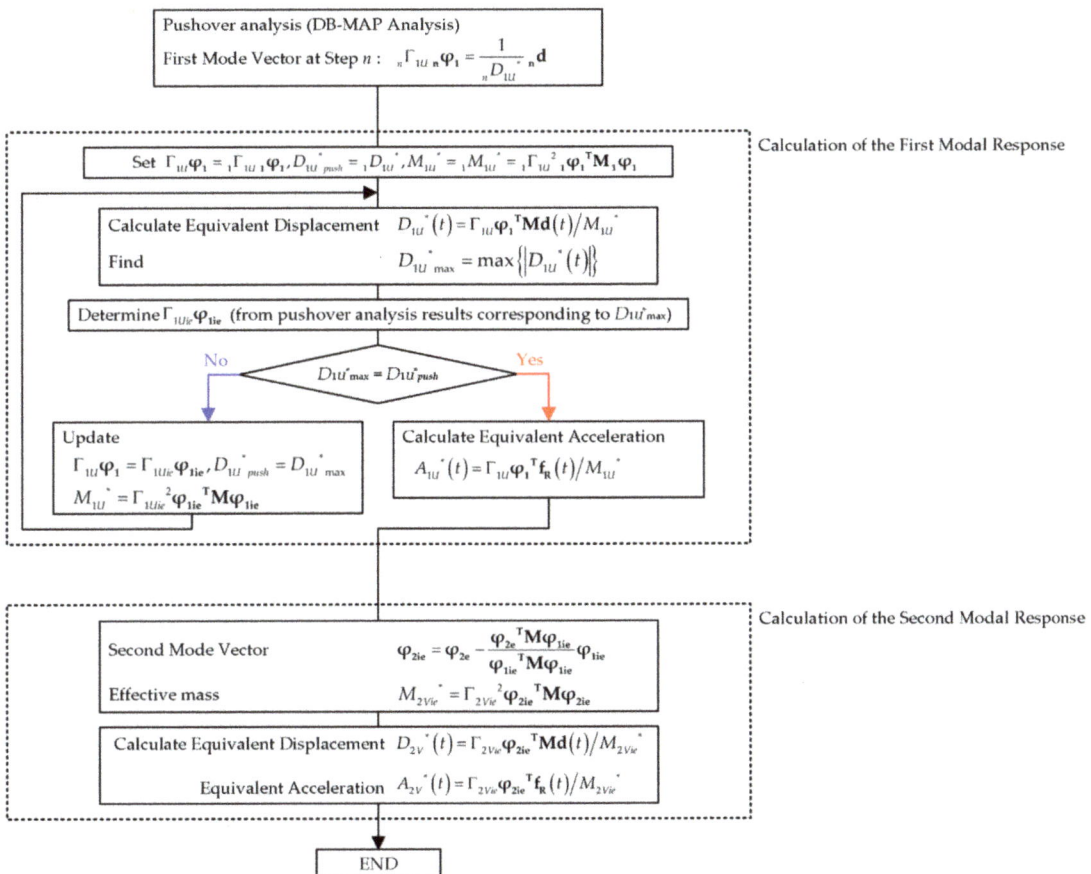

Figure 31. Calculation flow for the first and second modal responses.

In Equation (25), $\dot{D}_{1U}{}^*(t)$ is the derivative of $D_{1U}{}^*(t)$ with respect to t, $a_{gU}(t)$ is the ground acceleration component of the U-axis (the principal axis of the first modal response corresponds to $D_{1U}{}^*{}_{max} = \max\{|D_{1U}{}^*(t)|\}$), and Δt is the duration of a half cycle of the first modal response. The maximum momentary input energy of the first modal response per unit mass ($\Delta E_{1U}{}^*{}_{max}/M_{1U}{}^*$) is the maximum value of $\Delta E_{1U}{}^*/M_{1U}{}^*$ calculated by Equation (25) over the course of the seismic event. Figure 32 shows the definition of the maximum momentary input energy of the first modal response per unit mass. As shown in this figure, the half cycle when the maximum equivalent displacement ($D_{1U}{}^*{}_{max}$) occurs corresponds to the half cycle when the maximum momentary energy input occurs.

The equivalent velocity of the maximum momentary input energy of the first modal response ($V_{\Delta E1U}{}^*$) and the response period of the first modal response (T_1') were calculated such that:

$$V_{\Delta E1U}{}^* = \sqrt{2\Delta E_{1U}{}^*{}_{max}/M_{1U}{}^*}, \quad (26)$$

$$T_1' = 2\Delta t. \quad (27)$$

5.2. Relationship between the Peak Equivalent Displacement and the Maximum Momentary Input Energy of the First Modal Response

Figures 33 and 34 show comparisons between the $V_{\Delta E\mu 1U}{}^* - D_{1U}{}^*$ curve and the $V_{\Delta E1U}{}^* - D_{1U}{}^*{}_{max}$ relationship obtained from the nonlinear time-history analysis, with the results of Models Tf34 and Tf44 shown in Figures 33 and 34, respectively. These figures confirm that the $V_{\Delta E\mu1U}{}^*$–$D_{1U}{}^*$ curve fits the plots obtained from the time-history analysis results very well. This is because the contribution of the first modal response to the whole response may be large: The effective modal mass ratio of the first mode with respect to the U-axis is large (more than 0.7) in both Models Tf34 and Tf44.

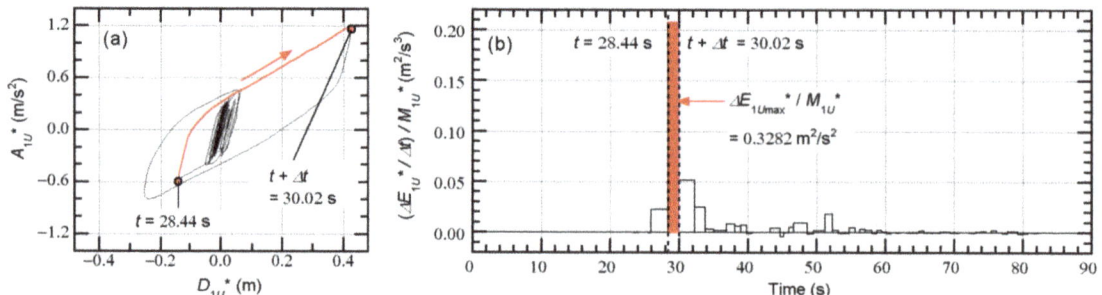

Figure 32. Definition of the maximum momentary input energy of the first modal response per unit mass (structural model: Model-Tf44, ground motion: TCU, angle of incidence of seismic input: $\psi = -30°$): (**a**) Hysteresis of the first modal response and (**b**) time-history of the momentary energy input.

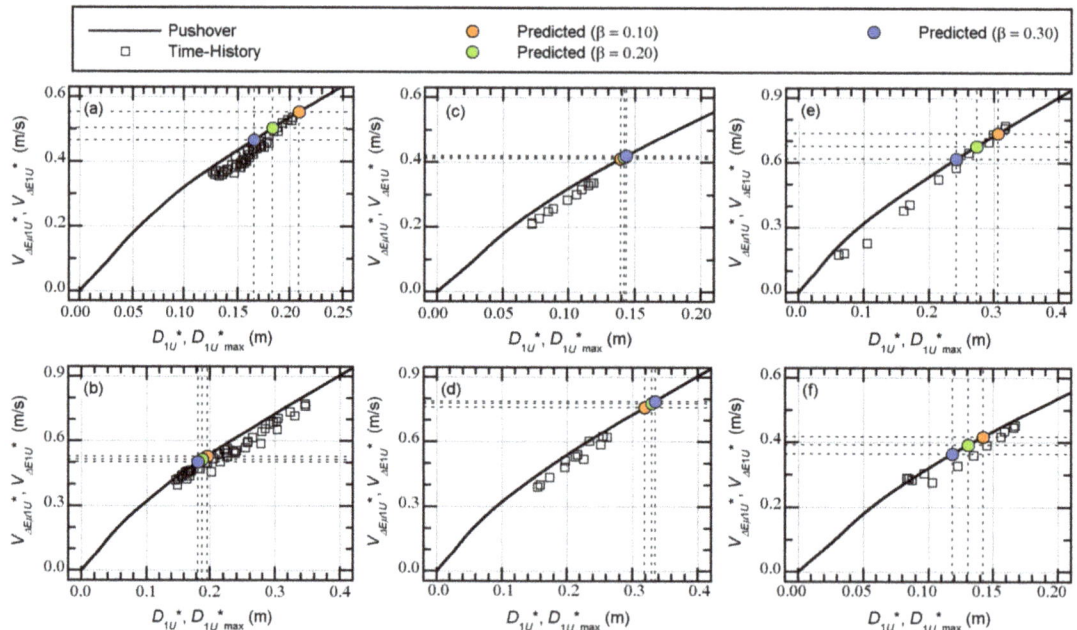

Figure 33. Comparison between the $V_{\Delta E\mu1U}{}^*$–$D_{1U}{}^*$ curve and the $V_{\Delta E1U}{}^*$–$D_{1U}{}^*{}_{max}$ relationship obtained from the time-history analysis (Model-Tf34): (**a**) Art-1 series; (**b**) Art-2 series; (**c**) UTO0414; (**d**) UTO0416; (**e**) TCU; (**f**) YPT.

Therefore, the accuracy of the predicted $D_{1U}{}^*{}_{max}$ relies on the accuracy of the predicted $V_{\Delta E1U}{}^*$, which is discussed in the next subsection.

5.3. Comparison of the Maximum Momentary Input Energy and the Bidirectional Momentary Input Energy Spectrum

Figure 35 shows the prediction of $V_{\Delta E1U}{}^*$ from the bidirectional $V_{\Delta E}$ spectrum for Model-Tf34. The plots shown in this figure indicate the nonlinear time-history analysis results.

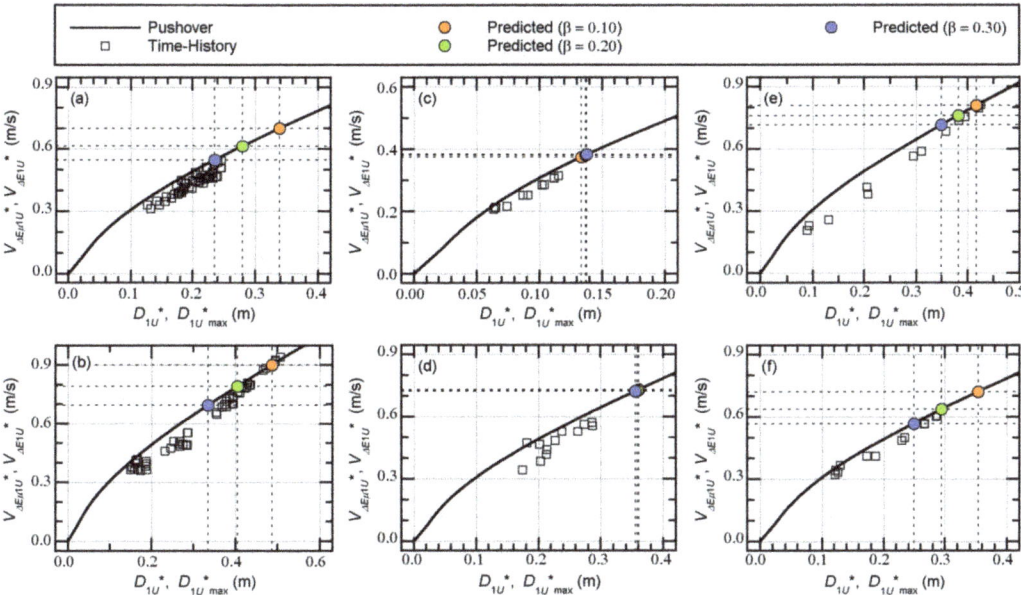

Figure 34. Comparison between the $V_{\Delta E\mu 1U}{}^*$–$D_{1U}{}^*$ curve and the $V_{\Delta E1U}{}^*$–$D_{1U}{}^*{}_{max}$ relationship obtained from the time-history analysis (Model-Tf44): (**a**) Art-1 series; (**b**) Art-2 series; (**c**) UTO0414; (**d**) UTO0416; (**e**) TCU; (**f**) YPT.

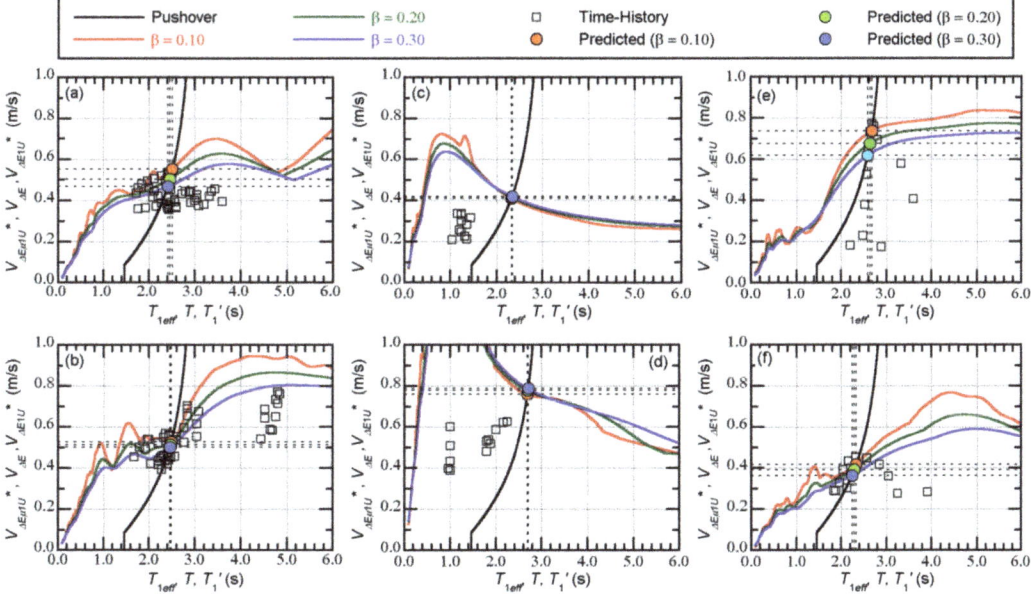

Figure 35. Prediction of $V_{\Delta E1U}{}^*$ from the bidirectional $V_{\Delta E}$ spectrum and its accuracy (Model-Tf34): (**a**) Art-1 series; (**b**) Art-2 series; (**c**) UTO0414; (**d**) UTO0416; (**e**) TCU; (**f**) YPT.

In most cases, the bidirectional $V_{\Delta E}$ spectrum with complex damping $\beta = 0.10$ approximated the upper bound of the plot of the time-history analysis results, as shown in Figure 35a,b,e,f. However, in the other cases shown in Figure 35c,d, the plots of the time-history analysis results were below those of the bidirectional $V_{\Delta E}$ spectrum.

From the comparisons between the predicted $V_{\Delta E1U}{}^*$ and that obtained from the time-history analysis results, the predicted $V_{\Delta E1U}{}^*$ provided a conservative estimation, except in the case of the Art-2 series shown in Figure 35b. This is because the predicted response points correspond to the "valley" of the bidirectional $V_{\Delta E}$ spectrum, therefore making the predicted $V_{\Delta E1U}{}^*$ smaller.

Figure 36 shows the prediction of $V_{\Delta E1U}{}^*$ from the bidirectional $V_{\Delta E}$ spectrum for Model-Tf44. Similar observations to those made for Model-Tf34 can be made for Model-Tf44.

Based on the above results, it can be concluded that that bidirectional $V_{\Delta E}$ spectrum can approximate the upper bound of the equivalent velocity of the maximum momentary input energy of the first mode $V_{\Delta E1U}{}^*$. As shown in Equation (25), $\Delta E_{1U}{}^*/M_{1U}{}^*$ was calculated from the ground acceleration component of U-axis, while the contribution of V-axis was none. Therefore, it is easily expected that the upper bound of the maximum momentary input energy of the first mode can be approximated by the bidirectional maximum momentary input energy. Comparisons between the unidirectional and bidirectional $V_{\Delta E}$ spectra can be found in Appendix B. Therefore, the upper bound of $V_{\Delta E1U}{}^*$ can be properly predicted using the bidirectional $V_{\Delta E}$ spectrum.

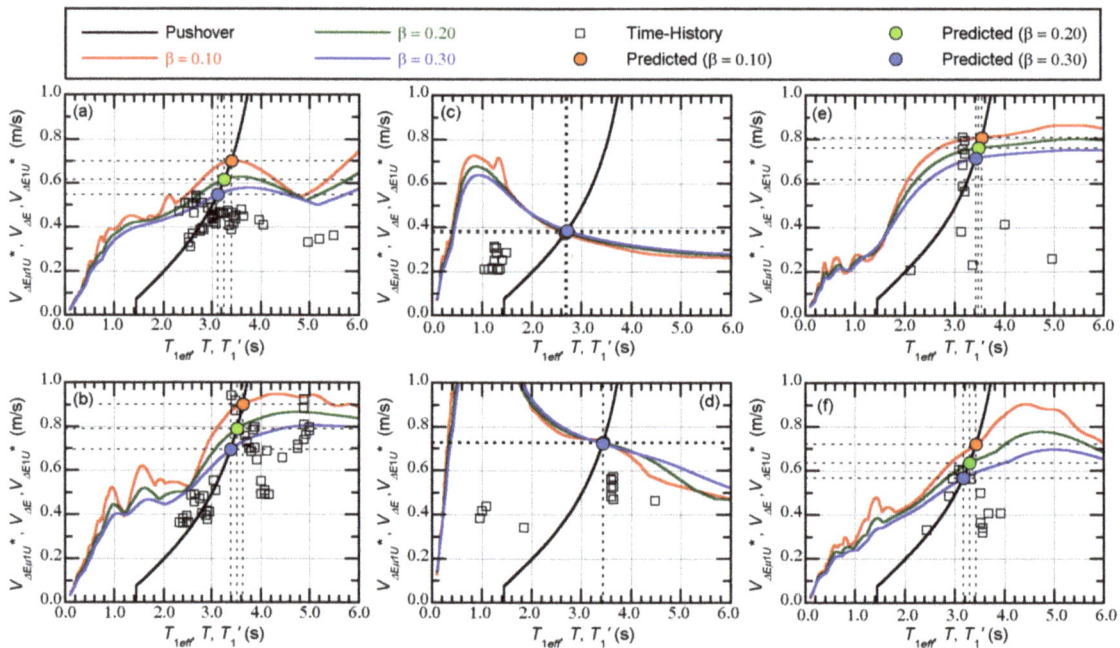

Figure 36. Prediction of $V_{\Delta E1U}{}^*$ from the bidirectional $V_{\Delta E}$ spectrum and its accuracy (Model-Tf44): (**a**) Art-1 series; (**b**) Art-2 series; (**c**) UTO0414; (**d**) UTO0416; (**e**) TCU; (**f**) YPT.

5.4. Accuracy of the Predicted Peak Equivalent Displacements of the First and Second Modal Responses

Figure 37 shows the accuracy of the predicted peak equivalent displacement of the first and second modal responses, $D_{1U}{}^*{}_{max}$ and $D_{2V}{}^*{}_{max}$, respectively, for Model-Tf34. The predicted peak in the case of complex damping $\beta = 0.10$ approximated the upper bound of the time-history analysis results, except when the ground motion dataset was the Art-2 series (Figure 37b) or YPT (Figure 37f).

Figure 38 shows the accuracy of the predicted $D_{1U}^*{}_{max}$ and $D_{2V}^*{}_{max}$ for Model-Tf44. The predicted peak in the case of $\beta = 0.10$ approximated the upper bound of the time-history analysis results in all cases.

It is also observed from Figures 37c–f and 38c–f that the largest $D_{1U}^*{}_{max}$ and $D_{2V}^*{}_{max}$ did not occur simultaneously. To understand this phenomena, Figures 39 and 40 show $D_{1U}^*{}_{max}$ and $D_{2V}^*{}_{max}$ for varying angles of incidence of the seismic input (ψ) for both models. As shown here, the angle where the largest $D_{2V}^*{}_{max}$ occurred was different from the angle where the largest $D_{1U}^*{}_{max}$ occurs, with the difference between the two angles being approximately 90°. This was clearly observed in the cases of TCU and YPT for both models.

Based on the above discussion, the upper bound of the peak equivalent displacements of the first and second modal responses, $D_{1U}^*{}_{max}$ and $D_{2V}^*{}_{max}$, can be predicted by the updated MABPA.

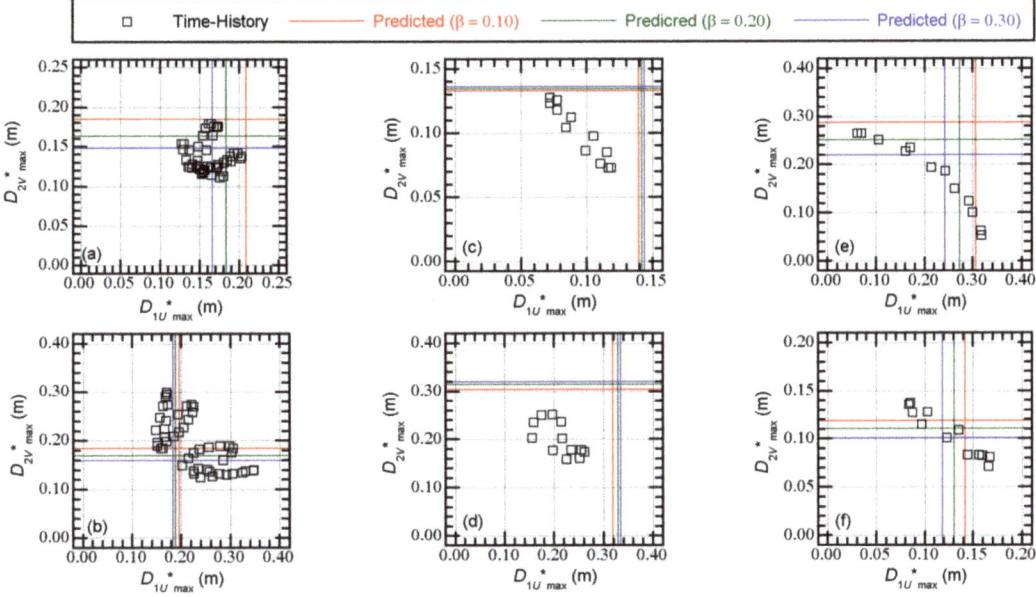

Figure 37. Accuracy of the predicted peak equivalent displacements of the first two modes (Model-Tf34): (**a**) Art-1 series; (**b**) Art-2 series; (**c**) UTO0414; (**d**) UTO0416; (**e**) TCU; (**f**) YPT.

The variation in $D_{1U}^*{}_{max}$ and $D_{2V}^*{}_{max}$ due to the angles of incidence of the seismic input ψ may explain the variation in the peak displacement of the isolatiors shown in Figures 29 and 30. As the example, the response of Model-Tf44 subjected to YPT ground motion was focused on. From Figure 40c, the angle (ψ) at which the largest $D_{1U}^*{}_{max}$ occurred was $-30°$. While the peak displacement at isolator $X_{1A}Y_6$ shown in Figure 30c, the peak displacement at the angle (ψ) of $-30°$ was close to the largest peak value. On the contrary, the peak displacement at isolator $X_{6A}Y_1$ at the angle (ψ) of $-30°$ was close to the smallest peak value.

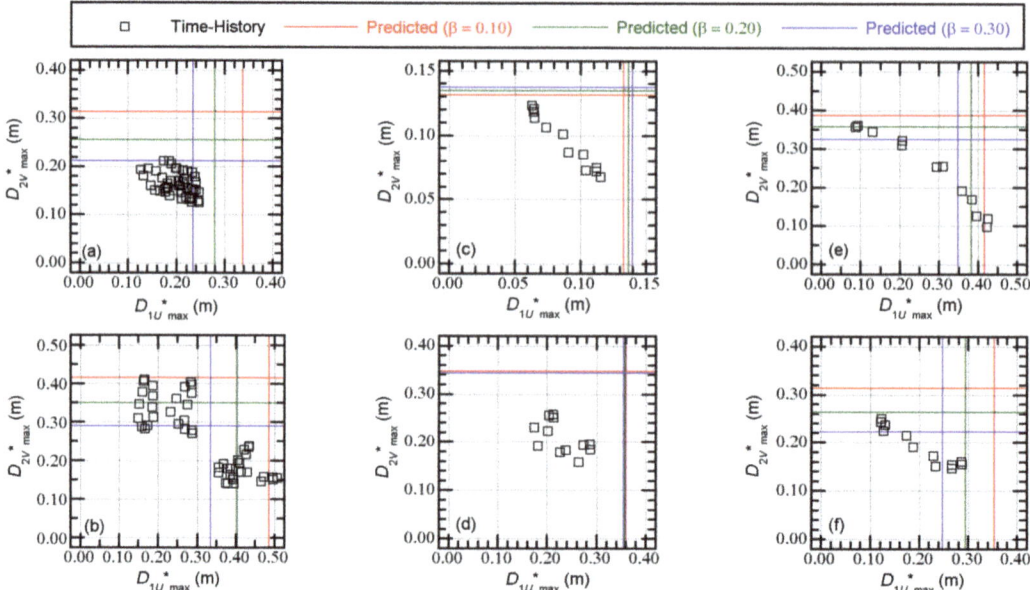

Figure 38. Accuracy of the predicted peak equivalent displacements of the first two modes (Model-Tf44): (**a**) Art-1 series; (**b**) Art-2 series; (**c**) UTO0414; (**d**) UTO0416; (**e**) TCU; (**f**) YPT.

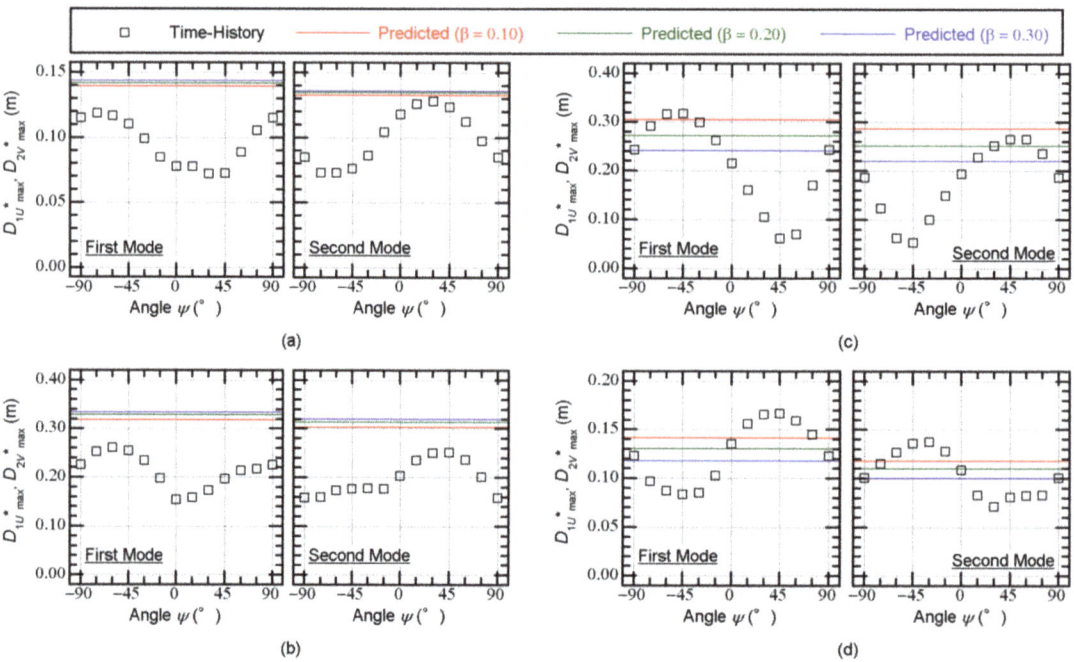

Figure 39. Peak equivalent displacements of the first two modes for various directions of seismic input (Model-Tf34): (**a**) UTO0414; (**b**) UTO0416; (**c**) TCU; (**d**) YPT.

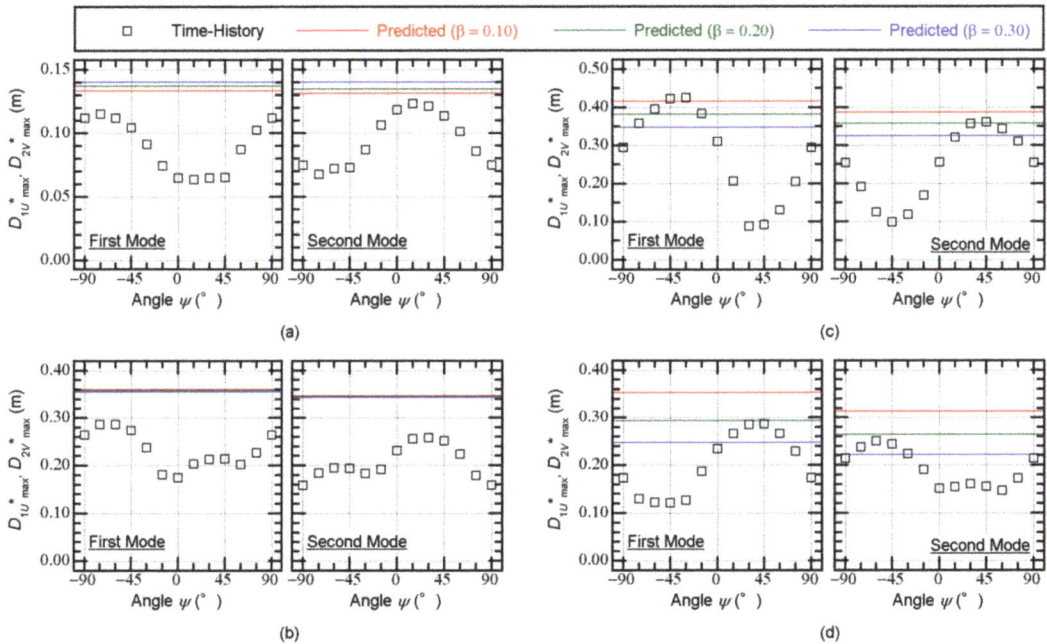

Figure 40. Peak equivalent displacements of the first two modes for various directions of seismic input (Model-Tf44): (**a**) UTO0414; (**b**) UTO0416; (**c**) TCU; (**d**) YPT.

5.5. Contribution of the Higher Mode to the Displacement Response at the Edge of Level 0

As discussed in Section 4.2.2, the accuracy of the predicted horizontal distribution of the peak response at level 0 depends on the ground motion dataset. According to Model-Tf34, the predicted largest peak displacement in the Y direction occurred at X_{1A} (the flexible side in the Y direction), while the envelope of the nonlinear time-history analysis results indicates that the largest peak occurred at X_{6A} (the stiff side in the Y direction), the opposite side to X_{1A} in the case of UTO0414, as shown in Figure 26a. Conversely, in the case of TCU shown in Figure 26b, the envelope of the time-history analysis results indicates that the largest peak occurred at X_{1A}, which is consistent with the predicted results. In this subsection, the modal response at level 0 is calculated and discussed.

From the time-history of the displacement at the center of mass of level 0, the horizontal displacement in the Y direction at point j on level 0 can be calculated as:

$$d_{Y0j}(t) = y_0(t) - L_{X0j}\theta_0(t). \tag{28}$$

In Equation (28), L_{X0j} is the location of point j in the X direction from the center of mass of level 0. Therefore, the modal response of the horizontal displacement at point j can be calculated as:

$$d_{Y0j1}(t) = \Gamma_{1U}\left(\phi_{Y01} - L_{X0j}\phi_{\Theta 01}\right)D_{1U}^*(t), \tag{29}$$

$$d_{Y0j2}(t) = \Gamma_{2V}\left(\phi_{Y02} - L_{X0j}\phi_{\Theta 02}\right)D_{2V}^*(t), \tag{30}$$

$$d_{Y0jh}(t) = d_{Y0j}(t) - \{d_{Y0j1}(t) + d_{Y0j2}(t)\}. \tag{31}$$

Figure 41 shows comparisons of the modal responses at the edge of level 0. The structural model shown in this figure is Model-Tf34, the input ground motion dataset is UTO0414, and the angle of incidence of the seismic input (ψ) is $-75°$, the angle at which the largest peak response at X_{1A} occurs. Note that "All modes" is the response originally obtained from the time-history analysis results ($d_{Y0j}(t)$)), "First mode" and "Second mode" are the first and second modal responses ($d_{Y0j1}(t)$ and $d_{Y0j2}(t)$, respectively), and "Higher mode" is the higher (residual) modal response calculated from Equation (31) ($d_{Y0jh}(t)$).

In the response of X_{1A}, shown in Figure 41a, the contribution of the higher modal response was non-negligible, even though the contribution of the first modal response was predominant. In addition, the sign of the higher modal response at the time the peak response occurred at X_{1A} was opposite to that of the "All mode" response, with the contribution of the higher mode reducing the peak response at X_{1A}.

In the response of X_{6A}, shown in Figure 41b, the contribution of the first modal response was negligibly small and those of the second and higher modal responses were noticeable. In addition, the sign of the higher modal response at the time the peak response occurred at X_{6A} was the same as that of the "All mode" response, with the contribution of the higher mode increasing the peak response at X_{6A}.

This indicates that the reason why the largest peak response in the envelope of the time-history analysis results occurred at X_{6A} (*not* at X_{1A}) in the case of UTO0414 can be explained by the contributions of the higher modal response. In the case of UTO0414, the contribution of the higher modal response was non-negligibly large.

Another comparison is made in Figure 42 for the structural model Model-Tf34, the input ground motion dataset TCU, and an angle of incidence of the seismic input (ψ) of 60°, where the angle of the largest peak response at X_{1A} occurs. In the response of X_{1A}, shown in Figure 42a, the contribution of the first modal response was predominant, while that of the higher modal response was small. Meanwhile, in the response of X_{6A}, shown in Figure 42b, the contribution of the first modal response was negligibly small and those of the second and higher modal responses were noticeable.

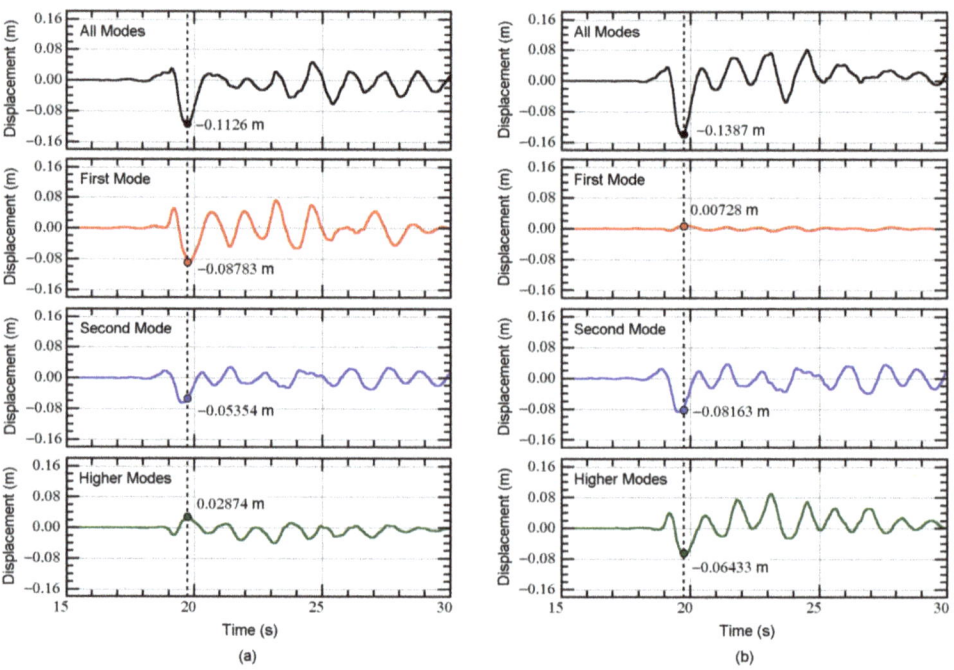

Figure 41. Comparisons of the modal responses at the edge of level 0 (structural model: Model-Tf34, ground motion: UTO0414, angle of incidence of seismic input: $\psi = -75°$): (**a**) X_{1A} and (**b**) X_{6A}.

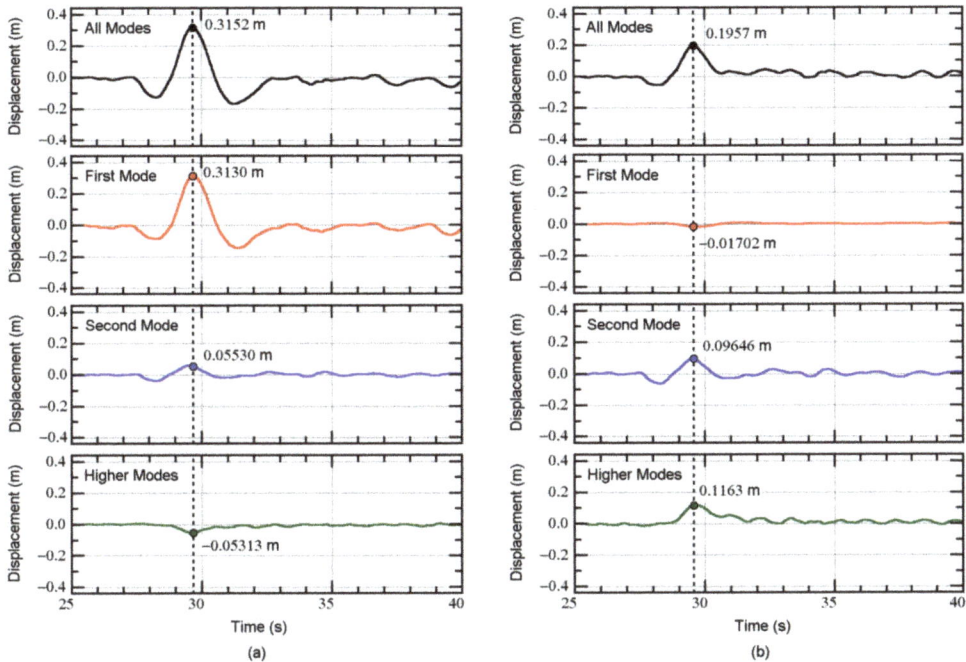

Figure 42. Comparisons of the modal responses at the edge of level 0 (structural model: Model-Tf34, ground motion: TCU, angle of incidence of seismic input: $\psi = 60°$): (**a**) X_{1A} and (**b**) X_{6A}.

Therefore, the accuracy of the predicted horizontal distribution of the peak response at level 0 relies on the contribution of the higher modal response. The envelope of the time-history analysis results was notably different from the predicted results in the case of UTO0414, because the contribution of the higher modal response was significant. Meanwhile, the predicted results were close to the envelope of the time-history analysis results in the case of TCU because the contribution of the higher modal response was small. To confirm this, Table 6 lists the equivalent velocities of the maximum momentary input energy predicted from the bidirectional $V_{\Delta E}$ spectrum (complex damping ratio (β) of 0.10) for the two cases. In this table, the values of the first and second modal responses are those predicted using MABPA, while the value of the third mode was predicted assuming that the effective period (T_{3eff}) equals the natural period in the elastic range (T_{3e}). This table confirms that the contribution of the third mode may be noticeable in the case of UTO0414, while it may be small in the case of TCU.

Table 6. Equivalent velocities of the maximum momentary input energy predicted from the bidirectional $V_{\Delta E}$ spectrum ($\beta = 0.100$).

Ground Motion Set	First Mode $V_{\Delta E}(T_{1eff})$ (m/s)	Second Mode $V_{\Delta E}(T_{2eff})$ (m/s)	Third Mode $V_{\Delta E}(T_{3e})$ (m/s)	Ratio (2nd/1st)	Ratio (3rd/1st)
UTO0414	0.4108	0.4232	0.7151	1.030	1.741
TCU	0.7575	0.7319	0.2519	0.9662	0.3325

5.6. Summary of the Discussions

In this section, the discussions are focused on the four points that are important to discuss the accuracy of MABPA. The summary of the discussions are as follows.

The relationship between the maximum momentary input energy and the peak equivalent displacement of the first modal response was discussed in Section 5.2. The results confirmed that the $V_{\Delta E\mu 1U}{}^*$–$D_{1U}{}^*$ curve obtained from pushover analysis results fit the plots obtained from the time-history analysis results very well. This is because the contribution of the first modal response to the whole response may be large in both Models Tf34 and Tf44.

Then, the predictability of the maximum momentary input energy of the first modal response was discussed in Section 5.3. The results confirmed that bidirectional $V_{\Delta E}$ spectrum can approximate the upper bound of the equivalent velocity of the maximum momentary input energy of the first mode $V_{\Delta E1U}{}^*$. This is because that the maximum momentary input energy of the first mode was calculated from the unidirectional input in U-axis, while the effect of simultaneous bidirectional input was included in the calculation of bidirectional $V_{\Delta E}$ spectrum automatically.

Since (a) the $V_{\Delta E\mu 1U}{}^*$–$D_{1U}{}^*$ curve can be properly predicted from the pushover analysis results, and (b) the upper bound of the equivalent velocity of the maximum momentary input energy can be predicted via the bidirectional $V_{\Delta E}$ spectrum, the largest peak equivalent displacement should be predicted accurately. The accuracy of the predicted equivalent displacements of the first and second modal responses was discussed in Section 5.4. The results confirmed that the upper bound of the peak equivalent displacement of the first and second modal responses can be predicted accurately.

Although the upper bound of the peak equivalent displacement of the first and second modal responses can be predicted accurately, the accuracy of the predicted horizontal distributions of the peak displacement at level 0 depends on the ground motion dataset. Section 5.5 discussed the contribution of the higher mode to the displacement response at the edge of level 0. The results showed that higher modal responses may not be negligible for the prediction of the peak displacement at the edge of level 0. In the MABPA prediction, only the contributions of the first and second modal responses were considered. Therefore, a discrepancy of the predicted results from the time-history analysis may occur because of the lack of a contribution from the higher modal responses.

6. Conclusions

In this study, the main building of the former Uto City Hall was investigated as a case study of the retrofitting of an irregular reinforced concrete building using the base-isolation technique. The nonlinear peak responses of two retrofitted building models subjected to horizontal bidirectional ground motions were predicted by MABPA, and the accuracy of the method was evaluated. The main conclusions and results are as follows:

- The predicted peak response according to the updated MABPA agreed satisfactorily with the envelope of the time-history analysis results. The peak relative displacement at $X_{3A}Y_3$ at each floor can be satisfactorily predicted. The predicted distribution of the peak displacement at level 0 (just above the isolation layer) approximated the envelope of the nonlinear time-history analysis results, even though in some cases, the predicted distributions differed from the envelope of the nonlinear time-history analysis. A discrepancy between the predicted results and nonlinear time-history analysis may occur because of the lack of a contribution from the higher modal responses.
- The relationship between the equivalent velocity of the maximum momentary input energy of the first modal response ($V_{\Delta E1U}{}^*$) and the peak equivalent displacement of the first modal response ($D_{1U}{}^*{}_{max}$) can be properly evaluated from the pushover analysis results. The plots obtained from the nonlinear time-history analysis results fit the evaluated curve from the pushover analysis results well.
- The upper bound of the peak equivalent displacements of the first two modal responses can be predicted using the bidirectional $V_{\Delta E}$ spectrum [52]. Comparisons between the predicted peak equivalent displacements and those calculated from the nonlinear time-history analysis results showed that the predicted peak approximated

the upper bound of the nonlinear time-history analysis results. The upper bound of $V_{\Delta E1U}{}^*$ can be approximated by the bidirectional $V_{\Delta E}$ spectrum.

Based on the above findings, the updated MABPA appears to predict the peak responses of irregular base-isolated buildings with accuracy. However, MABPA still has two shortcomings. The first shortcoming is the limitation of the applicability of MABPA. As discussed in a previous study [43], the application of MABPA is limited to buildings classified as torsionally stiff buildings. The current (updated) MABPA has the same restriction. This limitation can be avoided if the torsional resistance of the isolation layer is sufficiently provided, as shown in this study. The second shortcoming involves the contributions of the higher modal responses. In the original MABPA for non-isolated buildings, only the first two modes were considered for the prediction. Therefore, the prediction was less accurate for cases when the response in the stiff-side perimeter was larger than that in the flexible-side perimeter. The contributions of the third and higher modal responses need to be investigated.

Another aspect of this study to be emphasized is the application of the bidirectional $V_{\Delta E}$ spectrum for the prediction of the peak response of a base-isolated building. The results shown in this study imply that the bidirectional $V_{\Delta E}$ spectrum [52] is a promising candidate for a seismic intensity parameter for the peak response. As discussed in a previous study [53], one of the biggest advantages of the bidirectional momentary input energy is that it can be directly calculated from the Fourier amplitude and phase angle of the ground motion components using a time-varying function of the momentary energy input, without knowing the time-history of the ground motion. This means that researchers can eliminate otherwise unavoidable fluctuations from the nonlinear time-history analysis results. Therefore, the pushover analysis and the bidirectional $V_{\Delta E}$ spectrum are an optimal combination to understand the fundamental characteristics of both base-isolated and non-isolated asymmetric buildings.

The optimal distribution of the hysteresis dampers according to the design of the isolation layer needed to minimize the torsional response was not discussed in this study. However, the updated MABPA presented here can help in the optimization of the damper distribution. The next update of MABPA for base-isolated irregular buildings with other kinds of dampers (e.g., linear and nonlinear oil dampers, viscous mass dampers, and other kings of "smart passive dampers" [19]) is also an important issue that will be investigated in subsequent studies.

Author Contributions: Conceptualization, K.F. and T.M.; data curation, K.F. and T.M.; formal analysis, K.F. and T.M.; funding acquisition, K.F.; investigation, K.F. and T.M.; methodology, K.F.; project administration, K.F.; resources, K.F. and T.M.; software, K.F.; supervision, K.F.; validation, K.F.; visualization, K.F.; writing—original draft, K.F.; writing—review and editing, K.F. All authors have read and agreed to the published version of the manuscript.

Funding: This research received no external funding.

Institutional Review Board Statement: Not applicable.

Informed Consent Statement: Not applicable.

Data Availability Statement: The data presented in this study are available on request from the corresponding author. The data are not publicly available because they are not part of ongoing research.

Acknowledgments: The ground motions used in this study were obtained from the website of the National Research Institute for Earth Science and Disaster Resilience (NIED) (http://www.kyoshin.bosai.go.jp/kyoshin/, last accessed on 14 December 2019) and the Pacific Earthquake Engineering Research Center (PEER) (https://ngawest2.berkeley.edu/, last accessed on 14 December 2019). The contributions during the beginning stage of this study made by Ami Obikata, a former undergraduate student at the Chiba Institute of Technology, are greatly appreciated.

Conflicts of Interest: The authors declare no conflict of interest.

Appendix A. Time-Histories of the Recorded Ground Motions Used in This Study

Table A1 shows the date of event, magnitude (Meteorological Agency Magnitude M_J, or moment magnitude M_W), location of the epicenter, distance, and station name of each record. Figures A1–A4 show the time-histories and orbits of the original ground motion records.

Table A1. Event date, magnitude, location of the epicenter, distance, and station name of each record.

ID	Event Date	Magnitude	Distance	Station Name	Direction of Components	
					ξ-Dir	ζ-Dir
UTO0414	14 April 2016	$M_J = 6.5$	15 km	K-Net UTO (KMM008)	EW	NS
UTO0416	16 April 2016	$M_J = 7.3$	12 km	K-Net UTO (KMM008)	EW	NS
TCU	20 September 1999	$M_W = 7.6$	0.89 km *	TCU075	Major **	Minor **
YPT	17 August 1999	$M_W = 7.5$	4.83 km *	Yarimca	Major **	Minor **

* This distance is the closest distance from the rupture plane defined in the Pacific Earthquake Engineering Research Center (PEER) database, while those from the Japanese database are the epicentral distances. ** Horizontal major and minor axes were determined following the works of Arias [63] and Penzien and Watabe [64].

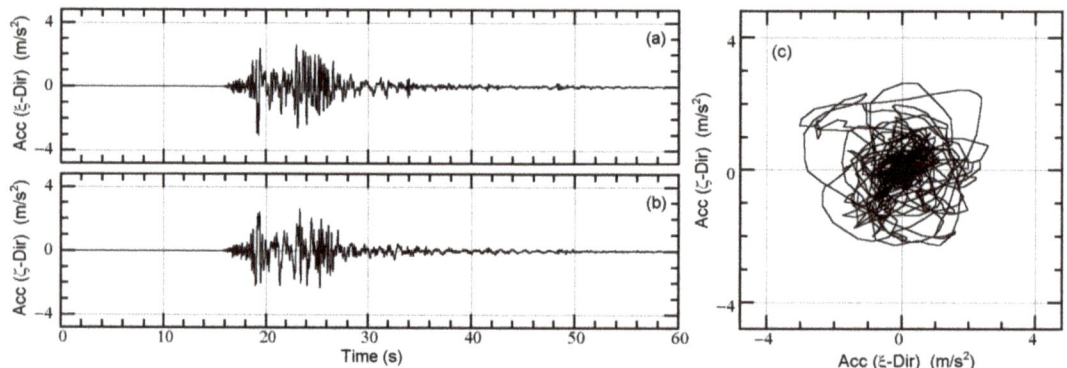

Figure A1. Two components of the recorded ground motion (UTO0414): (**a**) ξ direction; (**b**) ζ direction; (**c**) orbit.

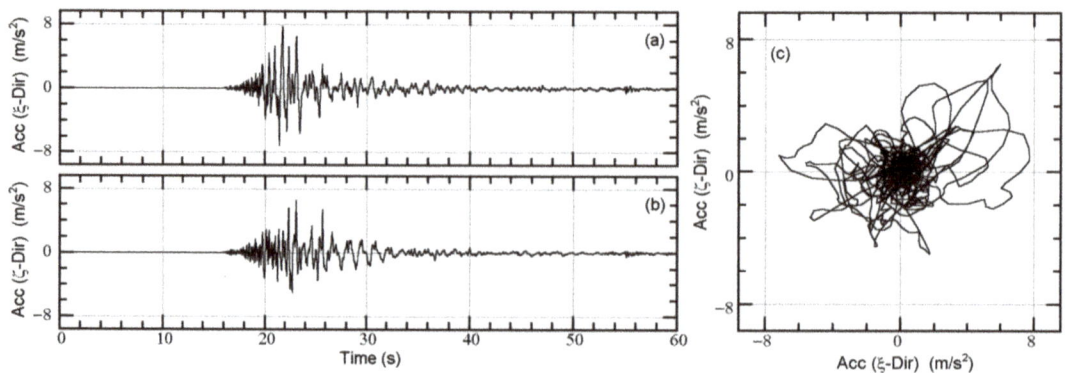

Figure A2. Two components of the recorded ground motion (UTO0416): (**a**) ξ direction; (**b**) ζ direction; (**c**) orbit.

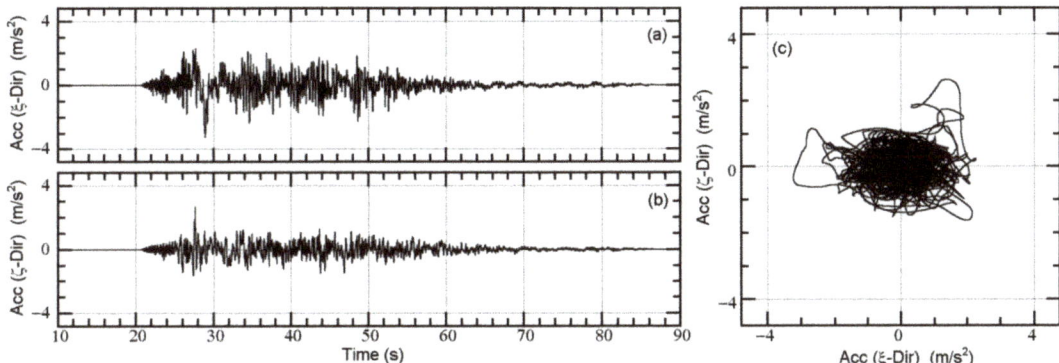

Figure A3. Two components of the recorded ground motion (TCU): (**a**) ξ direction; (**b**) ζ direction; (**c**) orbit.

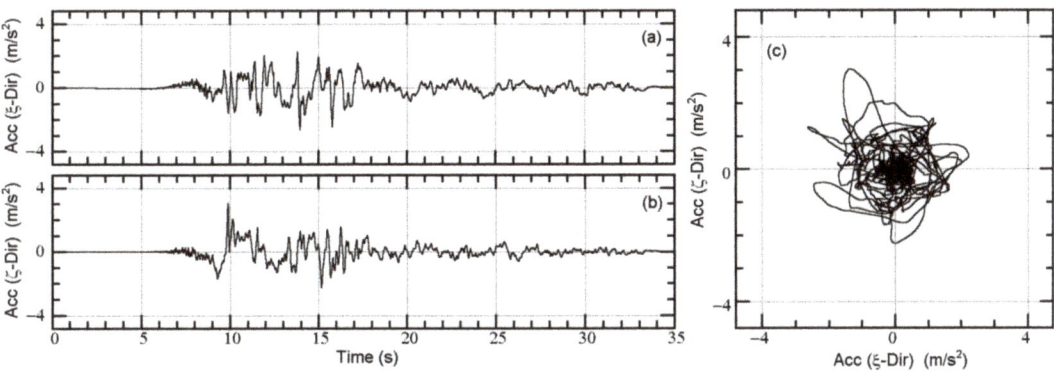

Figure A4. Two components of the recorded ground motion (YPT): (**a**) ξ direction; (**b**) ζ direction; (**c**) orbit.

Appendix B. Comparisons of the Unidirectional and Bidirectional $V_{\Delta E}$ Spectra

Figure A5 shows the comparisons of the unidirectional and bidirectional $V_{\Delta E}$ spectra for Art-1 and Art-2 series. In this figure, the unidirectional $V_{\Delta E}$ spectrum was calculated as the maximum obtained from the linear time-history analysis using all possible rotations between 0° and 360° degrees with intervals of 5°. Viscous damping (damping ratio 0.10) was considered, while the bidirectional $V_{\Delta E}$ spectrum was calculated using the time-varying function described in Section 2.2.1. Two damping models, complex and viscous damping (damping ratio 0.10), were considered. Note again that the bidirectional $V_{\Delta E}$ spectrum calculated from time-varying function is independent of phase-shift. As shown in this figure, the bidirectional $V_{\Delta E}$ spectrum approximates the maximum of the unidirectional $V_{\Delta E}$ spectrum.

Figure A6 shows comparisons of the unidirectional and bidirectional $V_{\Delta E}$ spectra for recorded ground motion datasets. In this figure, the maximum, minimum, and medium of the set of geometrical means obtained using all possible rotations between 0° and 90° (GMRotD50) defined by Boore et al. [65] of unidirectional $V_{\Delta E}$ spectra are shown. As shown in this figure, the bidirectional $V_{\Delta E}$ spectrum approximates the maximum of the unidirectional $V_{\Delta E}$ spectrum.

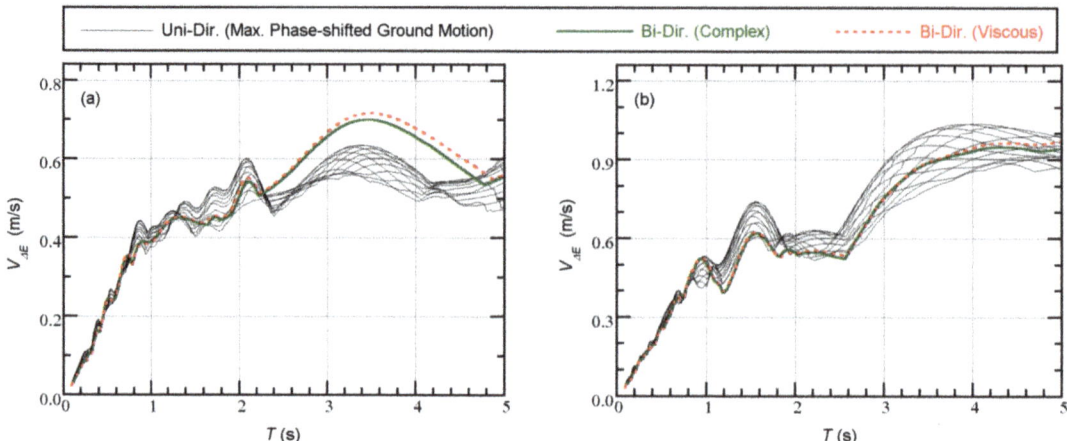

Figure A5. Comparisons of the unidirectional and bidirectional $V_{\Delta E}$ spectra for artificial ground motion datasets: (**a**) Art-1 series and (**b**) Art-2 series.

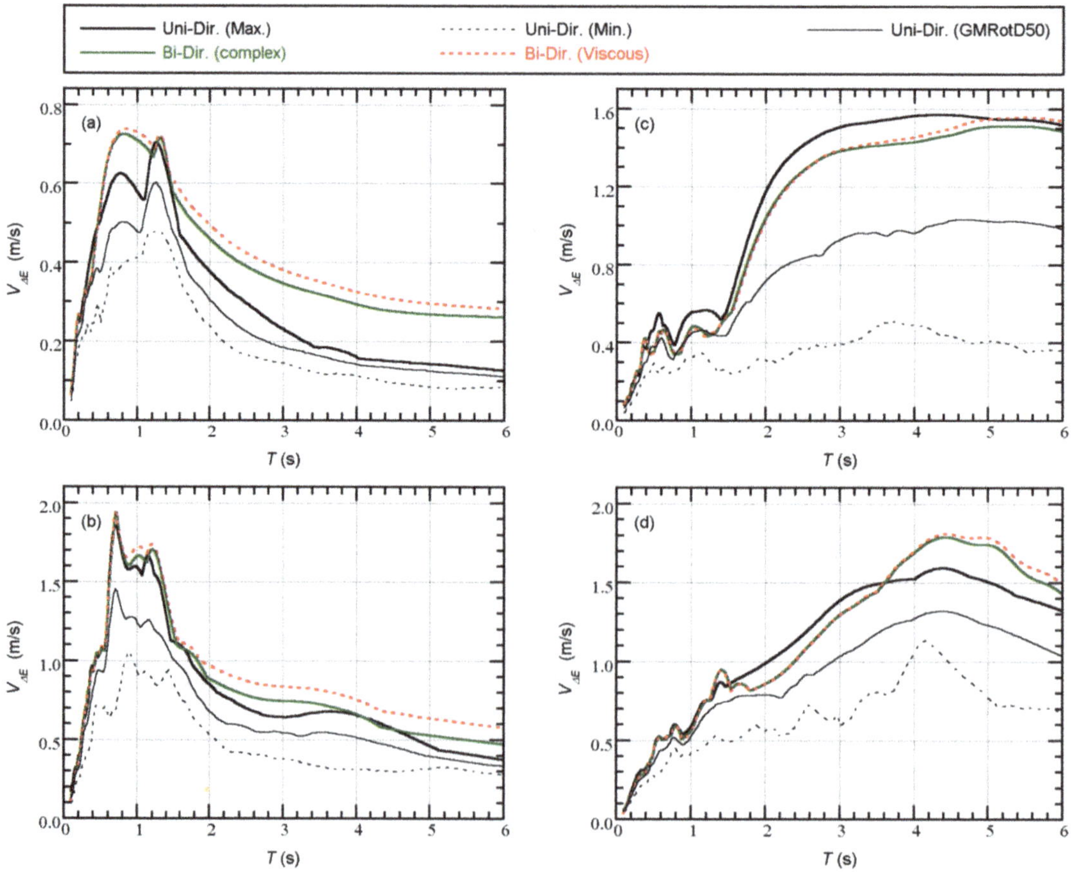

Figure A6. Comparisons of the unidirectional and bidirectional $V_{\Delta E}$ spectra for recorded ground motion datasets: (**a**) UTO0414; (**b**) UTO0416; (**c**) TCU; (**d**) YPT.

References

1. Charleson, A.; Guisasola, A. *Seismic Isolation for Architects*; Routledge: London, UK; New York, NY, USA, 2017.
2. Architectural Institute of JAPAN (AIJ). *Design Recommendations for Seismically Isolated Buildings*; Architectural Institute of Japan: Tokyo, Japan, 2016.
3. ASCE. *Seismic Evaluation and Retrofit of Existing Buildings*; ASCE Standard, ASCE/SEI41-17; American Society of Civil Engineering: Reston, VA, USA, 2017.
4. Seki, M.; Miyazaki, M.; Tsuneki, Y.; Kataoka, K. A Masonry school building retrofitted by base isolation technology. In Proceedings of the 12th World Conference on Earthquake Engineering, Auckland, New Zealand, 30 January–4 February 2000.
5. Kashima, T.; Koyama, S.; Iiba, M.; Okawa, I. Dynamic behaviour of a museum building retrofitted using base isolation system. In Proceedings of the 14th World Conference on Earthquake Engineering, Beijing, China, 12–17 October 2008.
6. Clemente, P.; De Stefano, A. *Application of seismic isolation in the retrofit of historical buildings. Earthquake Resistant Engineering Structures VIII*; WIT Press: Southampton, UK, 2011; Volume 120, pp. 41–52.
7. Gilani, A.S.; Miyamoto, H.K. Base isolation retrofit challenges in a historical monumental building in Romania. In Proceedings of the 15th World Conference on Earthquake Engineering, Lisbon, Portugal, 24–28 September 2012.
8. Nakamura, H.; Ninomiya, T.; Sakaguchi, T.; Nakano, Y.; Konagai, K.; Hisada, Y.; Seki, M.; Ota, T.; Yamazaki, Y.; Mochizuki, S.; et al. Seismic isolation retrofit of Susano City Hall situated above lava tubes. In Proceedings of the 15th World Conference on Earthquake Engineering, Lisbon, Portugal, 24–28 September 2012.
9. Sorace, S.; Terenzi, G. A viable base isolation strategy for the advanced seismic retrofit of an r/c building. *Contemp. Eng. Sci.* **2014**, *7*, 817–834. [CrossRef]
10. Ferraioli, M.; Mandara, A. Base isolation for seismic retrofitting of a multiple building structure: Evaluation of equivalent linearization method. *Math. Probl. Eng.* **2016**, *2016*, 8934196. [CrossRef]
11. Terenzi, G.; Fuso, E.; Sorace, S.; Costoli, I. Enhanced seismic retrofit of a reinforced concrete building of architectural interest. *Buildings* **2020**, *10*, 211. [CrossRef]
12. Vailati, M.; Monti, G.; Bianco, V. Integrated solution-base isolation and repositioning-for the seismic rehabilitation of a preserved strategic building. *Buildings* **2021**, *11*, 164. [CrossRef]
13. Usta, P. Investigation of a base-isolator system's effects on the seismic behavior of a historical structure. *Buildings* **2021**, *11*, 217. [CrossRef]
14. Nishizawa, T.; Suekuni, R. Outline of structural design for Kyoto City Hall. In Proceedings of the 17th World Conference on Earthquake Engineering, Sendai, Japan, 20 September–2 November 2021.
15. Hashimoto, T.; Fujita, K.; Tsuji, M.; Takewaki, I. Innovative base-isolated building with large mass-ratio TMD at basement for greater earthquake resilience. *Future Cities Environ.* **2015**, *1*, 9. [CrossRef]
16. Ikenaga, M.; Ikago, K.; Inoue, N. Development of a displacement-dependent damper for base isolated structures. In Proceedings of the Tenth International Conference on Computational Structures Technology, Valencia, Spain, 14–17 September 2010.
17. Ikenaga, M.; Ikago, K.; Inoue, N. Seismic displacement control design of base isolated structures by magneto-rheological dampers based on a pseudo-complex-damping rule. In Proceedings of the Thirteenth International Conference on Computational Structures Technology, Crete, Greece, 6–9 September 2011.
18. Nakaminami, S.; Ikago, K.; Inoue, N.; Kida, H. Response characteristics of a base-isolated structure incorporated with a force-restricted viscous mass damper. In Proceedings of the 15th World Conference on Earthquake Engineering, Lisbon, Portugal, 24–28 September 2012.
19. Inoue, N.; Ikago, K. Displacement control design concept for long-period structures. In Proceedings of the 16th World Conference on Earthquake Engineering, Santiago, Chile, 9–13 January 2017.
20. Anajafi, H.; Medina, R.A. Comparison of the seismic performance of a partial mass isolation technique with conventional TMD and base-isolation systems under broad-band and narrow-band excitations. *Eng. Struct.* **2018**, *158*, 110–123. [CrossRef]
21. Luo, H.; Ikago, K.; Chong, C.; Keivan, A.; Phillips, B.M. Performance of low-frequency structures incorporated with rate-independent linear damping. *Eng. Struct.* **2019**, *181*, 324–335. [CrossRef]
22. Tariq, M.A.; Usman, M.; Farooq, S.H.; Ullah, I.; Hanif, A. Investigation of the structural response of the MRE-based MDOF isolated structure under historic near- and far-fault earthquake loadings. *Appl. Sci.* **2021**, *11*, 2876. [CrossRef]
23. Nagarajaiah, S.; Reinhorn, A.M.; Constantinou, M.C. Torsion in base-isolated structures with elastomeric isolation systems. *J. Struct. Eng. ASCE* **1992**, *119*, 2932–2951. [CrossRef]
24. Tena-Colunga, A.; Gómez-Soberón, L. Torsional response of base-isolated structures due to asymmetries in the superstructure. *Eng. Struct.* **2002**, *24*, 1587–1599. [CrossRef]
25. Tena-Colunga, A.; Zambrana-Rojas, C. Dynamic torsional amplifications of base-isolated structures with an eccentric isolation system. *Eng. Struct.* **2006**, *28*, 72–83. [CrossRef]
26. Tena-Colunga, A.; Escamilla-Cruz, J.L. Torsional amplifications in asymmetric base-isolated structures. *Eng. Struct.* **2008**, *29*, 237–247. [CrossRef]
27. Seguín, C.E.; De La Llera, J.C.; Almazán, J.L. Base–structure interaction of linearly isolated structures with lateral–torsional coupling. *Eng. Struct.* **2008**, *30*, 110–125. [CrossRef]
28. Di Sarno, L.; Chioccarelli, E.; Cosenza, E. Seismic response analysis of an irregular base isolated building. *Bull. Earthq. Eng.* **2011**, *9*, 1673–1702. [CrossRef]

29. Mazza, F.; Mazza, M. Nonlinear seismic analysis of irregular r.c. framed buildings base-isolated with friction pendulum system under near-fault excitations. *Soil Dyn. Earthq. Eng.* **2016**, *90*, 299–312. [CrossRef]
30. Cancellara, D.; De Angelis, F. Assessment and dynamic nonlinear analysis of different base isolation systems for a multi-storey RC building irregular in plan. *Comput. Struct.* **2017**, *180*, 74–88. [CrossRef]
31. Mazza, F. Seismic demand of base-isolated irregular structures subjected to pulse-type earthquakes. *Soil Dyn. Earthq. Eng.* **2018**, *108*, 111–129. [CrossRef]
32. Volcev, R.; Postolov, N.; Todorov, K.; Lazarov, L. Base isolation as an effective tool for plan irregularity reduction. In *Seismic Behaviour and Design of Irregular and Complex Civil Structures III*; Köber, D., De Stefano, M., Zembaty, Z., Eds.; Springer Nature Switzerland: Cham, Switzerland, 2020; pp. 377–389.
33. Köber, D.; Semrau, P.; Weber, F. Design approach for an irregular hospital building in Bucharest. In Proceedings of the 9th European Workshop on the Seismic Behaviour of Irregular and Complex Structures, Online, 15–16 December 2020.
34. Reyes, J.C. Seismic behaviour of torsionally-weak buildings with and without base isolators. In Proceedings of the 9th European Workshop on the Seismic Behaviour of Irregular and Complex Structures, Online, 15–16 December 2020.
35. Koren, D.; Kilar, V. Seismic behaviour of asymmetric base isolated structures with various distributions of isolators. *Eng. Struct.* **2009**, *31*, 910–921.
36. Koren, D.; Kilar, V. The applicability of the N2 method to the estimation of torsional effects in asymmetric base-isolated buildings. *Earthq. Eng. Struct. Dyn.* **2011**, *40*, 867–886. [CrossRef]
37. Peruš, I.; Fajfar, P. On the inelastic torsional response of single-storey structures under bi-axial excitation. *Earthq. Eng. Struct. Dyn.* **2005**, *34*, 931–941. [CrossRef]
38. Fajfar, P.; Marušić, D.; Peruš, I. Torsional effects in the pushover-based seismic analysis of buildings. *J. Earthq. Eng.* **2005**, *9*, 831–854. [CrossRef]
39. Fujii, K. Nonlinear static procedure for multi-story asymmetric frame buildings considering bi-directional excitation. *J. Earthq. Eng.* **2011**, *15*, 245–273. [CrossRef]
40. Fujii, K. Prediction of the largest peak nonlinear seismic response of asymmetric buildings under bi-directional excitation using pushover analyses. *Bull. Earthq. Eng.* **2014**, *12*, 909–938. [CrossRef]
41. Fujii, K. Application of the pushover-based procedure to predict the largest peak response of asymmetric buildings with buckling-restrained braces. In Proceedings of the 5th ECCOMAS Thematic Conference on Computational Methods in Structural Dynamics and Earthquake Engineering (COMPDYN), Crete Island, Greece, 25–27 May 2015.
42. Fujii, K. Assessment of pushover-based method to a building with bidirectional setback. *Earthq. Struct.* **2016**, *11*, 421–443. [CrossRef]
43. Fujii, K. Prediction of the peak seismic response of asymmetric buildings under bidirectional horizontal ground motion using equivalent SDOF model. *Jpn. Archit. Rev.* **2018**, *1*, 29–43. [CrossRef]
44. Fujii, K. Pushover-based seismic capacity evaluation of Uto City Hall damaged by the 2016 Kumamoto Earthquake. *Buildings* **2019**, *9*, 140. [CrossRef]
45. Building Center of Japan (BCJ). *The Building Standard Law of Japan on CD-ROM*; The Building Center of Japan: Tokyo, Japan, 2016.
46. Güneş, N.; Ulucan, Z.Ç. Nonlinear dynamic response of a tall building to near-fault pulse-like ground motions. *Bull. Earthq. Eng.* **2019**, *17*, 2989–3013. [CrossRef]
47. Akiyama, H. *Earthquake–Resistant Limit–State Design for Buildings*; University of Tokyo Press: Tokyo, Japan, 1985.
48. Nakamura, T.; Hori, N.; Inoue, N. Evaluation of damaging properties of ground motions and estimation of maximum displacement based on momentary input energy. *J. Struct. Constr. Eng. AIJ* **1998**, *513*, 65–72. (In Japanese) [CrossRef]
49. Inoue, N.; Wenliuhan, H.; Kanno, H.; Hori, N.; Ogawa, J. Shaking Table Tests of Reinforced Concrete Columns Subjected to Simulated Input Motions with Different Time Durations. In Proceedings of the 12th World Conference on Earthquake Engineering, Auckland, New Zealand, 30 January–4 February 2000.
50. Hori, N.; Inoue, N. Damaging properties of ground motion and prediction of maximum response of structures based on momentary energy input. *Earthq. Eng. Struct. Dyn.* **2002**, *31*, 1657–1679. [CrossRef]
51. Fujii, K.; Kanno, H.; Nishida, T. Formulation of the time-varying function of momentary energy input to a SDOF system by Fourier series. *J. Jpn. Assoc. Earthq. Eng.* **2019**, *19*, 247–266. (In Japanese) [CrossRef]
52. Fujii, K.; Murakami, Y. Bidirectional momentary energy input to a one-mass two-DOF system. In Proceedings of the 17th World Conference on Earthquake Engineering, Sendai, Japan, 20 September–2 November 2021.
53. Fujii, K. Bidirectional seismic energy input to an isotropic nonlinear one-mass two-degree-of-freedom system. *Buildings* **2021**, *11*, 143. [CrossRef]
54. Strong-Motion Seismograph Networks (K-NET, KiK-net). Available online: https://www.kyoshin.bosai.go.jp/ (accessed on 14 October 2021).
55. Bridgestone Corporation. Seismic Isolation Product Line-Up. Version 2017. Volume 1. Available online: https://www.bridgestone.com/products/diversified/antiseismic_rubber/pdf/catalog_201710.pdf (accessed on 7 September 2021).
56. Bridgestone Corporation. Kenchiku-Menshin-yo Sekisou-Gomu-Seihin Shiyou-Ichiran (Seismic Isolation Product Line-Up). Version 2021. Volume 1. Available online: https://www.bridgestone.co.jp/products/dp/antiseismic_rubber/product/pdf/product_catalog_202106.pdf (accessed on 7 September 2021).

57. Nippon Steel Engineering Co. Ltd. Men-Shin NSU Damper Line-Up. Available online: https://www.eng.nipponsteel.com/steelstructures/product/base_isolation/damper_u/lineup_du/ (accessed on 3 August 2019). (In Japanese)
58. Fujii, K. Prediction of the Maximum Seismic Member Force in a Superstructure of a Base-Isolated Frame Building by Using Pushover Analysis. *Buildings* **2019**, *9*, 201. [CrossRef]
59. Wada, A.; Hirose, K. Elasto-plastic dynamic behaviors of the building frames subjected to bi-directional earthquake motions. *J. Struct. Constr. Eng. AIJ* **1989**, *399*, 37–47. (In Japanese)
60. Almansa, F.L.; Weng, D.; Li, T.; Alfarah, B. Suitability of seismic isolation for buildings founded on soft soil. Case study of a RC building in Shanghai. *Buildings* **2020**, *10*, 241. [CrossRef]
61. Ordaz, M.; Huerta, B.; Reinoso, E. Exact Computation of Input-Energy Spectra from Fourier Amplitude Spectra. *Earthq. Eng. Struct. Dyn.* **2003**, *32*, 597–605. [CrossRef]
62. Kuramoto, H. Earthquake response characteristics of equivalent SDOF system reduced from multi-story buildings and prediction of higher mode responses. *J. Struct. Constr. Eng. AIJ* **2004**, *580*, 61–68. (In Japanese) [CrossRef]
63. Arias, A. A measure of seismic intensity. In *Seismic Design for Nuclear Power Plant*; MIT Press: Cambridge, MA, USA, 1970; pp. 438–483.
64. Penzien, J.; Watabe, M. Characteristics of 3-dimensional earthquake ground motions. *Earthq. Eng. Struct. Dyn.* **1975**, *3*, 365–373. [CrossRef]
65. Boore, D.M.; Watson-Lamprey, J.; Abrahamson, N.A. Orientation-Independent Measures of Ground Motion. *Bull. Seismol. Soc. Am.* **2007**, *96*, 1502–1511. [CrossRef]

Article

Application of Genetic Algorithm to Optimize Location of BRB for Reinforced Concrete Frame with Curtailed Shear Wall

Taufiq Ilham Maulana, Patricia Angelica de Fatima Fonseca and Taiki Saito *

Department of Architecture and Civil Engineering, Toyohashi University of Technology, Toyohashi 441-8580, Japan; taufiq.ilham.maulana.mt@tut.jp (T.I.M.); patricia.angelica.de.fatima.fonseca.fo@tut.jp (P.A.d.F.F.)
* Correspondence: saito.taiki.bv@tut.jp

Abstract: The shear walls are essential seismic elements to increase buildings bearing capacity against earthquakes. In mid- and high-rise buildings, shear walls are subjected to predominant bending deformation under earthquakes, and the responses in upper floors increase. In order to utilize the shear walls appropriately, previous studies proposed to install shear walls until a certain building level, referred to as the curtailed wall. However, the upper frame structure without shear walls suffered significant deformation during earthquakes compared to the lower stories. Therefore, the objective of this study is to present structural configuration for buildings with curtailed shear walls by installing buckling-restrained braces (BRBs) in the upper frame to reduce its deformation under earthquakes. Firstly, the analysis accuracy was verified by simulating the experimental results of four sets of scaled frames with curtailed walls tested on a shaking table. Then, ten- and twenty-story plane frames with the different heights of curtailed walls were created, and their nonlinear responses to earthquake ground motions were evaluated. The genetic algorithm was applied to establish the optimum BRB locations to satisfy the design criteria. It was proved that using BRBs at specific locations in upper frames can significantly improve the seismic response of buildings with curtailed walls.

Keywords: seismic response; optimization; curtailed shear wall; buckling-restrained brace; genetic algorithm

Citation: Maulana, T.I.; Fonseca, P.A.d.F.; Saito, T. Application of Genetic Algorithm to Optimize Location of BRB for Reinforced Concrete Frame with Curtailed Shear Wall. *Appl. Sci.* **2022**, *12*, 2423. https://doi.org/10.3390/app12052423

Academic Editors: Pier Paolo Rossi and Melina Bosco

Received: 25 January 2022
Accepted: 23 February 2022
Published: 25 February 2022

Publisher's Note: MDPI stays neutral with regard to jurisdictional claims in published maps and institutional affiliations.

Copyright: © 2022 by the authors. Licensee MDPI, Basel, Switzerland. This article is an open access article distributed under the terms and conditions of the Creative Commons Attribution (CC BY) license (https:// creativecommons.org/licenses/by/ 4.0/).

1. Introduction

1.1. RC Frame with Curtailed Wall

In the earthquake-prone area, the reinforced concrete (RC) shear wall is used together with the RC frame to increase the bearing capacity and rigidity against earthquakes. For low-rise to mid-rise buildings, it is appropriate to install the RC shear walls up to the top floor. Studies by Estekanchi et al. [1], Xia et al. [2], Bhatta et al. [3], and Bhat et al. [4] suggested that installing RC shear walls with the full height of the high-rise building is not effective. The reason is that the multi-story shear walls in a high-rise building are subjected to predominant bending deformation under earthquakes, and the inter-story deformation of the building increases in the upper floors.

On the other hand, in order to utilize the shear walls, a design method was proposed using shear walls until a certain level of the building. This shear wall is generally called the curtailed wall. Studies by Rathi et al. [5], Nollet et al. [6,7], and Atik et al. [8,9] suggested that the optimum height of the curtailed shear walls adopted for RC frames depends on several parameters such as the ratio of wall flexural rigidity to frame shear rigidity, axial stiffness coefficient, and the top deflection of the building. However, these studies [5–9] use only a linear approach without considering the nonlinear dynamic behavior of the building. Based on the results of nonlinear dynamic analysis of RC frames with curtailed shear walls, Costa et al. [10] and Paulay and Priestley [11] pointed out that the structure above the curtailed shear wall has a significant inter-story drift. It is a challenge to reduce this notable large inter-story drift.

Therefore, the objective of this study is to propose a new structural system for the buildings with curtailed shear walls by installing buckling-restrained braces (BRBs) in the upper frame to reduce its deformation under earthquakes.

1.2. Implementation of BRBs

Adding BRBs to RC frames was proven to be effective in improving the seismic response of buildings, especially in reducing inter-story drift. There were many previous studies regarding the optimum usage and location of BRBs. One approach to optimizing BRB usage in frame structures is by using a Genetic Algorithm (GA). The study by Farhat et al. [12] used the total weight, volume, or cost produced after adding BRB as the fitness function of GA. Oxborrow and Richards [13] and Oxborrow [14] utilized the parameters of building responses such as the building displacement, inter-story drift, and ductility response against earthquake ground motions for the fitness function. Park et al. [15] adopted the damage cost expected during the structure's life cycle for the parameter. Similar studies were performed by Mohammadi et al. [16], Tu et al. [17], Fujishita et al. [18], and Terazawa and Takeuchi [19,20].

In this study, the optimum location of BRBs is discussed based on the fitness function considering the damage index of structural members defined by the Park–Ang [21,22].

1.3. Numerical Analysis to Model Seismic Behavior of RC Structures

In order to conduct the study, a numerical approach is required to model the seismic behavior of reinforced concrete structures appropriately. Several studies were performed by previous researchers using several types of numerical techniques, such as Finite Element Method (FEM) and Applied Element Method (AEM). The FEM subdivides a large system into smaller and simpler parts, assembles the parts into the global system, and solves the problem by minimizing an associated error function. The AEM takes a different approach by simulating the structure by virtually dividing it into discrete elements linked by normal and shear springs at precise contact locations on the elements' surfaces [23]. AEM application examples in nonlinear dynamic analyses are available, such as the study of seismic rehabilitation of masonry buildings using FRP by Fathalla and Salem [24] and the study of seismic debris field for collapsed RC frame buildings by Sediek et al. [25].

In this study, nonlinear structural seismic response analyses are employed using STERA_3D, a software developed by the co-author [26]. In the previous study by Maulana et al. [27], STERA_3D was used to conduct seismic response analyses for RC setback buildings, and the results were considerably matched with experimental shaking table tests.

1.4. Objective of This Study

This study is conducted according to the following procedures:

a. In order to verify the accuracy of the nonlinear analysis using the STERA_3D (Structural Earthquake Response Analysis 3D) [26] software, four sets of RC frame–wall specimens tested by a shaking table [28] were analyzed, and the nonlinear responses were compared with the test results.
b. Two 5-bays two dimensional (2D) RC frames with 10 stories and 20 stories using three different coverage percentages of RC shear walls were analyzed to examine the damage concentration at the upper part of the structure without walls.
c. A new structural system for RC frames with curtailed walls by installing BRBs in the upper stories was proposed, and the optimum location of BRBs was obtained by employing the Genetic Algorithm. Three parameters were adopted to assess the genetic algorithm's fitness function, namely the number of BRBs, the inter-story drift response, and the damage indices of beam elements.

2. Simulation of RC Frames with Curtailed Walls Tested by Shaking Table

The accuracy of the software STERA_3D was examined by analyzing four frame specimens with curtailed walls of different heights tested on a shaking table by Moehle and Sozen [28]. The experiment was reported in 1980. It was stated that the overall test structure configuration was determined by the equipment limitations at that time. The structural elements were chosen so that the wall height effect on the seismic response of multi-story RC wall–frame structures could be inspected.

2.1. Simulation Method of RC Frame Analysis

The experiment was conducted to investigate the response of RC frames with different wall heights against strong earthquakes. Figures 1–3 present the structural element sections, the elevation view, and the plan view with connection details of the specimen. The specimen has nine stories and three bays with curtailed walls in the center. Shaking table experiments were conducted in the x-direction for three specimens with different heights of the curtailed walls. Names of the specimens were FNW, FSW, FHW, and FFW, where FNW is the frame with no wall, FSW is the frame with a one-story wall, FHW is the frame with a half wall (four-story wall), and FFW is the frame with a full-height wall.

The design concrete compressive strength was 38 MPa, and the tensile strength of the steel rebar was 399 MPa. The total weight of the specimens was 40.87 kN, and each story weight is about 4.55 kN. The experimental raw data, which consists of the specimens' dimensions, strengths, input seismic motions, displacement responses, and acceleration responses, were obtained from Datacenterhub [29].

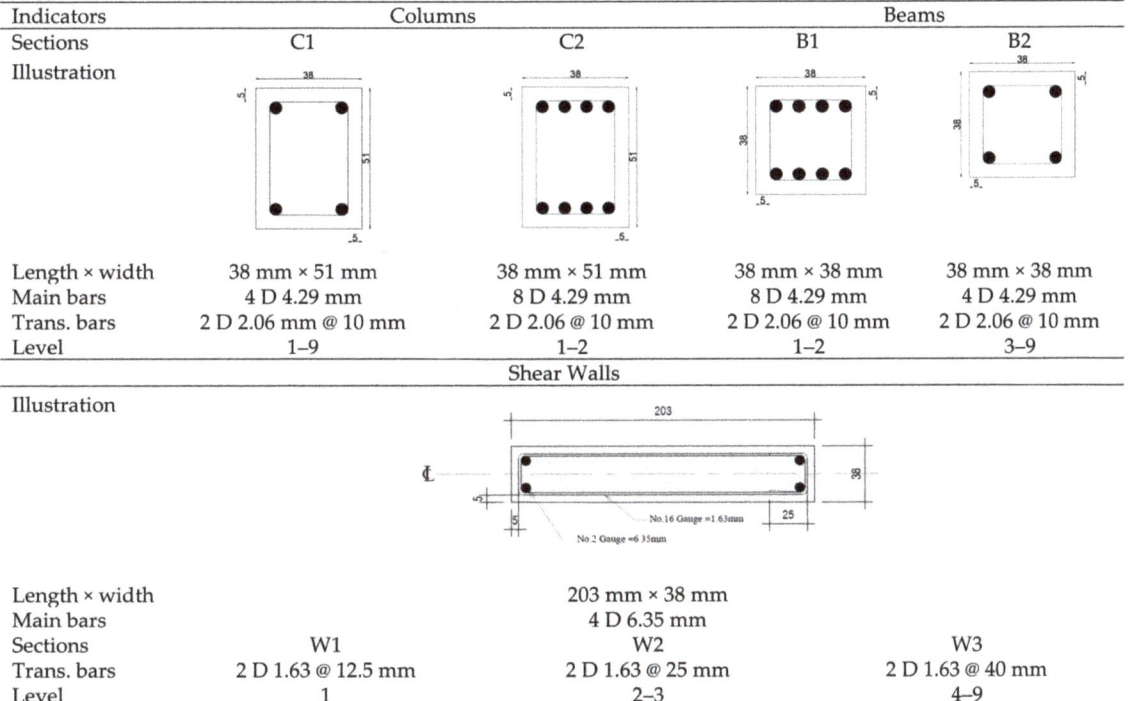

Figure 1. Beam, column, and wall element sections used in experimental tests by Moehle and Sozen (dimension in mm).

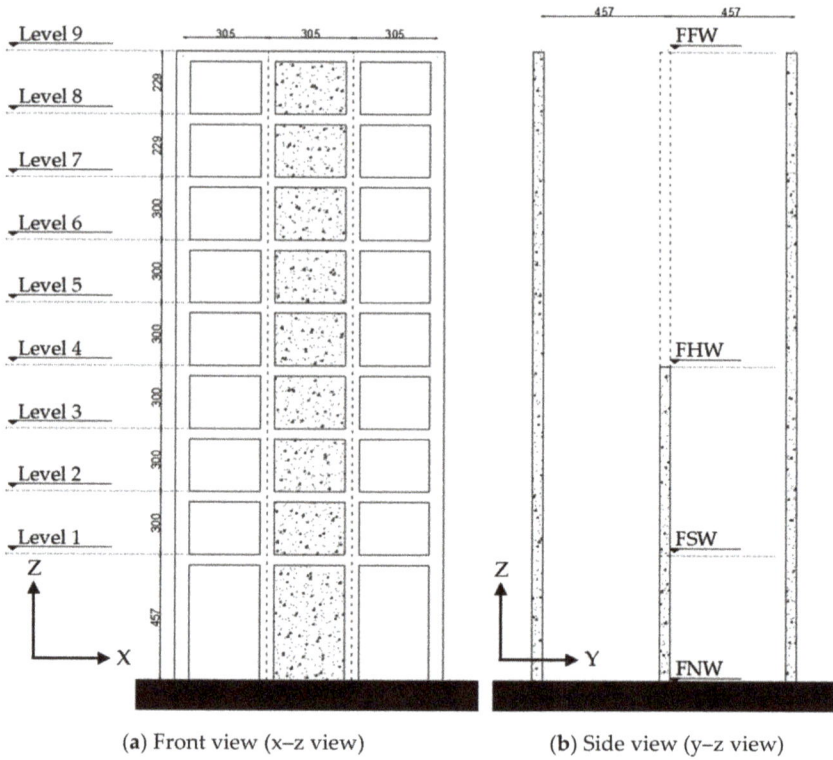

Figure 2. Elevation view of the test specimen by Moehle and Sozen (dimension in mm).

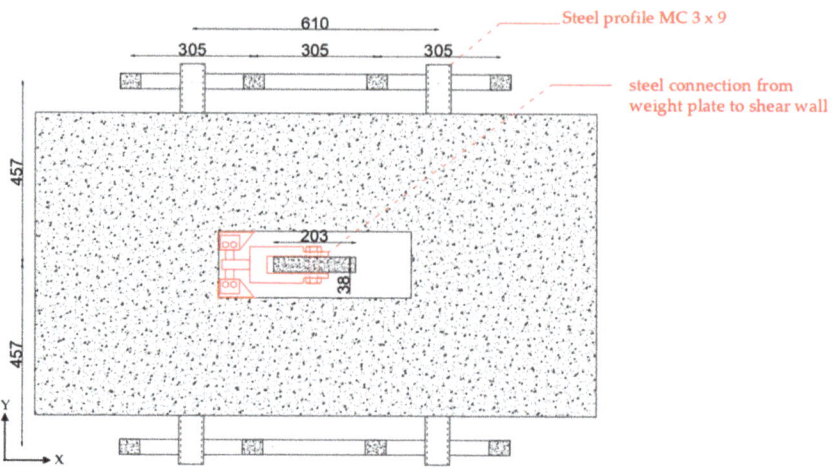

Figure 3. Plan view of the test specimen by Moehle and Sozen; showing connection between RC frames, weighting plates, and RC shear walls (dimension in mm).

2.2. Input Ground Motion

The specimens were tested on the shaking table subjected to the El Centro 1940 Imperial Valley earthquake, and the original time interval was divided by 2.5, considering the scale factor of the specimen. In order to attain a required amount of inelastic response, the peak acceleration was amplified to roughly 0.4 g. The recorded ground motion accelerations at the shaking table are shown in Figure 4 for each specimen.

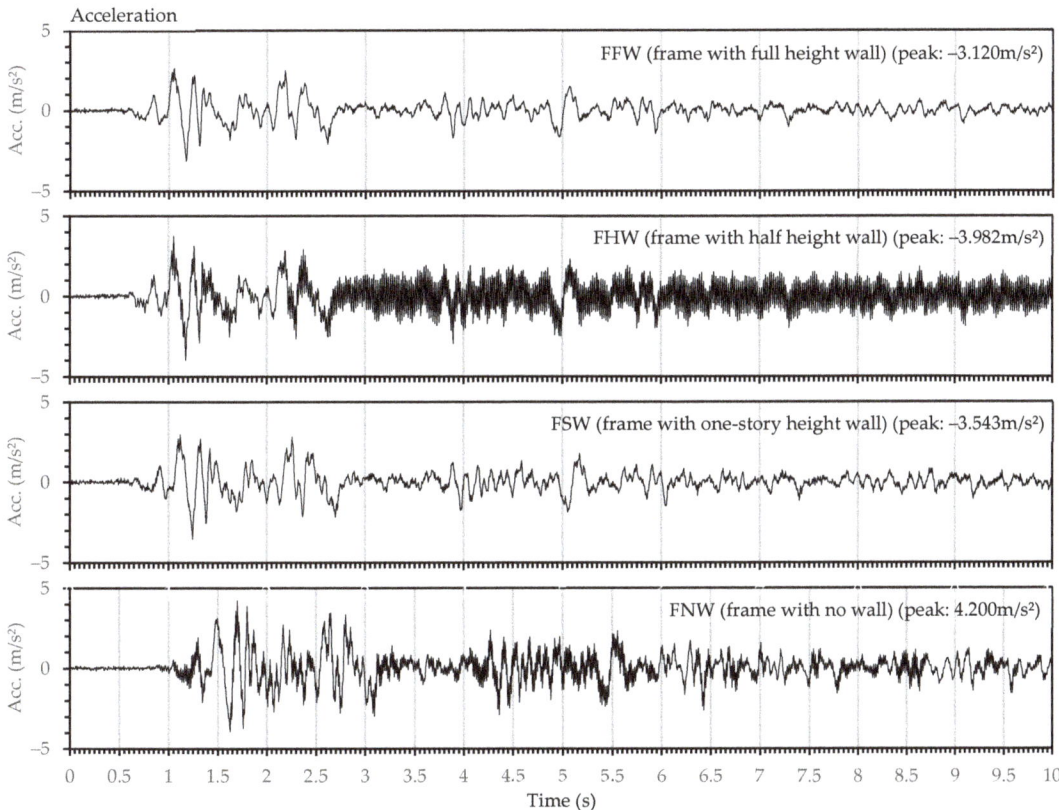

Figure 4. Four accelerations recorded at the shaking table test during experiments.

2.3. Simulation Method of RC Frame-Wall Analysis

Nonlinear dynamic analyses of the specimens were performed using the STERA_3D version 10.8, a program developed by one of the co-authors [26]. STERA_3D is a three-dimensional software for seismic analysis of buildings with a variety of structures developed for educational and research purposes. STERA_3D features a graphical interface for quickly and easily creating building models and visualizing the results. The beam and column elements are modeled as line elements with two nonlinear bending springs at both ends and one nonlinear shear spring at the middle, as shown in Figure 5. The end displacement vector was obtained from Equation (1) as the sum of the displacement vector of each component. The RC column section was modeled by using the multi springs model, originally proposed by Lai, Will, and Otani in 1984 [30], as shown in Figure 6. The wall was modeled as a line element with a nonlinear shear spring and a nonlinear bending spring in the middle. The nonlinearity in the model is considered in both flexural spring and shear spring, as illustrated in Figures 7 and 8. Figure 9 shows the model for the RC wall element, where k_s is the stiffness of the nonlinear shear spring, k_b is the stiffness of the nonlinear bending spring, and k_n is the stiffness of the axial spring. The floor is assumed to be rigid for in-plane deformation; therefore, the two RC frames and RC wall are deforming conjunctively. The viscous damping was modeled as the proportional damping using the spontaneous stiffness matrix with a 1% damping factor. All the details of implemented modeling technique are available in STERA_3D Technical Manual [26]. The analytical models of the specimens in the STERA_3D are illustrated in Figure 10a–d.

$$\left\{ \begin{array}{c} \theta_A \\ \theta_B \\ \delta_x \end{array} \right\} = \left\{ \begin{array}{c} \tau_A \\ \tau_B \\ \delta_x \end{array} \right\} + \left\{ \begin{array}{c} \phi_A \\ \phi_B \\ 0 \end{array} \right\} + \left\{ \begin{array}{c} \eta_A \\ \eta_B \\ 0 \end{array} \right\} \quad (1)$$

where:

θ: is the total rotation at the element joint;
δ_x: is the element deformation at direction x;
τ: is the elastic element rotation;
ϕ: is the nonlinear element rotation due to bending;
η: is the nonlinear element rotation due to shear.

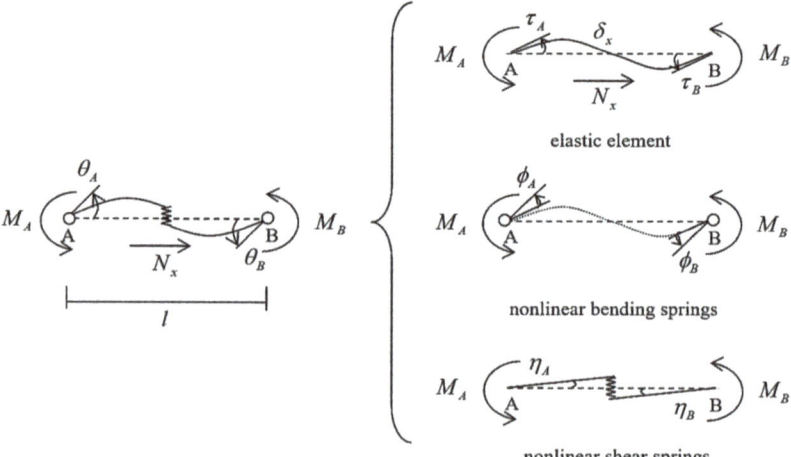

Figure 5. Elastic, nonlinear bending, and nonlinear shear springs for elements modeled by STERA_3D [26].

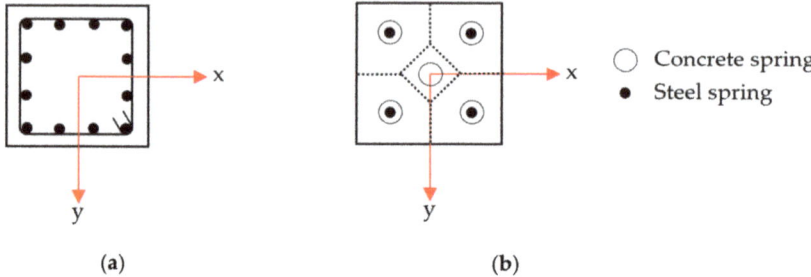

Figure 6. RC Column section modelling using multi-spring models: (**a**) original column section, and (**b**) multi-spring model idealization [26].

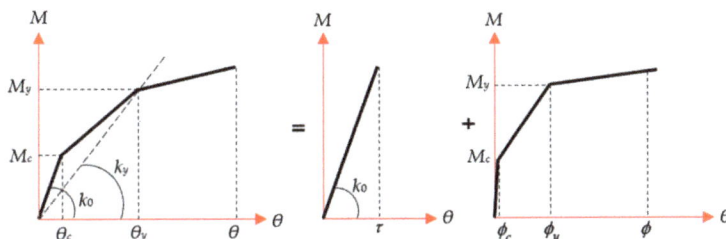

Figure 7. Moment-rotation relationship at bending spring for nonlinearity consideration in structural elements [26].

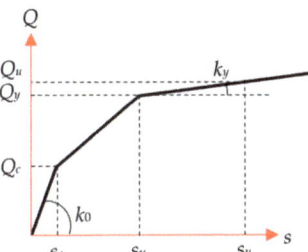

Figure 8. Force-deformation relationship at shear spring for nonlinearity consideration in structural elements [26].

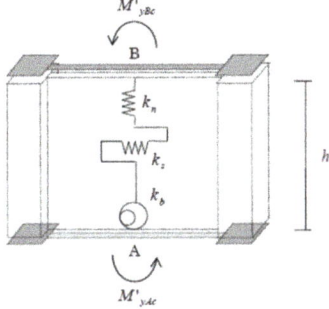

Figure 9. Wall element idealization in STERA_3D [26].

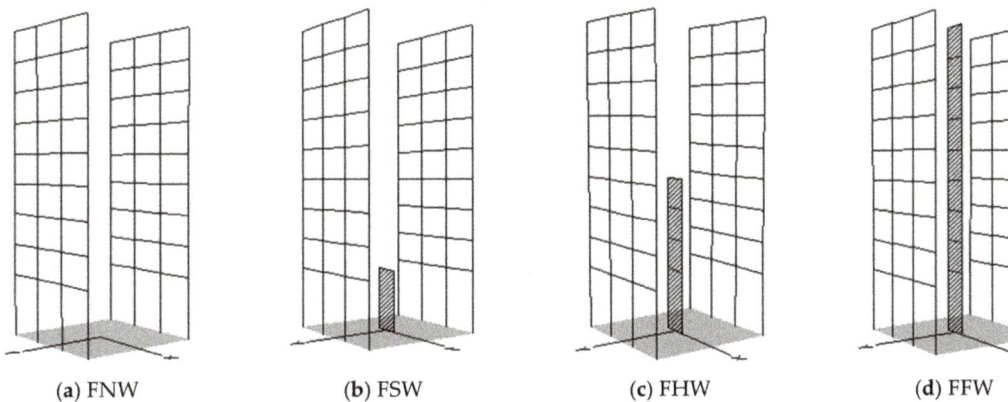

(a) FNW **(b)** FSW **(c)** FHW **(d)** FFW

Figure 10. STERA_3D models of test specimens.

2.4. Results of Comparison

The responses of acceleration and displacement at the top floor were calculated and compared with the experimental data, as shown in Figures 11 and 12. Since the experimental and numerical results only had a slight difference, it is considered adequate to use the STERA_3D for further nonlinear response analyses of buildings with curtailed walls.

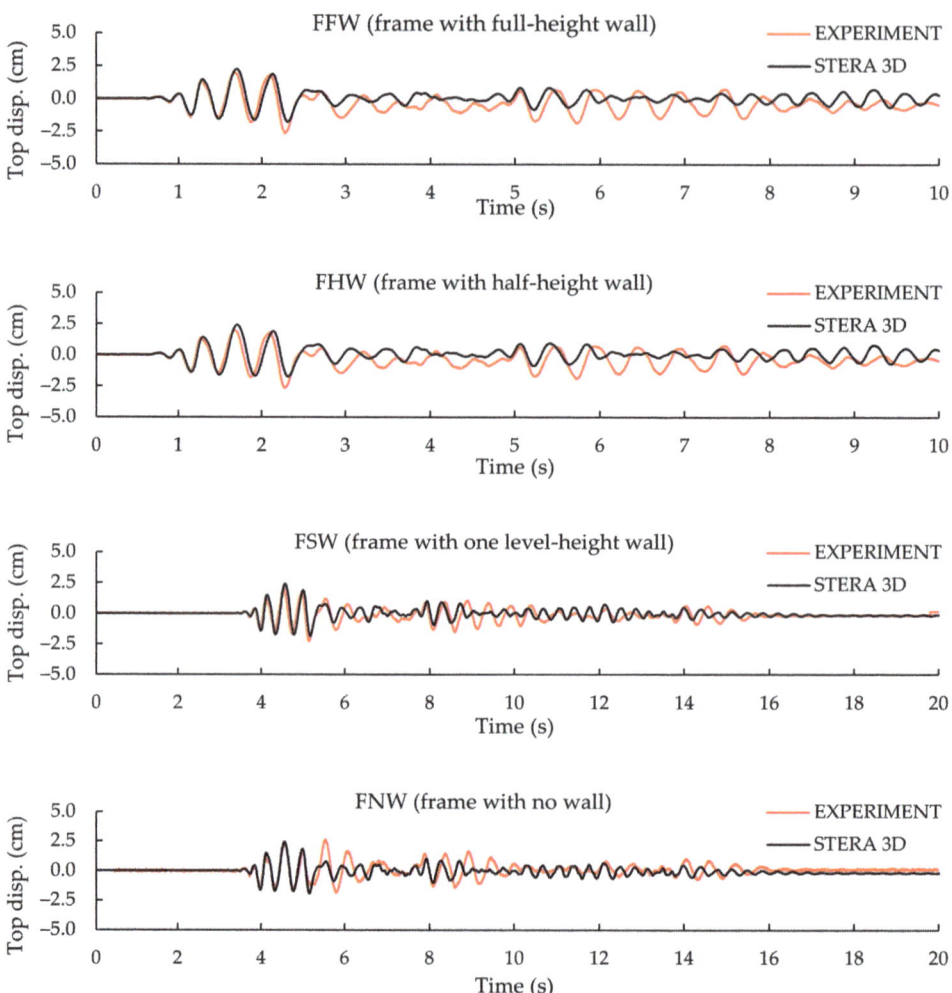

Figure 11. Comparison of displacement response between Experiment and STERA_3D analysis.

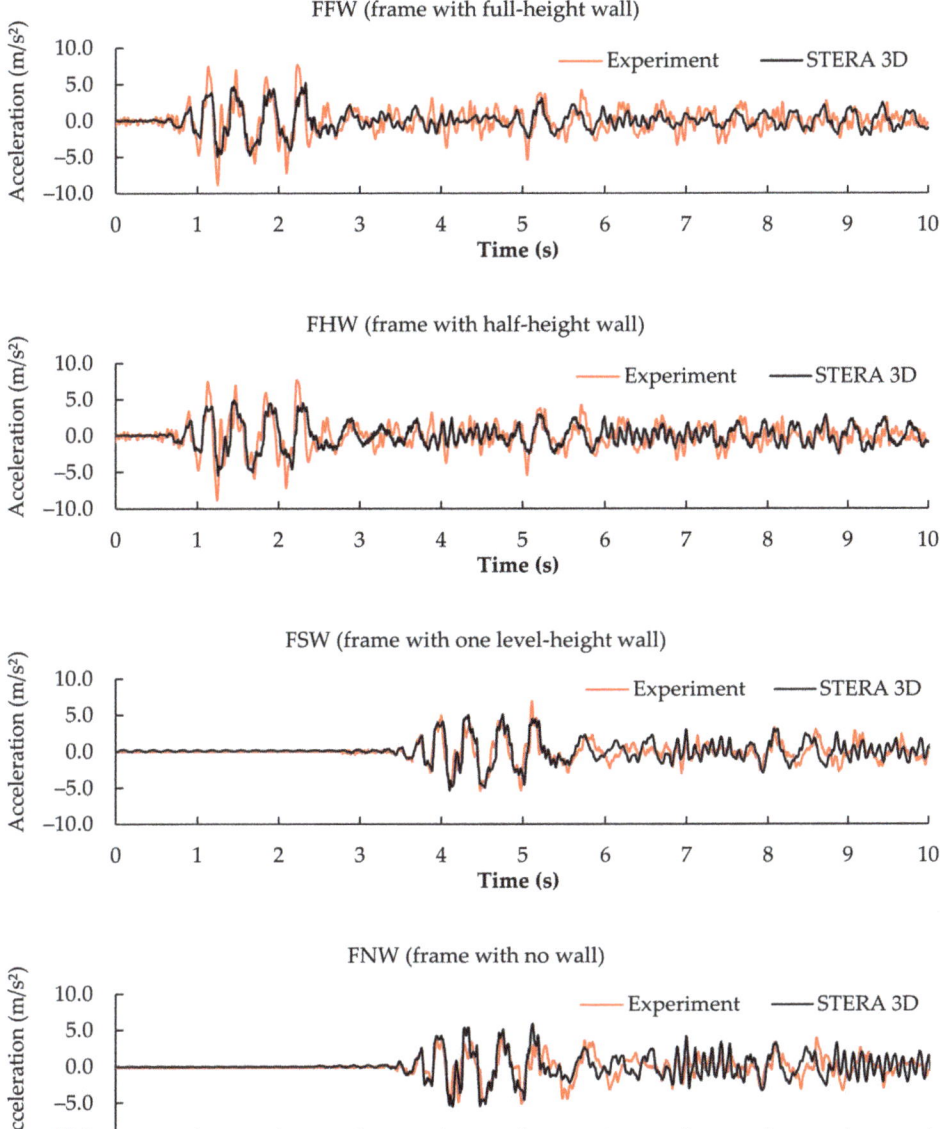

Figure 12. Comparison of acceleration response between Experiment and STERA_3D analysis.

3. Response Characteristic of Frame Buildings with Curtailed Walls

3.1. Models of Frames with Curtailed Shear Walls

An analytical study was conducted for the reinforced concrete frames of ten and twenty-story with curtailed shear walls. Three different ratios of the curtailed wall heights were selected as 30%, 50%, and 70% of the total height of the structures. Each structure had five bays, and the span length was 6400 mm. The story height was 3000 mm, and the story weight was 3000 kN. In total, there are six types of building models. Each model was then named with the unique code based on its total height and wall percentage coverage: 1030, 1050, 1070, 2030, 2050, and 2070. Figure 13 shows the illustration of the models. Tables 1–3 show the detail of building dimensions and the concrete strengths. The tensile strength of main and shear reinforcement rebars for all members are 490 MPa and 295 MPa, respectively. The model was designed to satisfy the criteria of Japanese design standards.

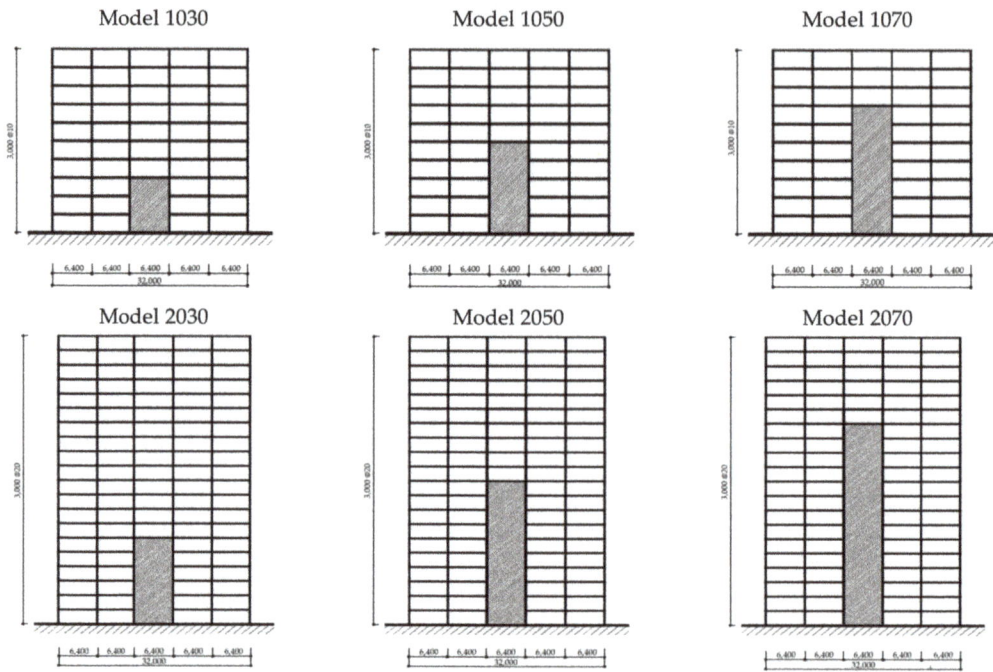

Figure 13. Ten-story (**top**) and twenty-story (**bottom**); 30%, 50%, and 70 % of curtailed shear wall (in mm).

Table 1. Structural parameters of RC columns.

Story	B × D (mm)	Main Rebar	Shear Rebar	Concrete Strength, f_c (MPa)	Rebar Strength, f_y (MPa)
20–11	850 × 800	16D32	2D13 @ 100 mm	36	
10–6	900 × 900	16D38	2D13 @ 100 mm	42	
	900 × 850	16D35	2D13 @ 100 mm	42	490
5–1	900 × 900	16D38	2D13 @ 100 mm	48	
	900 × 850	24D38	2D13 @ 100 mm	48	

Table 2. Structural parameters of RC beams.

Story	B × D (mm)	Top Bar	Bottom Bar	Shear Bar	f_c (MPa)	f_y (MPa)
20–16	600 × 800	4D22	4D22	4D13 @ 100 mm	36	
15–11	600 × 800	4D32	4D32	4D13 @ 100 mm	36	
10–6	600 × 800	4D32	4D32	4D13 @ 100 mm	42	490
		4D25	4D25	4D13 @ 100 mm		
5–1	600 × 800	6D32	6D32	4D13 @ 100 mm	48	
		4D29	4D29	4D13 @ 100 mm		

Table 3. Structural parameters of RC walls.

Story	Thickness (mm)	Shear Reinforcement	f_c (MPa)
20–11			36
10–6	150	2D13 @ 150 mm	42
5–1			48

3.2. Input Earthquake Ground Motions

Ten input motions were selected and scaled so that the maximum inter-story drift response is equal to 1/75 of the inter-story height. Table 4 is the list of earthquake ground motions, and Figures 14 and 15 show their acceleration response spectra and the wave forms before scaling. Table 5 shows the scale factors to obtain the same maximum inter-story drift ratio of 1/75.

Table 4. List of earthquake ground motions.

No	Event	Year	Station	Component	Original Max. Acc. (m/s^2)	Code
1	Imperial Valley	1940	El Centro	NS	3.41	ELC
2	Kern County	1952	Taft	EW	1.52	TAF
3	Chi-chi	1999	CHY080	360 DEG	8.36	CHI
4	Northridge	1994	Arleta-Nordhoff Ave Fire Station	90 DEG	3.37	NOR
5	Loma Prieta	1989	Saratoga-Aloha Ave	0 DEG	4.94	LOM
6	Valparaiso, Chile	1985	Vina del Mar	200 DEG	3.55	CHILE
7	Villita, Mexico	1985	Guerrero Array Stn VIL	N00W	1.25	MEX
8	Cape Mendocino	1992	Petrolia	0 DEG	5.78	CAPE1
9	Cape Mendocino	1992	Rio Dell-101/Painter St. Overpass	270 DEG	3.78	CAPE2
10	Kobe	1995	JMA	NS	8.17	KOB

Table 5. Amplification scale of input motions to produce maximum inter-story drift of 1/75 building inter-story height.

Earthquake	Scale 1030	Scale 1050	Scale 1070	Scale 2030	Scale 2050	Scale 2070
ELC	1.158	0.891	0.968	0.957	0.945	1.176
KOB	0.998	0.724	1.150	1.731	1.620	1.451
TAF	1.162	1.229	1.074	1.221	1.328	1.668
CHI	1.003	0.665	1.002	1.268	1.200	1.217
NOR	0.938	0.977	1.375	1.798	1.512	1.805
LOM	0.846	0.992	1.341	1.020	0.808	1.059
CHILE	0.740	0.721	0.776	1.427	0.980	1.057
MEX	1.445	1.172	1.672	1.372	1.504	1.843
CAPE1	1.138	0.750	0.942	1.506	1.350	1.388
CAPE1	1.085	1.042	1.062	1.572	1.171	1.462

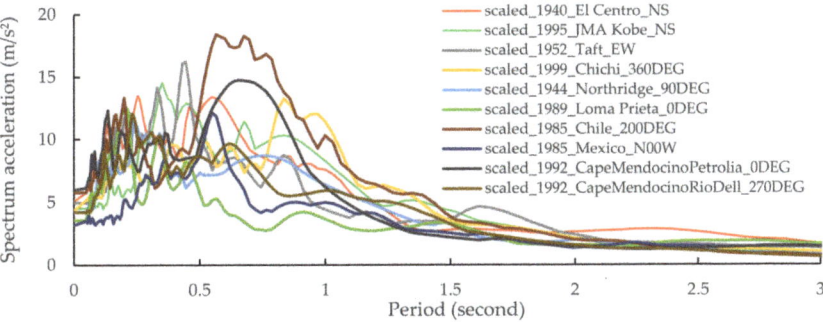

Figure 14. Acceleration response spectrum of original earthquake ground motions.

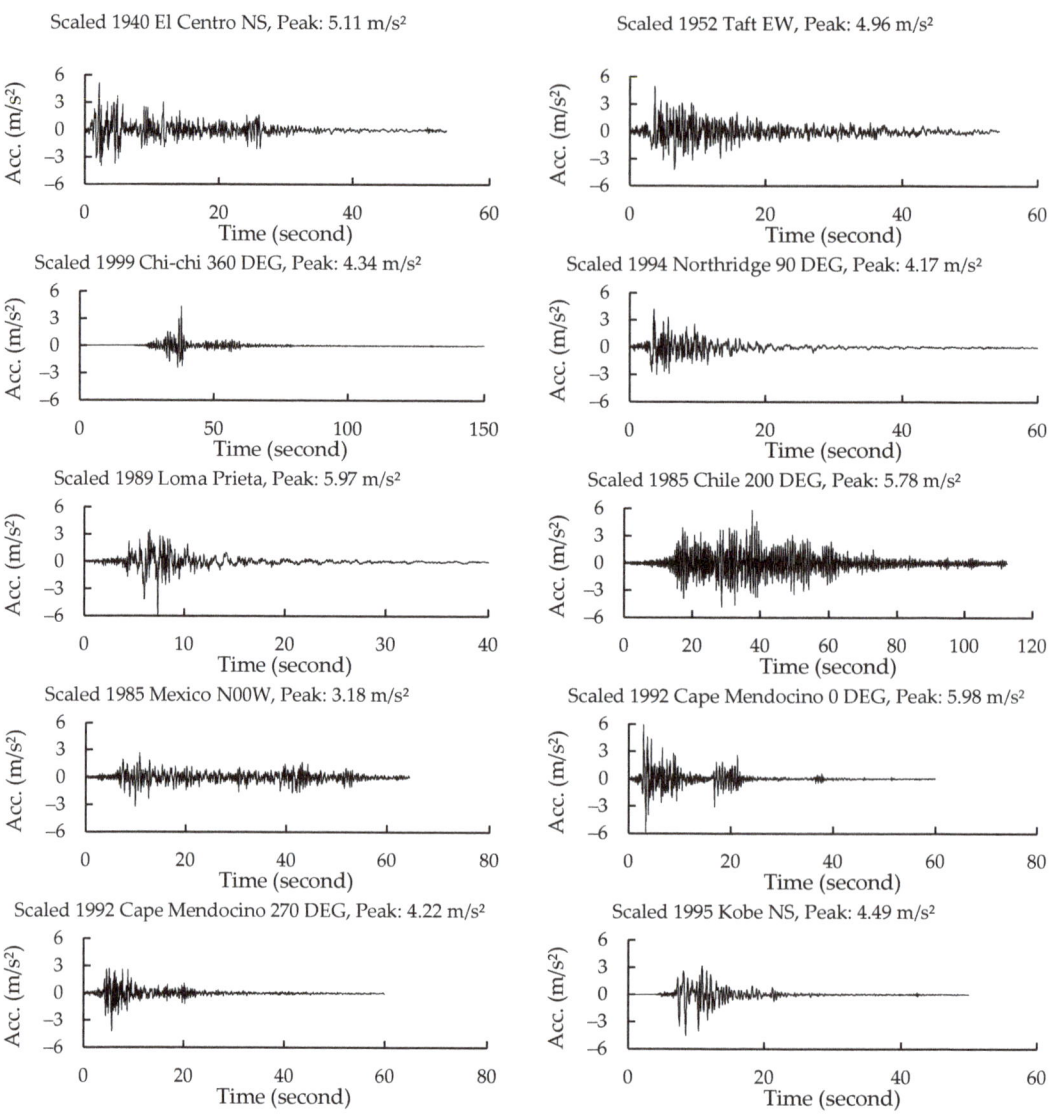

Figure 15. Original ten time history acceleration waves of earthquake motions.

3.3. The Response of RC Frame-Curtailed Walls

The inter-story drift responses of RC frame-curtailed walls under ten earthquake input motions are depicted in Figure 16. From all nonlinear analyses, it is observed that after the wall coverage is stopped, the drift is then significantly increasing, and it makes a big difference in story drift between the upper part frame without walls and the lower part frame with walls. It is necessary to make all specimens have the same state of inter-story drift before the optimization is performed. Although the location of the peak inter-story drift is different for each earthquake, the scale for input motions in Table 5 is successfully implemented to produce the same maximum inter-story drift of 4 cm, which corresponds to the story drift ratio of 1/75.

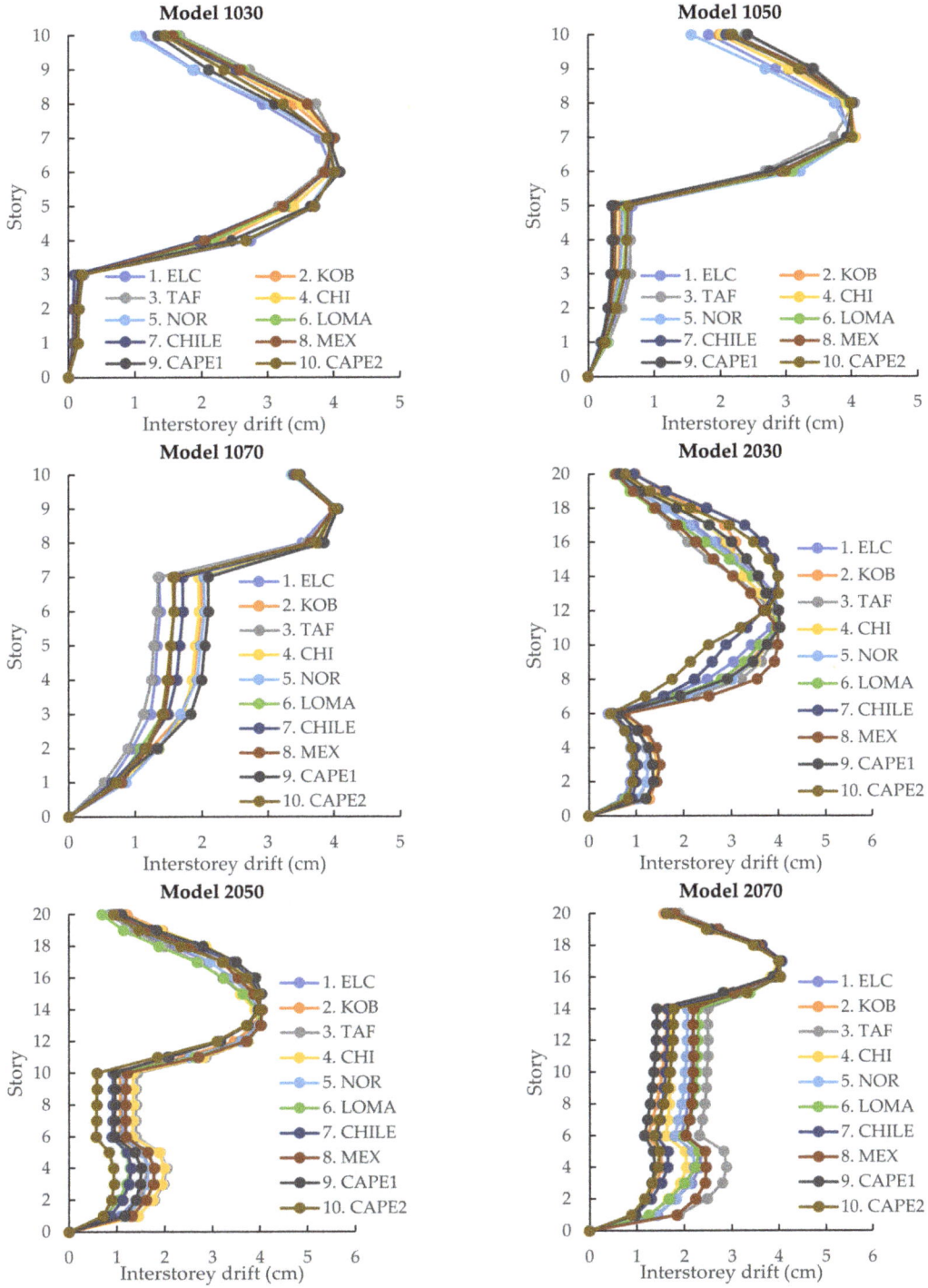

Figure 16. Inter-story drift of six specimens under ten selected input motions.

4. Optimum Arrangement of BRBs for RC Frames with Curtailed Walls

4.1. Implementation of Genetic Algorithm

The Genetic Algorithm (GA) is a form of optimization algorithm inspired by natural selection. It is a population-based search algorithm that makes use of the survival of the fittest concept [31]. In this algorithm, the population is subjected to a series of transformation processes, including mutation and crossover. Individuals in a population compete for survival by following a predetermined selection process to become the next generation. After a few generations, the best individual will represent the ideal solution.

The objective of this section is to decide the optimum locations of buckling-restrained braces (BRBs) to reduce the story drift in the upper part of the frames with curtailed walls. In order to optimize the best location of BRBs, the genetic algorithm (GA) is adopted in this study. In the GA, the strings represented by binary numbers are candidate solutions to the problem. These solutions are referred to as chromosomes, and the binaries are referred to as genes. A fitness function, formed as a formula to achieve, is predetermined to select a candidate solution's relative fitness, which the GA subsequently uses to guide the evolution of reasonable solutions.

In order to apply the GA in this study, the availability of BRB in the 2D frame structures is indicated using binary, which means the value will be zero if there is no BRB and one if there is a BRB. A set of binary numbers indicate the whole availability of BRB and their locations. These sets are associated with three main parameters to calculate the objective function. Then, six sets are initially generated randomly, and these sets are gathered to be an initial population during the GA process. The process is illustrated in Figure 17.

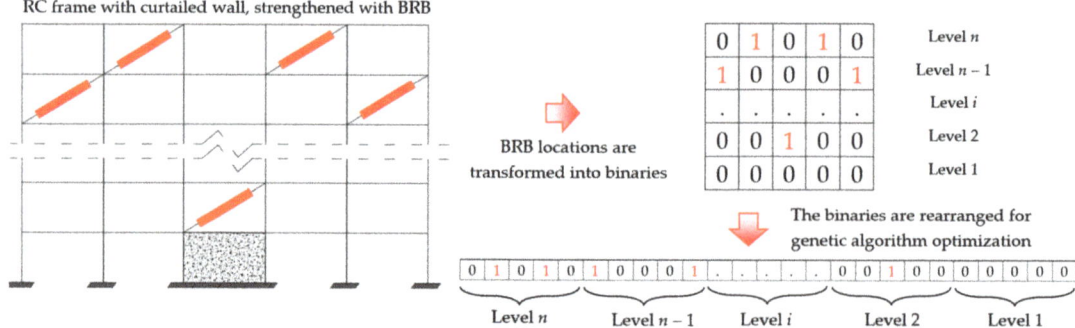

Figure 17. The diagram of BRBs location to binaries for Genetic Algorithm optimization process.

4.2. Procedures

The genetic algorithm consists of the following processes. The general process is shown in Figure 18.

a. Initialization: The initial population of candidate solutions is generated randomly across the search space. Each generation contains six solutions per population;
b. Evaluation: After the population is initialized, the fitness values of the candidate solutions are evaluated using the determined fitness function;
c. Selection: Selection is performed by ranking the solution based on the fitness function, and it imposes the survival-of-the-fittest mechanism on the candidate solutions;
d. Crossover: In this step, the first half of the first solution is combined with the last half of the second solution to produce a new population for the next generation;
e. Mutation: The modification is made by randomly changing one digit of the binaries;
f. Replacement: In this step, the new population in the newest generation will replace the other population if the fitness function value is closer to the determined target;
g. The steps b to f are repeated until one or more stopping criteria are met.

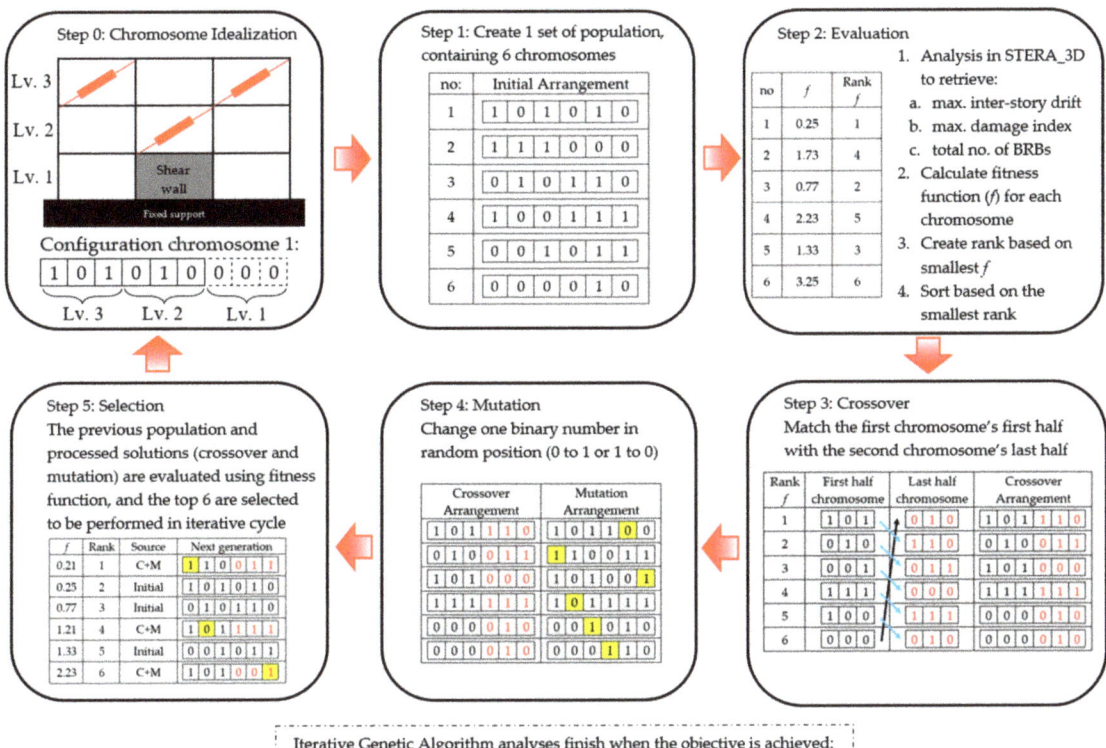

Figure 18. The general process of Genetic Algorithm sample.

4.3. Initial Population

A set of the initial population was created using the parameters such as number of spans (n_{bays}), number of stories (n_{story}), and number of shear wall story coverage (n_{wall}), with the number of solutions for each population (spp) is defined as the following Equation (2).

$$Population\_size = (spp, ((n_{bays} - 2) \times (n_{story} - n_{wall}))) \quad (2)$$

4.4. Determination of Objective and Fitness Function

The binaries number produced in step one is then transferred to the STERA_3D to recreate the frame with BRBs. The nonlinear dynamic response is then performed with ten ground motions to obtain the inter-story drift and the damage index of beams. These parameters were used to calculate the objective function, together with the total number of BRBs. The fitness function is defined as the sum of three conditions: the total number of BRBs, inter-story drift, and the damage index of the RC members, as shown in Equation (3).

$$Minimize: f = w_1\phi_1 + w_2\phi_2 + w_3\phi_3 \quad (3)$$

where:

w_1, w_2, w_3: Weighting factors, with $w_1 = 0.3$, $w_2 = 2$, and $w_3 = 0.2$ (based on the trial);
ϕ_1: Condition for total number of BRBs;
ϕ_2: Condition for story drift;
ϕ_3: Condition for damage index of RC members.

The parameter ϕ_1 is the contribution effect of the number of BRBs. The ϕ_1 is calculated using Equation (4).

$$\phi_1 = \frac{1}{5(N-k)} \sum_{i=k+1}^{N} n_{D,i} \qquad (4)$$

where:

$n_{D,i}$: Number of BRBs in i-th story;
N: Total number of stories;
k: Top story number of the curtailed wall.

The parameter ϕ_2 is the effect of the maximum inter-story drift. The ϕ_2 is defined by Equations (5) and (6), where it is zero if the maximum inter-story drift is less than the allowable drift δ_{allow}, which equals 1/100 of the inter-story height.

$$\phi_2 = \frac{1}{(N-k)} \sum_{i=k+1}^{N} \Omega_i \qquad (5)$$

$$\Omega_i = \begin{cases} 0, & if\ \delta_{max,i} < \delta_{allow} \\ \frac{\delta_{max,i}}{\delta_{allow}}, & if\ \delta_{max,i} \geq \delta_{allow} \end{cases} \qquad (6)$$

where:

$\delta_{max,i}$: The maximum story drift in i-th story;
δ_{allow}: Allowable story drift.

The parameter ϕ_3 is the effect of the damage index of the RC members. The ϕ_3 is defined by Equation (7). In this equation, the average damage index between the left side and the right side of the beam element is adopted.

$$\phi_3 = \frac{1}{5(N+1-k)} \sum_{i=k}^{N} \left(\sum_{j=1}^{5} \left(\frac{DI_{BL,i} + DI_{BR,i}}{2} \right) \right) \qquad (7)$$

where:

j: The inspected bays number;
$DI_{BL,i}$: Damage index of beam on the left joint;
$DI_{BR,i}$: Damage index of beam on the right joint.

The damage index (DI) is expressed by Equation (8) based on the structural deformation and the hysteretic energy response due to seismic excitations.

$$DI = \frac{u_m}{u_u} + \beta \frac{E_h}{F_y u_u} \qquad (8)$$

where:

u_m: Maximum displacement response of structure element due to earthquake;
u_u: Ultimate displacement capacity under monotonic loading;
E_h: Hysteretic energy dissipated by the structural element;
F_y: Yield force;
β: Non-negative parameter based on repeated loading effect.

4.5. BRBs Strengths

The BRB applied for the RC frame structure with the curtailed wall is chosen based on the study performed by Naqi and Saito [32]. This adoption consideration is taken on account of the similarity in the building's type, height, and dimensions. The applied BRB strength detail for this study are elaborated in Table 6 and Figure 19.

Table 6. Strength of BRB adopted in study 1.

	Structural Parameters of BRB Members		
Story	K0 (kN/mm)	K1/K0	Fy (kN)
20–16	80	0.02	520
15–11	80	0.02	520
10–6	100	0.02	650
5–4	120	0.02	780

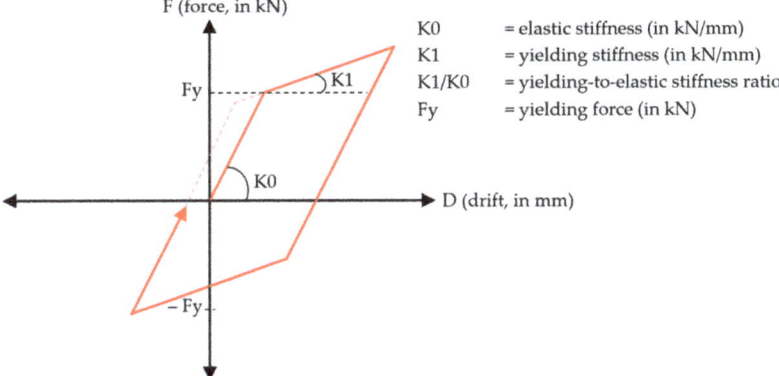

Figure 19. The adopted BRB's bilinear hysteresis.

4.6. Simple Probabilistic Method to Determine the Final BRB Arrangement

The optimum positions of BRBs under selected ten earthquake ground motions are different for each ground motion. Figures 20 and 21 show the distribution of BRBs location, added with the color, which indicates the probability of occurrence of BRBs location, based on the optimization from 10 different ground motions. The value in the panel is between 0 and 1, where 0 means there are no BRB and 1 means all 10 input motions determine that there is BRB in that panel. The final BRBs locations are then selected ascendingly based on the higher value of possibility.

4.7. Response of RC Frame-Wall with Final BRB Arrangement under the Scaled Motions

The optimized locations of BRBs for six RC frame structures with curtailed walls were selected based on the GA and the simple probabilistic method. The seismic performance improvement after the BRBs placement is presented in Figure 22 for models with 10 stories and Figure 23 for models with 20 stories. The red straight lines indicate the average inter-story drift of RC frame with different wall heights before adding the BRB, while the black straight lines are after placing BRBs. Especially at the upper region of walls, the inter-story drift is improved from 1/75 of inter-story height to 1/100.

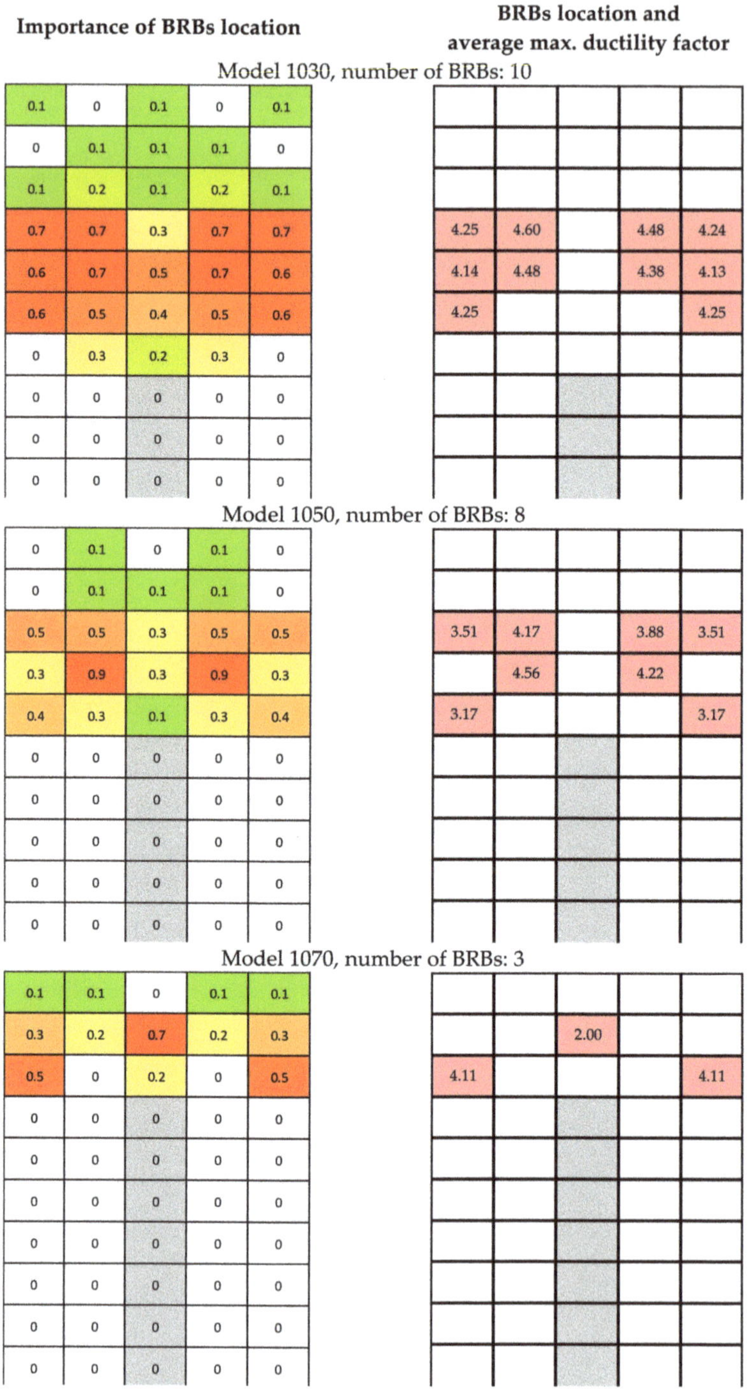

Figure 20. Implementation of BRBs placement for 10 stories based on optimization under 10 input motions.

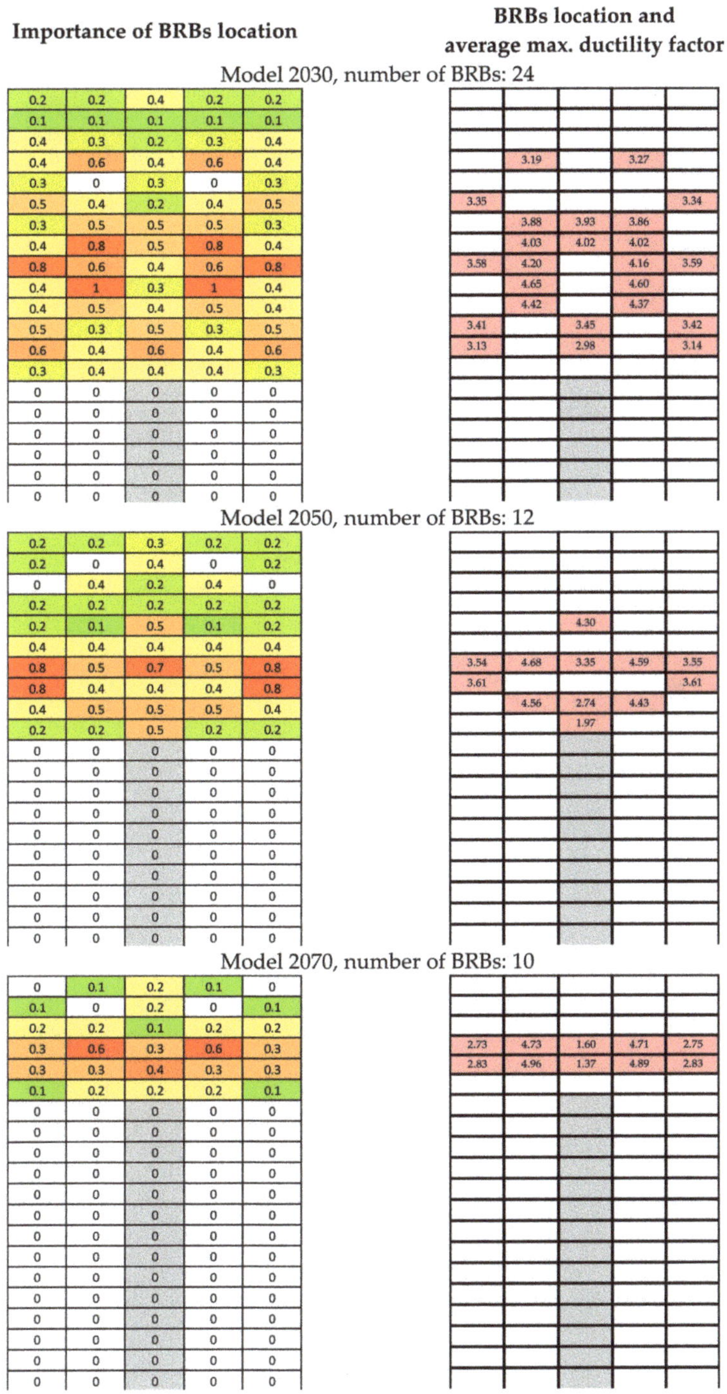

Importance of BRBs location | **BRBs location and average max. ductility factor**

Model 2030, number of BRBs: 24

C1	C2	C3	C4	C5		C1	C2	C3	C4	C5
0.2	0.2	0.4	0.2	0.2						
0.1	0.1	0.1	0.1	0.1						
0.4	0.3	0.2	0.3	0.4						
0.4	0.6	0.4	0.6	0.4			3.19		3.27	
0.3	0	0.3	0	0.3						
0.5	0.4	0.2	0.4	0.5		3.35				3.34
0.3	0.5	0.5	0.5	0.3			3.88	3.93	3.86	
0.4	0.8	0.5	0.8	0.4			4.03	4.02	4.02	
0.8	0.6	0.4	0.6	0.8		3.58	4.20		4.16	3.59
0.4	1	0.3	1	0.4			4.65		4.60	
0.4	0.5	0.4	0.5	0.4			4.42		4.37	
0.5	0.3	0.5	0.3	0.5		3.41		3.45		3.42
0.6	0.4	0.6	0.4	0.6		3.13		2.98		3.14
0.3	0.4	0.4	0.4	0.3						
0	0	0	0	0						
0	0	0	0	0						
0	0	0	0	0						
0	0	0	0	0						
0	0	0	0	0						
0	0	0	0	0						

Model 2050, number of BRBs: 12

C1	C2	C3	C4	C5		C1	C2	C3	C4	C5
0.2	0.2	0.3	0.2	0.2						
0.2	0	0.4	0	0.2						
0	0.4	0.2	0.4	0						
0.2	0.2	0.2	0.2	0.2						
0.2	0.1	0.5	0.1	0.2				4.30		
0.4	0.4	0.4	0.4	0.4						
0.8	0.5	0.7	0.5	0.8		3.54	4.68	3.35	4.59	3.55
0.8	0.4	0.4	0.4	0.8		3.61				3.61
0.4	0.5	0.5	0.5	0.4			4.56	2.74	4.43	
0.2	0.2	0.5	0.2	0.2				1.97		
0	0	0	0	0						
0	0	0	0	0						
0	0	0	0	0						
0	0	0	0	0						
0	0	0	0	0						
0	0	0	0	0						
0	0	0	0	0						
0	0	0	0	0						
0	0	0	0	0						
0	0	0	0	0						

Model 2070, number of BRBs: 10

C1	C2	C3	C4	C5		C1	C2	C3	C4	C5
0	0.1	0.2	0.1	0						
0.1	0	0.2	0	0.1						
0.2	0.2	0.1	0.2	0.2						
0.3	0.6	0.3	0.6	0.3		2.73	4.73	1.60	4.71	2.75
0.3	0.3	0.4	0.3	0.3		2.83	4.96	1.37	4.89	2.83
0.1	0.2	0.2	0.2	0.1						
0	0	0	0	0						
0	0	0	0	0						
0	0	0	0	0						
0	0	0	0	0						
0	0	0	0	0						
0	0	0	0	0						
0	0	0	0	0						
0	0	0	0	0						
0	0	0	0	0						
0	0	0	0	0						
0	0	0	0	0						
0	0	0	0	0						
0	0	0	0	0						
0	0	0	0	0						

Figure 21. Implementation of BRBs placement for 20 stories based on optimization under 10 input motions.

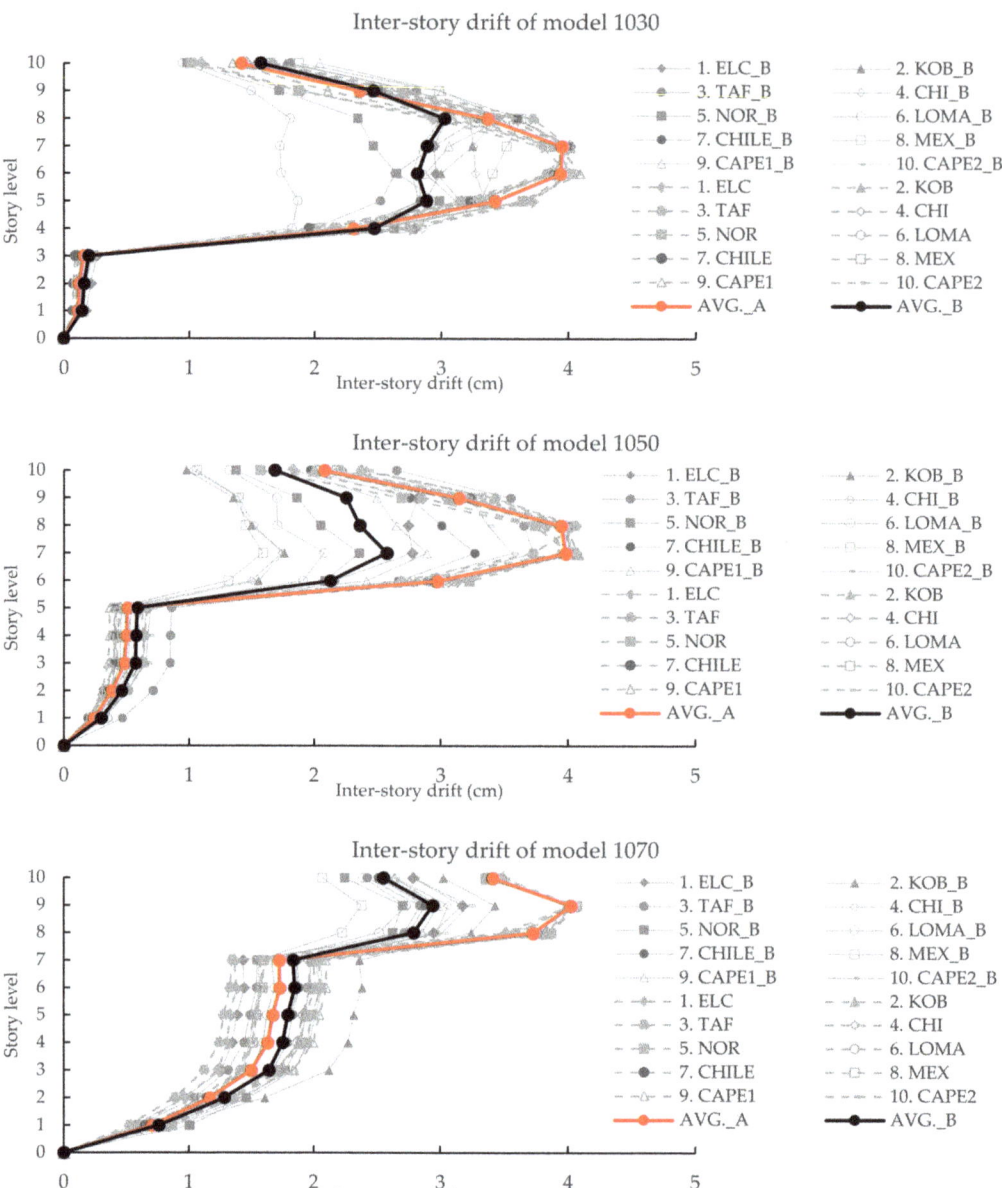

Figure 22. Comparison of inter-story drift distribution of frame-curtailed wall structure without BRBs and with BRBs for 10 stories.

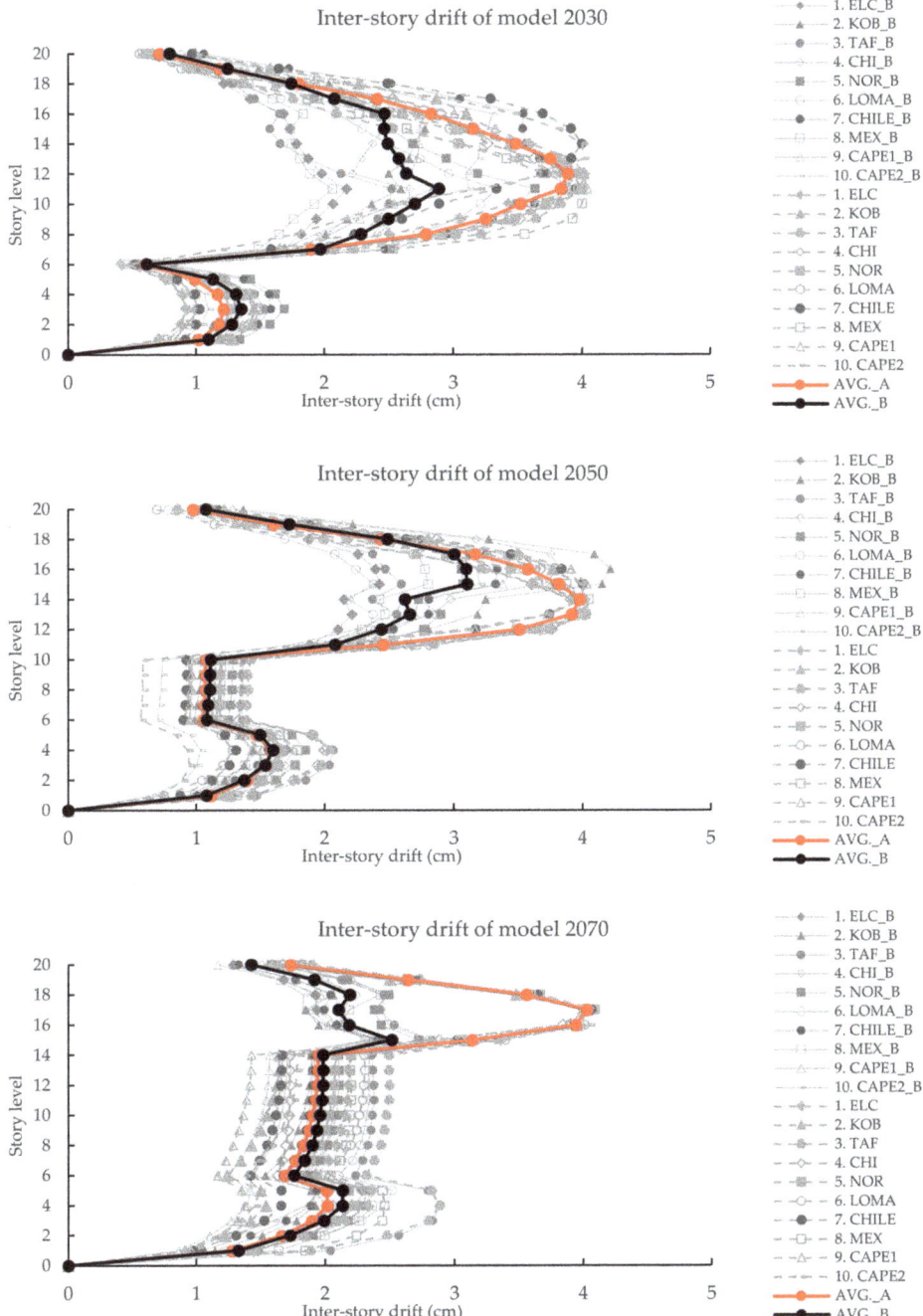

Figure 23. Comparison of inter-story drift distribution of frame-curtailed wall structure without BRBs and with BRBs for 20 stories.

5. Conclusions

The seismic performance of RC frame structures with different heights of shear walls was studied numerically under the nonlinear dynamic analysis.

First, four sets of previous experimental one-directional test models by Moehle and Sozen [28] comprising RC frames with different wall height coverages were modeled and analyzed by the STERA_3D. The simulation results matched well with the test results.

Then, 10-story and 20-story RC frames with 30%, 50%, and 70% wall coverages were generated and analyzed using the STERA_3D. The results show that the upper region where the wall coverage stops has a significant response of inter-story drift. In order to decrease the seismic response at the upper region of the frame without wall coverage, the installation of BRBs was proposed.

The genetic algorithm is applied to determine the optimum locations of BRBs by considering three main parameters in the fitness function: the inter-story drifts, the damage index of the beams, and the total number of BRBs. Under ten selected earthquake motions, the optimum locations of BRBs were found to be different for each motion. Thus, a simple probabilistic method is employed to select the final locations of BRBs. The method successfully reduced the average inter-story drifts from 1/75 to 1/100 and gave a better seismic response at the upper region.

Author Contributions: Conceptualization, T.S., P.A.d.F.F. and T.I.M.; methodology, P.A.d.F.F. and T.I.M.; software, T.S.; writing—original draft preparation, P.A.d.F.F. and T.I.M.; writing—review and editing, T.S.; supervision, T.S. All authors have read and agreed to the published version of the manuscript.

Funding: This research received no external funding.

Institutional Review Board Statement: Not applicable.

Informed Consent Statement: Not applicable.

Data Availability Statement: The data presented in this study are available on request from the corresponding author.

Conflicts of Interest: The authors declare no conflict of interest.

References

1. Estekanchi, H.E.; Harati, M.; Mashayekhi, M.R. An investigation on the interaction of moment-resisting frames and shear walls in RC dual systems using endurance time method. *Struct. Des. Tall. Spec. Build.* **2018**, *27*, e1489. [CrossRef]
2. Xia, G.; Shu, W.; Stanciulescu, I. Efficient analysis of shear wall-frame structural systems. *Eng. Comp.* **2019**, *36*, 2084–2110. [CrossRef]
3. Bhatta, B.D.; Vimalanandan, G.; Senthilselvan, S. Analytical study on effect of curtailed shear wall on seismic performance of high-rise building. *Int. J. Civ. Eng. Tech.* **2017**, *8*, 511–519.
4. Bhatt, G.; Titiksh, A.; Rajepandhare, P. Effect of Curtailment of Shear Walls for Medium Rise Structures. In Proceedings of the 2nd International Conference on Sustainable Computing Techniques in Engineering, Science and Management (SCESM-2017), Belagavi, India, 27–28 January 2017; pp. 501–507.
5. Rathi, N.; Muthukumar, G.; Kumar, M. Influence of Shear Core Curtailment on the Structural Response of Core-Wall Structures. In *Lecture Notes in Civil Engineering: Recent Advances in Structural Engineering*; Springer: Singapore, 2019; Volume 1, pp. 207–215.
6. Nollet, M.J.; Stafford Smith, B. Behavior of curtailed wall-frame structures. *J. Struct. Eng.* **1993**, *119*, 2835–2854. [CrossRef]
7. Nollet, M.J. Behaviour of Wall-Frame Structures: A Study of the Interactive Behaviour of Continuous and Discontinuous Wall-Frame Structures. Ph.D. Thesis, McGill University, Montréal, QC, Canada, 1991.
8. Atik, M.; Badawi, M.M.; Shahrour, I.; Sadek, M. Optimum level of shear wall curtailment in wall-frame buildings: The continuum model revisited. *J. Struct. Eng.* **2019**, *140*, 06013005. [CrossRef]
9. Atik, M. The Effect of Curtailed Walls in Wall-Frame Structures to Resist Lateral Loads. Master's Thesis, University of Aleppo, Aleppo, Syria, 2010.
10. Costa, A.G.; Oliveira, C.S.; Duarte, R.T. Influence of vertical irregularities on seismic response of buildings. In Proceedings of the 9th World Conference on Earthquake Engineering (9th WCEE), Tokyo, Japan, 2–9 August 1988; pp. 491–496.
11. Paulay, T.; Priestley, M.J.N. *Seismic Design of Reinforced Concrete and Masonry Buildings*, 1st ed.; Wiley-Interscience: New York, NY, USA, 1992; pp. 500–531.

12. Farhat, F.; Nakamura, S.; Takahashi, K. Application of genetic algorithm to optimization of buckling restrained braces for seismic upgrading of existing structures. *Comput. Struct.* **2009**, *87*, 110–119. [CrossRef]
13. Oxborrow, G.T.; Richards, P. Optimized distribution of strength in tall buckling-restrained brace frames. In *Behaviour of Steel Structures in Seismic Areas*; CRC Press: London, UK, 2009; Volume 1, pp. 819–824.
14. Oxborrow, G.T. Optimized Distribution of Strength in Buckling-Restrained Brace Frames in Tall Buildings. Master's Thesis, Brigham Young University, Provo, UT, USA, 2009.
15. Park, K.; Oh, B.K.; Park, H.S.; Choi, S.W. GA-based multi-objective optimization for retrofit design on a multi-core PC cluster. *Comput.-Aided Civ. Infrastruct Eng.* **2015**, *30*, 965–980. [CrossRef]
16. Mohammadi, R.K.; Garoosi, M.R.; Hajirasouliha, I. Practical method for optimal rehabilitation of steel frame buildings using buckling restrained brace dampers. *Soil Dyn. Earthq. Eng.* **2019**, *123*, 242–251. [CrossRef]
17. Tu, X.; He, Z.; Huang, G. Performance-based multi-objective collaborative optimization of steel frames with fuse-oriented buckling-restrained braces. *Struct. Multidiscipl. Optim.* **2020**, *61*, 365–379. [CrossRef]
18. Fujishita, K.; Sutcu, F.; Matsui, R.; Takeuchi, T. Optimization of Damper Arrangement with Hybrid GA using Elasto-plastic Response Analysis on Seismic Response Control Retrofit. *J. Struct. Constr. Eng.* **2016**, *81*, 537–546. [CrossRef]
19. Terazawa, Y.; Takeuchi, T. Generalized response spectrum analysis for structures with dampers. *Earthq. Spectra* **2018**, *34*, 1459–1479. [CrossRef]
20. Terazawa, Y.; Takeuchi, T. Optimal damper design strategy for braced structures based on generalized response spectrum analysis. *Japan Archit. Rev.* **2018**, *2*, 477–493. [CrossRef]
21. Park, Y.-J.; Ang, A.H.-S. Mechanistic seismic damage model for reinforced concrete. *J. Struct. Eng.* **1985**, *111*, 722–739. [CrossRef]
22. Park, Y.-J.; Ang, A.H.-S.; Wen, Y.K. Seismic damage analysis of reinforced concrete buildings. *J. Struct. Eng.* **1985**, *111*, 740–757. [CrossRef]
23. Tagel-Din, H.; Meguro, K. Applied Element Method for Dynamic Large Deformation Analysis of Structures. *Int. J. Jpn. Soc. Civ. Eng.* **2010**, *17*, 215s–224s. [CrossRef]
24. Fathalla, E.; Salem, H. Parametric Study on Seismic Rehabilitation of Masonry Buildings Using FRP Based upon 3D Non-Linear Dynamic Analysis. *Buildings* **2018**, *8*, 124. [CrossRef]
25. Sediek, O.A.; El-Tawil, S.; McCormick, J. Seismic Debris Field for Collapsed RC Moment Resisting Frame Buildings. *J. Struct. Eng.* **2021**, *147*, 04021045. [CrossRef]
26. Saito, T. Structural Earthquake Response Analysis, STERA_3D Version 10.8. Available online: http://www.rc.ace.tut.ac.jp/saito/software-e.html (accessed on 1 October 2020).
27. Maulana, T.I.; Enkhtengis, B.; Saito, T. Proposal of Damage Index Ratio for Low- to Mid-Rise Reinforced Concrete Moment-Resisting Frame with Setback Subjected to Uniaxial Seismic Loading. *Appl. Sci.* **2021**, *11*, 6754. [CrossRef]
28. Moehle, J.P.; Sozen, M. *Experiment to Study Earthquake Response of R/C Structures with Stiffness Interruptions*; National Science Foundation Report; University of Illinois at Urbana-Champaign: Champaign, IL, USA, 1980.
29. Moehle, J.; Sozen, M. Experiments to Study Earthquake Response of R/C Structures with Stiffness Interruptions (NEES-2011-1058). Available online: https://datacenterhub.org/deedsdv/publications/view/298 (accessed on 1 October 2020).
30. Lai, S.-S.; Will, G.T.; Otani, S. Model for Inelastic Biaxial Bending of Concrete Members. *J. Struct. Eng.* **1984**, *110*, 2563–2584. [CrossRef]
31. Michalewicz, Z. *Genetic Algorithms + Data Structures = Evolution Programs*; Springer: New York, NY, USA, 1992.
32. Naqi, A.; Saito, T. Performance of a BRB RC High-rise Buildings Under Successive Application of Wind-Earthquake Scenarios. In Proceedings of the 1st Croatian Conference on Earthquake Engineering (1st CroCEE), Zagreb, Croatia, 22–24 March 2021; pp. 909–919.

Article

Numerical Assessment of an Innovative RC-Framed Skin for Seismic Retrofit Intervention on Existing Buildings

Diego Alejandro Talledo [1], Irene Rocca [1], Luca Pozza [2], Marco Savoia [2] and Anna Saetta [1,*]

[1] Department of Architecture and Arts (DCP), University IUAV of Venice, 30135 Venezia, Italy; dtalledo@iuav.it (D.A.T.); irocca@iuav.it (I.R.)
[2] Department of Civil, Chemical, Environmental and Materials Engineering (DICAM), University of Bologna, 40126 Bologna, Italy; luca.pozza2@unibo.it (L.P.); marco.savoia@unibo.it (M.S.)
* Correspondence: saetta@iuav.it

Abstract: The seismic safety of existing building stock has become a very critical issue in recent years, mainly in earthquake-prone South Europe where most of the buildings were designed before the enforcement of seismic standards. Therefore, the concept, development and testing of efficient and cost-effective seismic retrofitting technologies are nowadays strongly needed, both for the society and for the scientific community. This study deals with the seismic assessment of a new RC-framed skin for retrofit intervention of existing buildings, evaluated through nonlinear static (pushover) analyses. A preliminary description of the proposed technology is provided, then numerical modeling of a typical RC existing building before and after retrofitting intervention is performed within the OpenSees framework. The results revealed that the proposed retrofitting technology improves the seismic performance of the RC building, also modifying the failure mode from a brittle soft-story mechanism to a more ductile one. The presented study, dedicated to the structural aspects of the system, is part of the TIMESAFE research project, where the thermo-hygrometric and acoustic performances achievable by the proposed RC-framed skin are also investigated.

Keywords: RC-framed skin; seismic assessment; nonlinear static analysis; existing RC buildings; retrofitting intervention

Citation: Talledo, D.A.; Rocca, I.; Pozza, L.; Savoia, M.; Saetta, A. Numerical Assessment of an Innovative RC-Framed Skin for Seismic Retrofit Intervention on Existing Buildings. *Appl. Sci.* **2021**, *11*, 9835. https://doi.org/10.3390/app11219835

Academic Editors: Pier Paolo Rossi and Melina Bosco

Received: 22 September 2021
Accepted: 18 October 2021
Published: 21 October 2021

Publisher's Note: MDPI stays neutral with regard to jurisdictional claims in published maps and institutional affiliations.

Copyright: © 2021 by the authors. Licensee MDPI, Basel, Switzerland. This article is an open access article distributed under the terms and conditions of the Creative Commons Attribution (CC BY) license (https://creativecommons.org/licenses/by/4.0/).

1. Introduction

The seismic inadequacy of masonry and reinforced concrete (RC) existing building stock has been tragically highlighted by the strong earthquakes occurred in the last decade in South Europe. Actually, a number of these buildings, not designed according to seismic standards (they are commonly named gravity load designed, GLD, buildings), collapsed or were severely damaged causing the loss of human lives and significant economic losses. Moreover, most of these existing buildings have reached their nominal service life, exhibiting a significant decay of structural performance due to degradation phenomena, often not properly handled by an adequate maintenance program. Therefore, the need of efficient and cost-effective seismic retrofitting methods has become a priority for the society and for the scientific community in order to fulfill modern standards such as the Eurocodes (e.g., [1,2]).

In the last twenty years a number of systems have been proposed, developed, and experimentally tested for RC buildings retrofitting; see [3] for a thorough review carried out by subdividing the seismic upgrading techniques in two categories, i.e., local and global approaches depending on the level at which they operate: on single element or on the structure as a whole. Among others, in [4] externally anchored precast wall panels have been used for the seismic retrofitting of RC frame, with a significant reduction of the construction period and a limited invasiveness.

More recently, a number of technological systems developed for energy and seismic renovation intervention have been proposed, see [5] for a comprehensive state of the art of interventions techniques for RC-framed buildings. These integrated solutions are typically based on realization of an external seismic-resistant system. In this way, the interruption of the use of the building during the intervention is also avoided, which represents significant cost and social saving. Concerning intervention techniques for existing RC buildings, in [6] an innovative technique for the simultaneous seismic-energy retrofitting of old buildings has been proposed, by combining inorganic textile-based composites with thermal insulation. A technological double skin exoskeleton with dissipative elements has been proposed in [7], while a new RC infilled frames connected to the existing ones has been studied in [8]. An integrated solution based on an earthquake resistant exoskeleton constituted by steel braced frames has been proposed in [9]. Similarly, in [10] the idea of external steel exoskeletons has been investigated for multiple retrofitting alternatives. Finally, Cross Laminated Timber (CLT) panels have been also suggested for the external integrated energy-seismic retrofitting of existing RC buildings. As an example, in [11] the possibility of replacing existing masonry infill walls in RC frame buildings with CLT panels has been investigated. It is worth underling the high potential of such an intervention technique (e.g., low cost and invasiveness, low time of realization, and low carbon material use), even if the related theoretical study and experimental testing are still at a preliminary stage.

In this paper, the seismic performances of the new RC-framed skin for retrofit intervention on existing buildings developed within the TIMESAFE project [12] are analyzed by discussing the results of a number of nonlinear static (pushover) analyses carried out on both existing and retrofitted RC buildings.

The main features of the proposed RC-framed skin, briefly summarized in the next section, are thoroughly described in [13] where also its feasibility and sustainability are investigated by means of a holistic Life Cycle Assessment for environmental impact and Life Cycle Cost for an economic evaluation.

2. RC-Framed Skin for Retrofit Intervention on Existing Buildings

In this section, the main structural characteristics of the proposed RC-framed skin technology are briefly presented, together with a description of the in situ installation phases, which allow avoiding the interruption of use of the building and significantly reducing the invasiveness of the intervention.

Figure 1 shows the detail of the proposed retrofitting technology, which consists of a RC frame rigidly connected to the external façade of the existing building by means of anchor rods placed at every floor level. The anchor rods must be designed to guarantee the respect of the hierarchy of strength principle, i.e., the connections must be sized to withstand with adequate over-strength the shear actions, so behaving elastically for the target seismic action.

The RC-framed skin is casted on site by using prefabricated expanded polystyrene (EPS) modules as formwork, which ensures also the thermal enhancement of the building.

The proposed system is characterized by a modular geometry with variable dimensions for both beams and columns, variable RC reinforcement, and variable column interspace. All these parameters can be designed to achieve a specific target of seismic safety level. Actually, the geometric parameters of EPS modules, strictly related to both the cross sections and interspace of frame elements, can be optimized depending on the characteristics of the existing building and of the selected seismic action.

In detail, columns can be designed with a square cross-section in the range 150 to 300 mm, while the transversal beams with a rectangular cross-section with base equal to the side of the column plus 50 mm and height variable in the range 300 to 500 mm. Typical column interspace is 1200 mm, but some adjustments are usually required to fit the façade opening geometry. The beams are placed in contact and connected with the existing structure in order to maximize the effectiveness of the connection system.

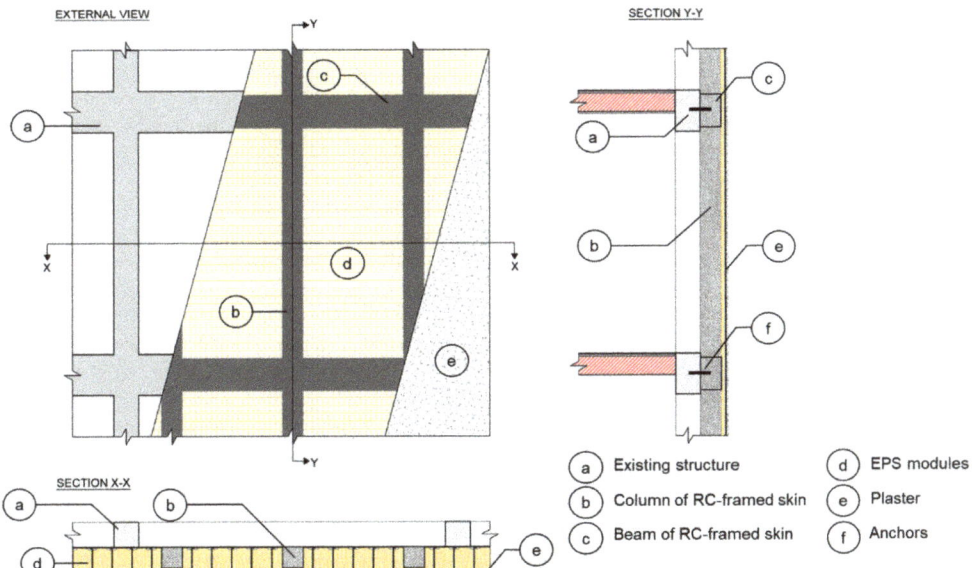

Figure 1. RC-framed skin.

Concerning the transversal reinforcement layout of the frame elements, continuous spirals are designed for the columns and rectangular stirrups for the beams. It is worth noting that, in order to ensure adequate confinement of the beam-column joint, the continuous spiral reinforcement of the column is provided across the joint itself.

The geometry of the system, and particularly of the EPS modules, is specifically shaped to realize an additional thick plaster on the external surface of the building. The external finishing plaster is reinforced with a steel mesh pre-assembled on the EPS modules and connected to the reinforcement of RC-framed skin by means of anchor rods.

The proposed retrofitting technology aims to limit invasiveness and interruption of use of the building; therefore, one of the main advantages of this technology is the possibility of working from the outside without internal intervention, by applying the RC-framed skin to the external façades.

The construction of the retrofitting system is of the platform type (i.e., floor by floor starting from the bottom) and consists of three main phases: (a) preparation of the façade and construction of the foundation beam; (b) installation of the prefabricated EPS modules and casting of the reinforcement frame; (c) realization of the external plaster and finishing of the façade, see [13] for the details.

It is worth noting that the existing foundation system may be inadequate for the new actions transmitted by the RC-framed skin, then, a simple connection to the existing foundation beams can be insufficient. In those cases, a retrofit of the foundation system may be needed. Nevertheless, the proposed technology provides for widespread interventions along the entire perimeter of the building, so generally leading to lower actions to the foundation if compared to the case of more localized retrofitted intervention, such as the addition of new shear walls.

3. Numerical Strategies

The structural behaviors of a typical GLD multi-story RC frame building constructed without any seismic detail, and of the same building retrofitted by using the proposed RC-framed skin technology, are investigated by means of nonlinear static analysis. The

Open System for Earthquake Engineering (OpenSees) framework [14] is adopted with the pre- and post-processor Scientific ToolKit for OpenSees (STKO) [15].

The modal analysis is performed to study the natural frequencies of both the existing and the retrofitted building, in order to compare the fundamental periods and evaluate the related displacement patterns, which are then assumed as the "modal" patterns within the nonlinear static analysis, according to Eurocode 8 [1].

3.1. Nonlinear Static Analyses

Concerning the nonlinear static analyses, after the application of the vertical loads, two vertical distributions of the lateral loads are applied, respectively: (i) a "uniform" pattern, based on lateral forces that are proportional to mass and (ii) a "modal" pattern, proportional to lateral forces consistent with first mode pattern obtained by the performed modal analysis. The structural performances of the existing and of the retrofitted building are discussed and compared in terms of capacity curves, i.e., the relations between base shear force and control node (top) displacement. In detail, peak shear force, ductility, and mode of failure are thoroughly evaluated and some considerations concerning the improvement obtained by retrofitting intervention are provided.

The numerical models are realized using a lumped plasticity approach with finite length of plastic hinges. Beams and columns are modelled as force-based nonlinear elements adopting the iterative flexibility formulation with the integration method proposed in [16], i.e., a modified two-point Gauss–Radau integration over each hinge region, so guaranteeing that localized deformations are integrated over the specified plastic hinge length. The length of the plastic hinge is evaluated according to the relation A.9 of EN1998-3 [2]:

$$L_{pl} = \frac{L_v}{30} + 0.2h + 0.11\frac{d_{bL}f_y}{\sqrt{f_c}} \quad (1)$$

where L_v is the shear span assumed equal to half the length of the element, h is the cross-section depth, d_{bL} is the mean diameter of the tension reinforcement, and f_c and f_y are the concrete compressive strength (MPa) and the steel yield strength (MPa), respectively. The P-Delta effect is considered for the columns in all the performed nonlinear static analyses.

The constitutive model adopted for concrete is the one proposed by Mander et al. [17] (based on the equation proposed by Popovics [18]) for both existing RC building and RC-framed skin, while the model proposed by Menegotto-Pinto model, as modified by Filippou et al. [19], is adopted for steel. In the concrete model, a zero tensile strength is assumed and the compressive stress is set equal to zero when the crushing strain is achieved.

The concrete model adopted in the column cross section's core accounts for the effects of passive confinement. To this aim, the formulation proposed in Mander et al. [17] is considered in this work both for existing columns (with rectangular stirrups) and for the columns of the reinforcing RC-framed skin (with circular spirals).

Concerning the implementation phase, for the columns of the reinforcing RC-framed skin, an ad-hoc GUI (Graphical User Interface) for STKO [15] has been specifically developed providing a utility for users to input easily the geometric and material characteristics and computing automatically the effect of confinement on core material. In particular, the plugin recognizes the class of constitutive model adopted for cover concrete in OpenSees (e.g., Concrete01, Concrete02, Concrete04, etc.) and computes the effect of confinement according to several formulations, including the one developed by Mander et al. [17], adopted in this work as recommended in EN1998-3 [2] and Italian National Standard [20]. Finally, the plugin creates on the fly (when the model runs) a new material for core concrete (confined) of the same class of the unconfined concrete. Of course, the user always has the possibility to specify a customized confined material law when he does not want to use the plugin automatism. The window of the GUI for the reinforcing column of the RC-framed skin is shown in Figure 2, where both the constitutive laws adopted in this work for confined and unconfined concrete are depicted in terms of stress-strain curve.

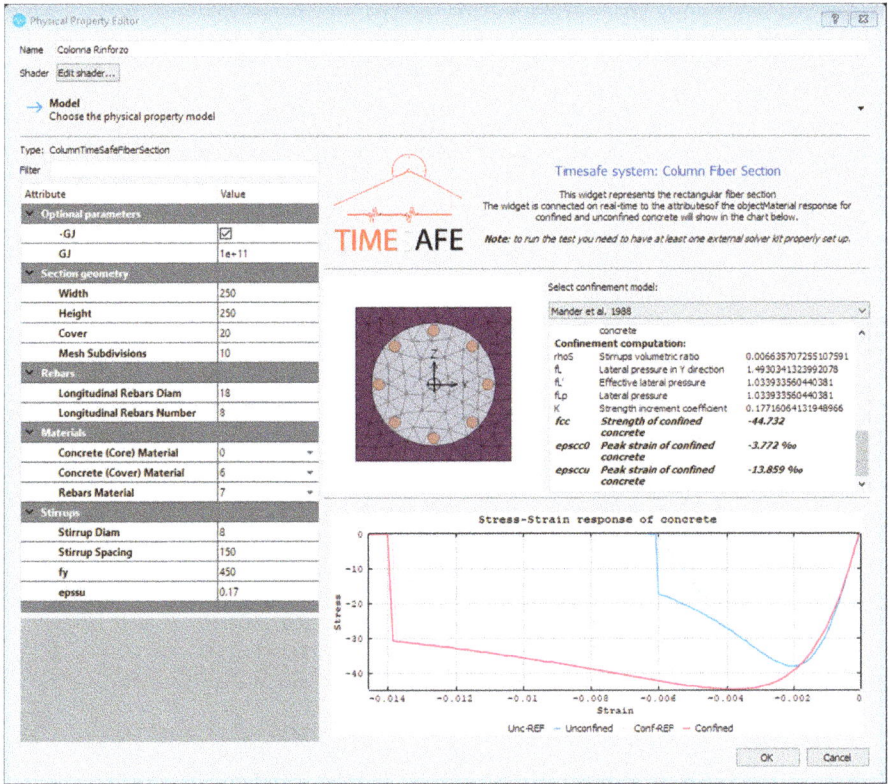

Figure 2. GUI developed for RC-framed skin column accounting automatically for effect of confinement.

3.2. Structural Capacity Assessment

After the nonlinear static analyses, both ductile and brittle mechanisms (i.e., shear) capacities are evaluated for all elements according to EN1998-3 [2].

In detail, for ductile mechanism, capacity checks are performed at the element level, evaluating the chord rotation demand and the corresponding capacity at the ends of each structural element (both beams and columns). The total chord rotation capacity at ultimate limit state (Near Collapse, NC) is evaluated according to the relation A.1 of EN1993-3 [2]:

$$\vartheta_{u,m} = \frac{1}{\gamma_{el}} 0.016 \cdot (0.3^\nu) \left[\frac{max(0.01; \omega')}{max(0.01; \omega)} f_c \right]^{0.225} \left(min\left(9; \frac{L_V}{h}\right) \right)^{0.35} 25^{\left(\alpha \rho_{sx} \frac{f_{yw}}{f_c}\right)} \left(1.25^{100\rho_d}\right) \quad (2)$$

where γ_{el} is equal to 1.5 for primary seismic elements and to 1.0 for secondary seismic elements; ν is the axial load ratio; ω, ω' are the mechanical reinforcement ratios of the tension (including the web reinforcement) and compression longitudinal reinforcement; α is the confinement effectiveness factor; ρ_{sx}, ρ_d are the ratios respectively of transverse steel parallel to the direction x of loading, and of diagonal reinforcement (if any), in each diagonal direction; f_{yw} is the stirrup yield strength (MPa). For RC existing frame designed without detailing for earthquake resistance, the value given by (1) must be divided by the factor 1.2.

Moreover, all the mean values of material strengths should be divided by the appropriate confidence factors based on the Knowledge Level, EN1998-3 [2]. The assessment with respect to Life Safety (LS) limit state refers to a chord rotation capacity equal to $3\vartheta_{u,m}/4$.

For brittle mechanism, the shear demand is compared with the cyclic shear capacity for all elements evaluated according to the relation A.12 of EN1993-3 [2] (with units: MN and meters):

$$V_R = \frac{1}{\gamma_{el,s}} \left[\frac{h-x}{2L_v} min(N; 0.55 A_c f_c) + \left(1 - 0.05 min\left(5; \mu_\Delta^{pl}\right)\right) \right.$$
$$\left. \cdot \left[0.16 max(0.5; 100 \rho_{tot}) \left(1 - 0.16 min\left(5; \frac{L_v}{h}\right)\right) \sqrt{f_c} A_c + V_w \right] \right] \quad (3)$$

where $\gamma_{el,s}$ is equal to 1.15 for primary seismic elements and to 1.0 for secondary seismic elements; x is the compression zone depth; N is the compressive axial force (positive, taken as being zero for tension); $A_c = b_w \cdot d$ with b_w the width and d the effective depth of the cross-section; ρ_{tot} is the total longitudinal reinforcement ratio; V_w is the contribution of transverse reinforcement to shear resistance. According to the Biskinis model [21], adopted in EN1998-3 [2], the decrease of shear strength is assumed as a function of the plastic part of ductility demand $\mu_\Delta^{pl} = \mu_\Delta - 1$ that can be calculated as the ratio between the plastic part of the chord rotation and the chord rotation at yielding. Note that shear resistance of primary seismic elements should be evaluated by adopting mean values of material strengths divided by the appropriate confidence factors based on the Knowledge Level and also by the partial factors for materials in accordance with EN1998-1 [22].

In the following analyses, a unitary confidence factor is assumed with the aim of evaluating the level of improvement attainable by adopting the proposed retrofitting technology in the limit case of maximum Knowledge Level for the existing building.

The attainment of ductile and brittle failures in the first elements, according to Equations (2) and (3), is marked in the capacity curves using for chord rotation capacity a yellow circle and for shear capacity a green triangle. Moreover, the type of elements where the limit is reached is highlighted by the letter C for column and B for beam (see also [23]).

Actually, the capacity curves are not interrupted when the first limit condition is reached but they continue up to a loss of at least 15% of the peak base shear capacity (according to Italian National Standard [20]), since the attainment of failure (whether ductile or brittle) in the first element is not assumed corresponding to failure of the whole structure, due to the 'conventional' nature associated with the definition of ultimate limit conditions.

For both RC existing and retrofitted building, in correspondence of specific steps of the loading history, the frame elements' state is represented to highlight where the limit conditions are reached, and the inter-story drift distribution is provided in order to support the interpretation of failure mechanisms and the validation of retrofitting measures.

4. Existing RC Frame

In this section the geometry, the loading conditions, and the material characteristics of a typical GLD multi-story RC frame representative of the Italian residential building stock of the 1970–1980s are described. Then, the main issues related to the numerical modeling approach adopted to assess the seismic performance of the selected building are illustrated and the results of static nonlinear analyses are presented.

4.1. Geometry and Loading Conditions

The selected building presents a rectangular plan with 14.30 m × 18.30 m dimensions and 3.30 to 3.50 m inter-story height, Figure 3. The geometrical characteristics of frame elements, as well as their reinforcement ratios, have been derived from a simulated design performed only for gravity load, according to the technical standards and design rules in force in the 1980s.

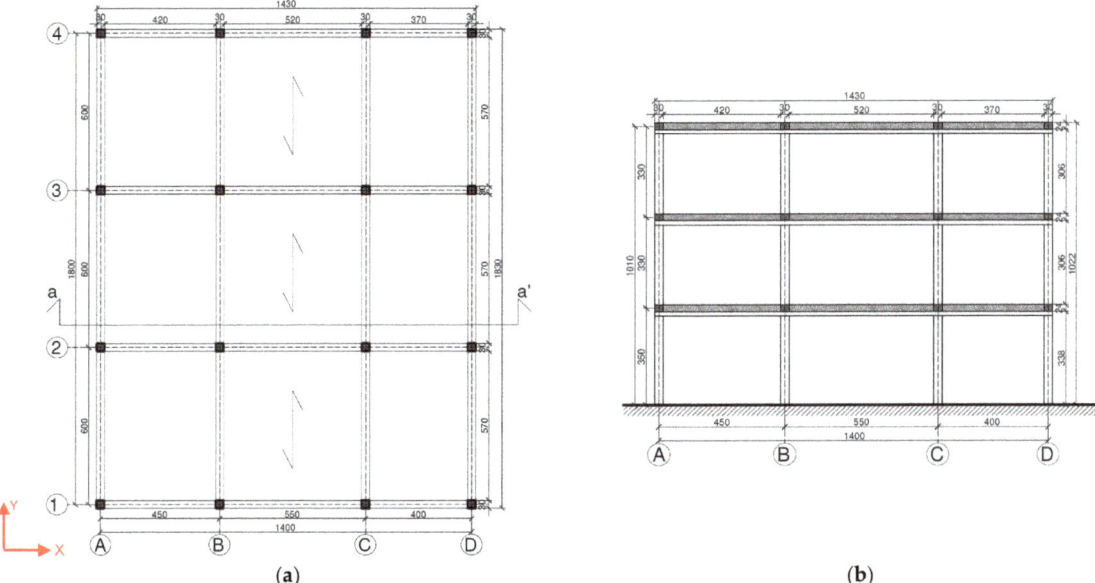

Figure 3. Three-story RC building: (**a**) Plan; (**b**) transversal view (units in cm).

The building has four lateral-load resisting frames along the transversal X direction, with constant beam cross-section 300 mm × 400 mm. Along the longitudinal Y direction, there are only two external frames where the columns are connected through a flat beam with cross section 300 mm × 240 mm. All the columns have the same cross-section, 300 mm × 300 mm. The reinforcement layouts of the typical cross-sections of the frame elements are shown in Figures 4 and 5. It is worth remarking that all RC elements are designed without any seismic detail.

Gravity loads are represented by a dead load of 5.00 kN/m² on the top floor and 6.0 kN/m² on the other floors. The live load is assumed 1.2 kN/m² on the top floor and 2.0 kN/m² on all the other floors corresponding to residential category (i.e., imposed load on floors, category A according to EN 1991-1-1 [24]). An average weight of 6.25 kN/m per unit length has been considered for external masonry infills.

4.2. Numerical Modeling

In this section, some issues related to the numerical modeling of the existing building are provided and the main results of pushover analyses are presented. All the displacements and rotations of the existing RC frame base nodes are fully restrained.

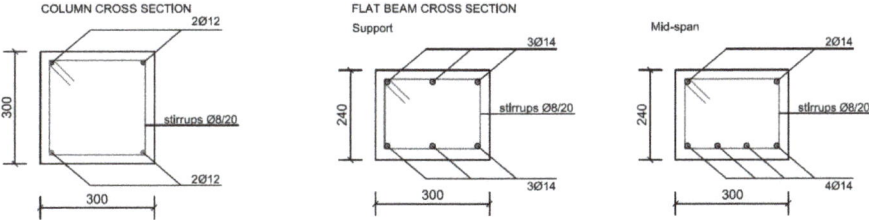

Figure 4. Three-story RC building. Reinforcement layout of RC element cross-sections: column and external flat beams in the weak Y direction (units in mm).

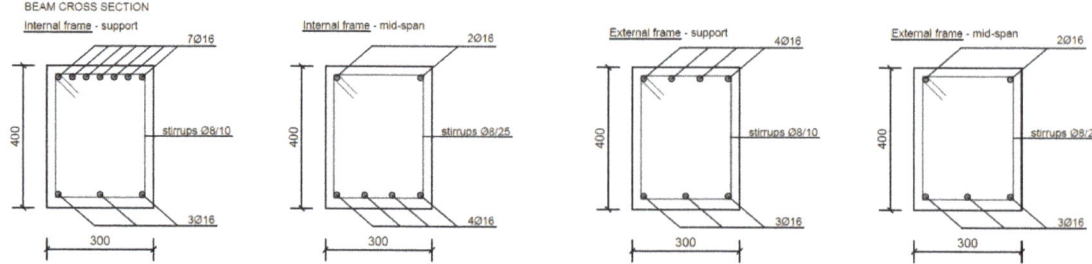

Figure 5. Three-story RC building. Reinforcement layout of RC element cross-sections: beams in the strong X directions (units in mm).

4.2.1. Material Properties

The material properties of both concrete and steel reinforcement of the existing building are summarized in Table 1. The concrete is assumed as a C25 grade according to Model Code 2010 [25], and the corresponding mean compressive value is evaluated as $f_{cm} = f_{ck} + \Delta f$ with $\Delta f = 8$ MPa [25]. The mean yield strength, the hardening ratio, and the ultimate strain of steel reinforcement are evaluated by means of STIL software developed by Verderame et al. [26], based on an extensive database of experimental tests on reinforcing bars collected from samples of the second half of the last century. In particular, almost 2500 samples of steel were selected within the period 1970 and 1990, computing the mean and standard deviation of yield strength, hardening ratio, and ultimate strain.

Table 1. Material properties for existing RC frame.

Material	Property	Value
Concrete: C25	Mean compressive strength	33 MPa
	Strain at peak stress	0.002
	Ultimate strain	0.006
Steel: FeB44k	Mean yield strength	510 MPa
	Hardening ratio [1]	1.52
	Ultimate strain	21.7%

[1] The hardening ratio is defined as ultimate strength over yield strength f_u/f_y.

4.2.2. Nonlinear Static Analysis

In this section, the main results in terms of capacity curves and distribution of plastic hinges at different time steps of the analysis are presented for both the strong and the weak directions, e.g., the X and Y direction, respectively.

As for the plastic hinge lengths, according to relation (1), for the existing frame they are in the range 300–330 mm for the deep beams, depending on their length, 290 mm for the flat beam, and 235 mm for the columns.

The results in terms of base shear vs. top displacement curves are reported for X (strong) direction in Figure 6a and for Y (weak) direction in Figure 6b. In both cases, the two distributions of the lateral loads are applied, according to the first mode and to a uniform pattern proportional to mass. The capacity in terms of base shear is, in the weak Y direction, about half with respect to X direction. The attainment of the LS chord rotation capacity in the first column of the existing frame occurs after about 30 mm from the peak in both pushover analyses (uniform and modal) in the X direction, while for the analyses in the Y direction the rotation capacity of the first column at the ground floor is achieved just after the beginning of the softening branch. Moreover, in Y direction the flat beams achieve their shear capacity close to the peak of the curve.

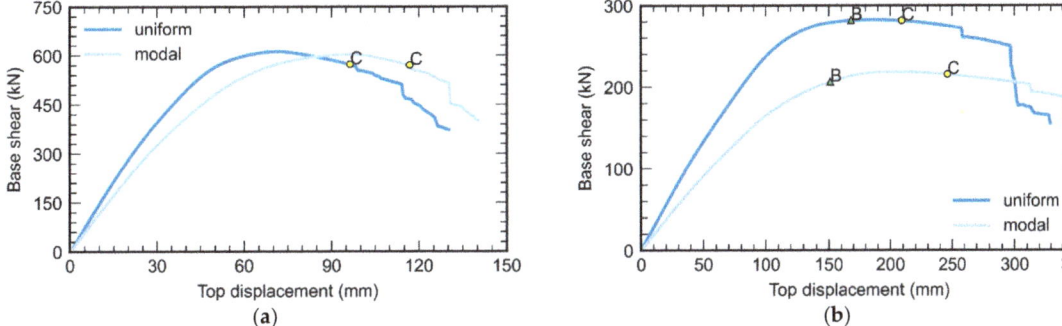

Figure 6. Capacity curves for RC existing building: (**a**) X direction; (**b**) Y direction. A yellow circle indicates the attainment of the LS chord rotation capacity and a green triangle the attainment of the shear capacity. The type of elements where the limit is reached is marked by the letter C for column and B for beam.

The failure mechanisms can be inferred by observing the evolution of the curvatures of the sections at the column and beam extremities during the analyses, reported in Figures 7 and 8 for loading in X and Y direction, respectively. Actually, in the X (strong) direction the first hinges occur at the base of the columns, with some hinges also at the top sections of the columns in the first and second floor, as shown in Figure 7a. Then, a soft-story mechanisms is clearly evidenced with plastic hinges in all of the columns of the ground floor at both base and top sections when the ultimate chord rotation capacity is attained, as depicted in Figure 7b.

As expected, the behavior in the Y (weak) direction is significantly different, with a much evident deformability (top displacement at maximum base shear almost three times greater than in the strong direction) and the formation of plastic hinges on the flat beams since the first steps of analysis and then a progressive formation of plastic hinges at the base of columns up to the structural failure (Figure 8).

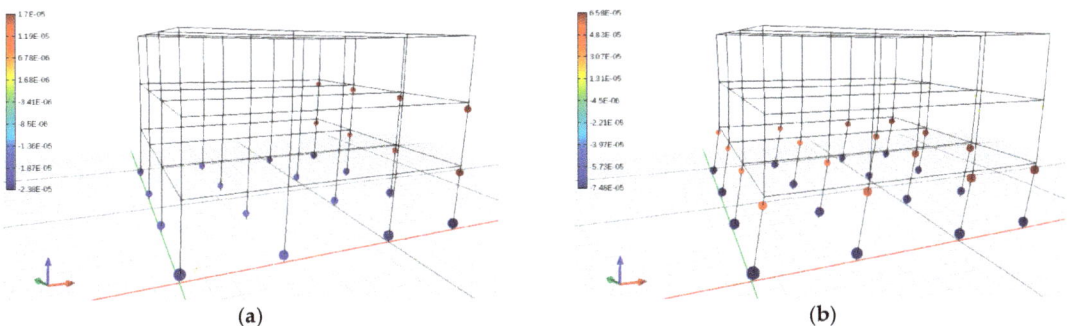

Figure 7. RC existing building, uniform pushover. Curvature distribution corresponding to a top displacement in X direction of: (**a**) 50 mm; (**b**) 95 mm (attainment of the LS chord rotation capacity). Red line X direction, green line Y direction.

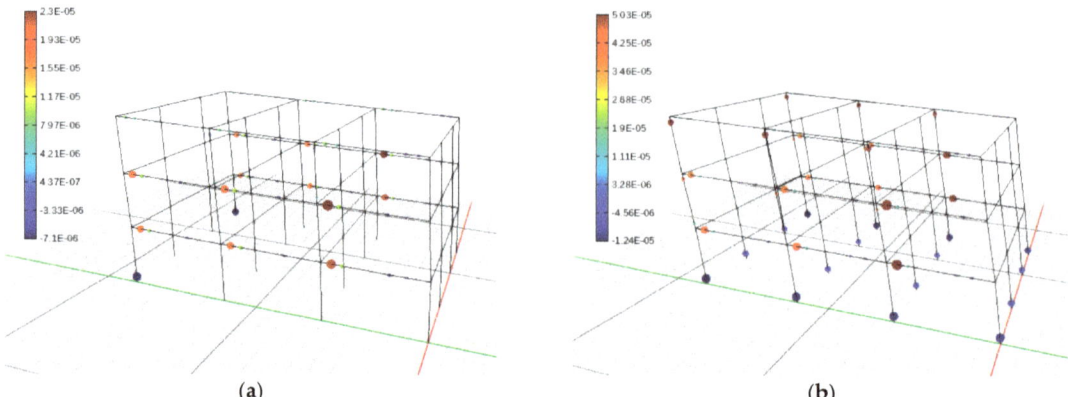

Figure 8. RC existing building, uniform pushover. Curvature distribution corresponding to a top displacement in Y direction of: (**a**) 100 mm; (**b**) 200 mm (attainment of the LS chord rotation capacity). Red line X direction, green line Y direction.

5. Retrofitted Building

In this section, the main characteristics of the existing building retrofitted by adopting the proposed new RC-framed skin are described, and the results of static nonlinear analyses are presented.

5.1. Geometry and Reinforcement Details

The proposed RC-framed skin is externally applied to the existing RC building described in the previous section and connected at each floor-level using anchor rods designed to guarantee a properly rigid connection as well as the respect of the hierarchy of strength principle. EPS modules 1200 mm wide and 325 mm thick are used to clad the structure. The adopted cross-sections for the RC-framed skin are 250 mm × 250 mm for the columns and 250 mm × 400 mm for the beams, respectively.

The geometry of the retrofitted building is depicted in Figure 9, where the spacing of the column is selected according to the opening layout of the existing building. Figure 10 shows the 3D view of retrofitted building, while Figure 11 shows the reinforcement layout of beam and column cross-sections of the RC-framed skin.

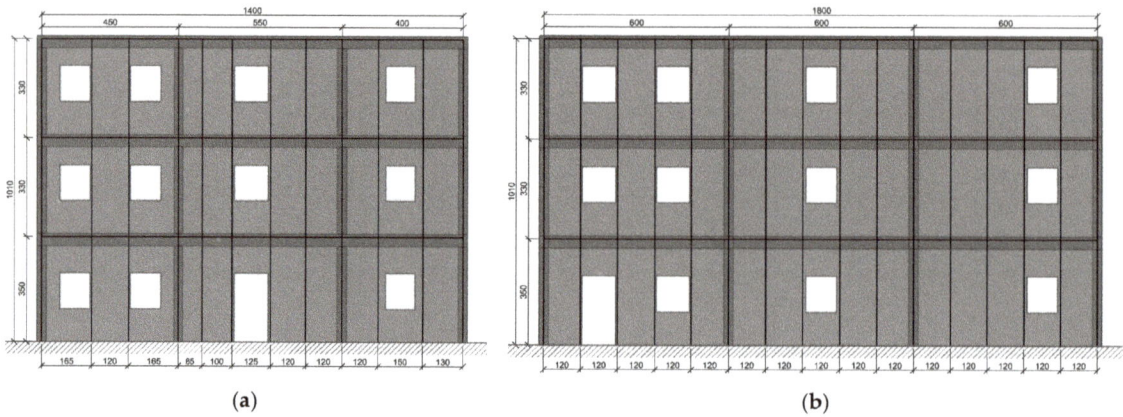

Figure 9. RC-framed skin applied to the existing RC building with evidenced the opening lay-out: (**a**) transversal view; (**b**) longitudinal view.

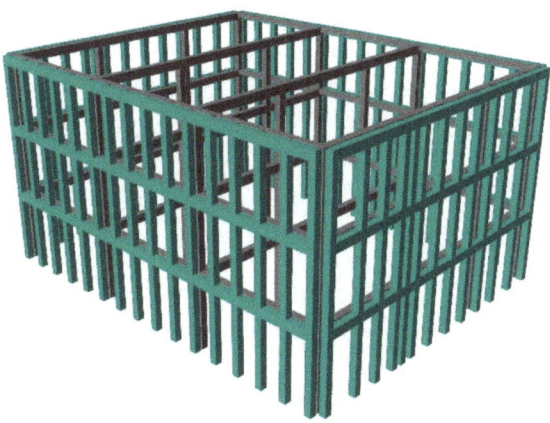

Figure 10. RC-framed skin as reinforcement for the existing RC building: 3D view.

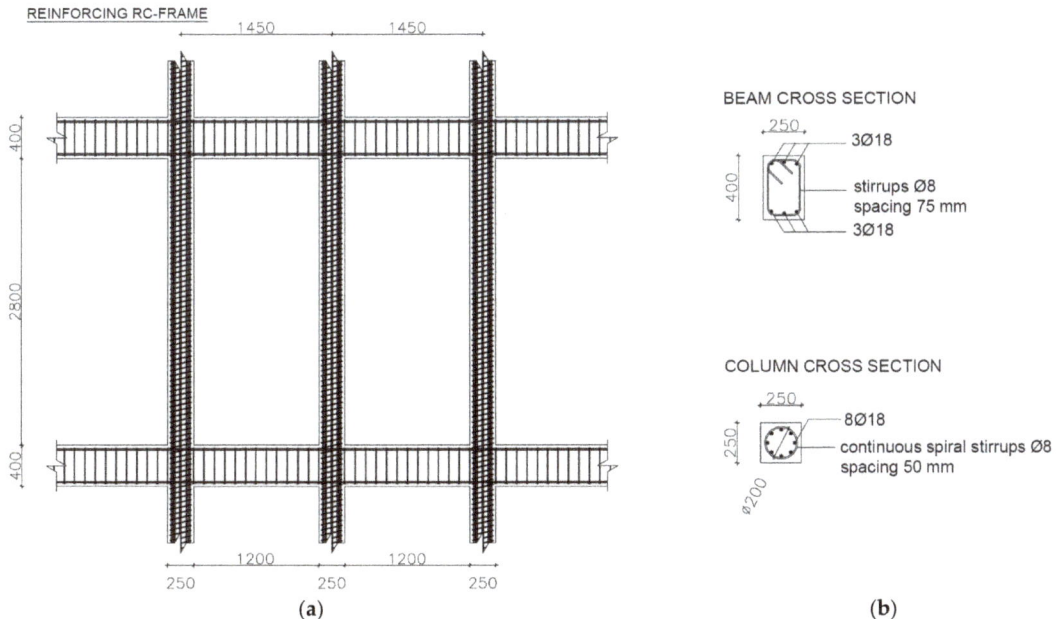

Figure 11. RC-framed skin: (**a**) reinforcement layout; (**b**) beam and column cross-sections.

5.2. Numerical Modeling

The RC-framed skin is modeled as connected to the existing structure by means of kinematic constraints imposing the equality of the horizontal and vertical displacement, while the base nodes are all fixed.

5.2.1. Material Properties

The material properties of both concrete and steel reinforcement of the RC-framed skin are summarized in Table 2. The concrete is assumed as a C30 grade according to Model Code 2010 [25], while the steel reinforcement is assumed equivalent to a B450C type.

Mean yield strength, hardening ratio, and ultimate strain of steel are taken from available experimental results [23].

Table 2. Material properties for RC-framed skin.

Material	Property	Value
Concrete: C30	Compressive strength	38 MPa
	Strain at peak stress	0.002
	Ultimate strain	0.006
Steel: B450C	Yield strength	514 MPa
	Hardening ratio [1]	1.19
	Ultimate strain	25.3%

[1] The hardening ratio is defined as ultimate strength over yield strength f_u/f_y.

5.2.2. Nonlinear Static Analysis

In this section, the main results in terms of capacity curves and distributions of plastic hinges at different time steps of the analysis are presented, for seismic action in X and Y directions.

The plastic hinge lengths, for the RC-framed skin elements according to relation (1), are about 280 mm for the beams and 285 mm for the columns.

The results in terms of base shear vs. top displacement curves are reported for X (strong) direction in Figure 12a and for Y (weak) direction in Figure 12b, for both the lateral forces distribution, i.e., uniform and modal. Due to the effect of the reinforcing RC-skin frame, the capacity in terms of base shear is significantly higher than the capacity of the existing building and the values for actions in X and Y directions are now comparable, leading to a structure without a weak direction. Actually, the capacity is slightly higher in the Y direction respect to that in the X direction, since there are more reinforcing columns along the sides parallel to Y direction. Moreover, it is possible to observe that, in both directions and for both pushover analyses (uniform and modal force distributions), the LS chord rotation capacity of the first column of the existing building is achieved just after the beginning of the softening branch.

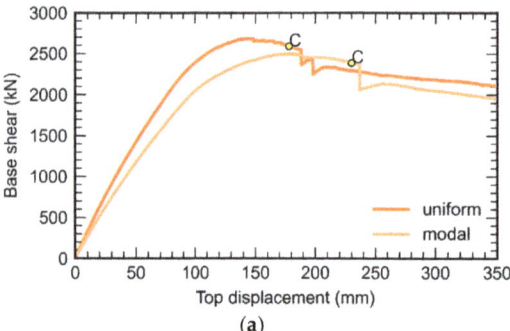
(a)

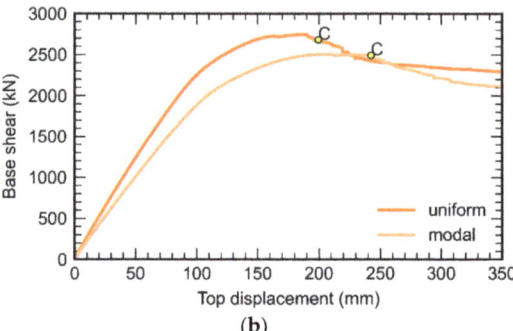
(b)

Figure 12. Capacity curves for the retrofitted building: (**a**) X direction; (**b**) Y direction. A yellow circle indicates the attainment of the LS chord rotation capacity and the type of elements where the limit is reached is marked by the letter C for column.

Figure 13 shows the curvature of the different sections at two steps of the pushover in X direction with uniform distribution of forces: Figure 13a refers to yielding corresponding to a top displacement of about 100 mm, while Figure 13b refers to the attainment of ultimate chord rotation capacity in the columns of the existing building. It is possible to observe the mechanism with the formation of plastic hinges at the base and the top section of all the existing columns of the first floor and some of the existing columns of the second floor, together with the formation of plastic hinges in the columns of the reinforcement RC-skin frame in the first and second floor.

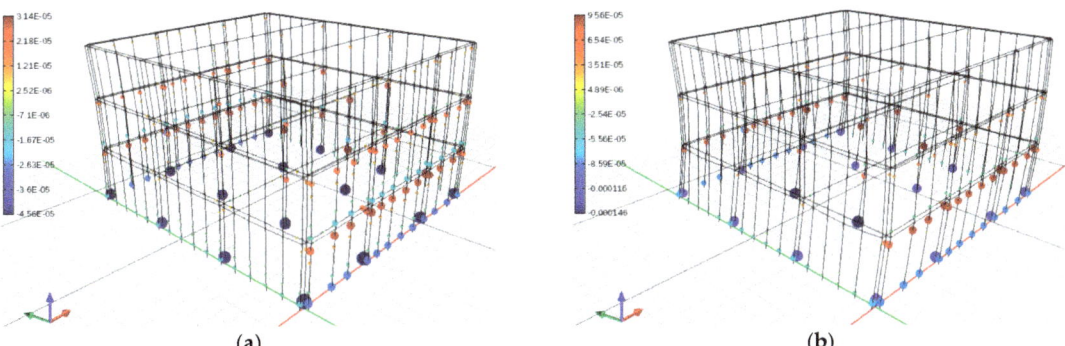

Figure 13. Retrofitted building, uniform pushover in X direction. Curvature distribution corresponding to a top displacement of about: (**a**) 100 mm; (**b**) 180 mm (attainment of the LS chord rotation capacity on a column of existing building).

Concerning the pushover in Y direction, Figure 14a depicts the curvature of the different sections for a top node displacement of about 100 mm: it is possible to observe that the plastic hinges occur at the bottom and top sections in the reinforcing columns at the first and second floor while, in the existing frame, the hinges are again at the base of the columns and on the flat beams. Figure 14b shows the curvature corresponding to the attainment of LS chord rotation capacity of the columns of the existing building. It is interesting to note that no shear failure was detected on the existing flat beams, as opposed to the pushover in Y direction of the existing building without reinforcement (see Section 4.2.2).

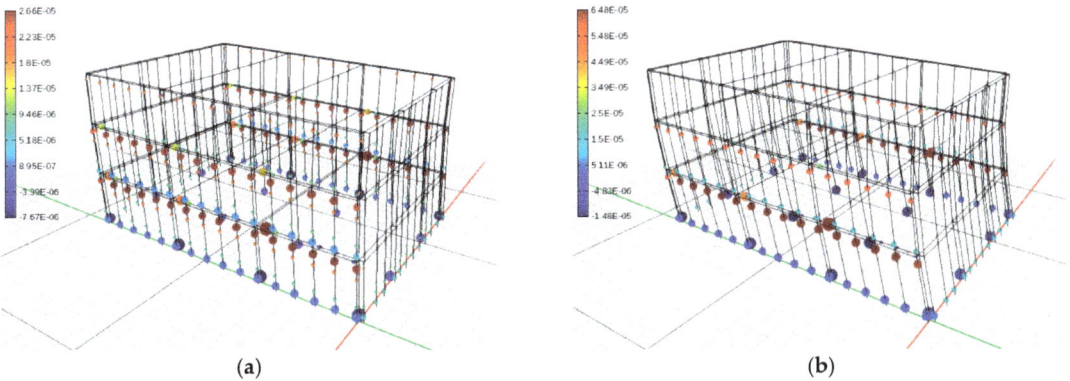

Figure 14. Retrofitted building, uniform pushover in Y direction. Curvature distribution corresponding to a top displacement of about: (**a**) 100 mm; (**b**) 200 mm (attainment of the LS chord rotation capacity on a column of existing building).

6. Discussion

As presented in the previous sections, a set of pushover analyses are carried out on the existing RC building and on the retrofitted building. In this section, a critical discussion of the results obtained is provided, also comparing the capacity curves of the unreinforced and reinforced building with the performance of the RC-framed skin only.

As an example, Figure 15 shows the three curves (unreinforced building, reinforced building, RC-framed skin only) in the case of uniform pushover analysis, highlighting the effect of the proposed intervention in terms of maximum shear capacity that increases respectively of about 4.5 times in X direction and more than 9 times in Y direction, i.e., the weak direction of the existing building.

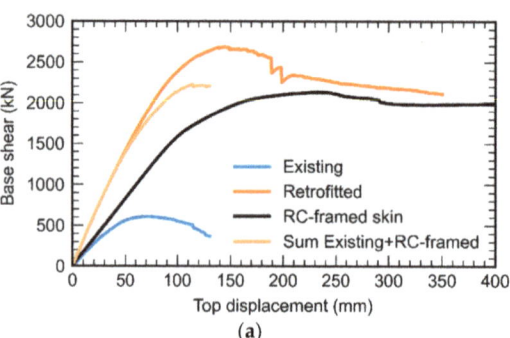 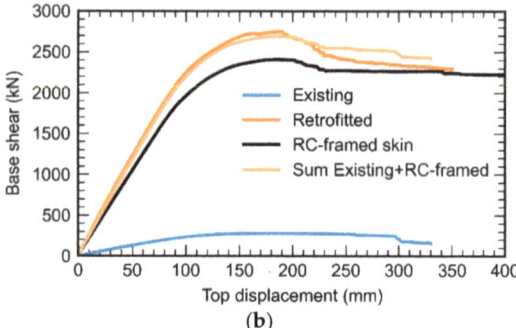

Figure 15. Uniform pushover analysis. Capacity curves of the existing RC building, the RC-framed skin, and the retrofitted building: (**a**) X direction; (**b**) Y direction.

It is possible also to see that the reinforcing frame increases significantly the stiffness of the building both in X and Y directions with an increase respectively of about 2 times and 10 times. The last very high value in the weak direction is due to the change of structural behavior from cantilever-like for the existing building to frame-like for the reinforced building. Of course, this aspect is important if the reinforcement is designed also to reduce the displacements and consequent damages to nonstructural components in the case of moderate earthquakes.

Concerning the displacement capacity, in X direction the displacement corresponding to the ultimate condition (attainment of ultimate chord rotation in columns) is increased of about 90%.

Indeed, comparing Figure 13a with Figure 7b it can be observed that, for a given value of top displacement, the reinforcement reduces the curvature at the bottom sections of the existing columns of almost 50%, so avoiding the formation of the soft-story mechanism, because the deformations are better distributed along the height of the existing structure. On the other hand, in Y direction the ultimate displacement remains substantially the same if one considers the attainment of ultimate chord rotation, while it increases of about 20% if one considers the attainment of shear capacity in the flat beams of the existing building that is avoided in the retrofitted building.

Moreover, it is worth noting that the maximum shear capacity of the retrofitted building in the strong direction reaches a value higher than the simple superposition of the existing building and RC-framed skin results of about 20%. This can be ascribed to the change in the mode of failure of the structure from a brittle soft-story collapse (for the existing building), to a more ductile one (for the retrofitted building), as described respectively in Section 4 (see also Figures 7 and 8) and in Section 5 (see also Figures 13 and 14). On the contrary, in the weak Y direction no difference between shear capacity of the retrofitted building and the simple sum of capacities of original structure and reinforcement is observed (Figure 15b).

The different failure-mode in the X direction for the retrofitted building with respect to the original building is also confirmed by the inter-story drift (ISD) distribution along the building height. Actually, Figures 16 and 17 show the evolution of the distributions of the ISD in the case of uniform pushover analysis for the existing RC building with and without retrofit measures. It is worth noting that the existing building exhibits a significant concentration of ISD in the ground floor story with seismic action in X direction, while the retrofitted building exhibits a more uniform distribution of drift, so demonstrating that a uniform distribution of both stiffness and strength is attained through the retrofit intervention.

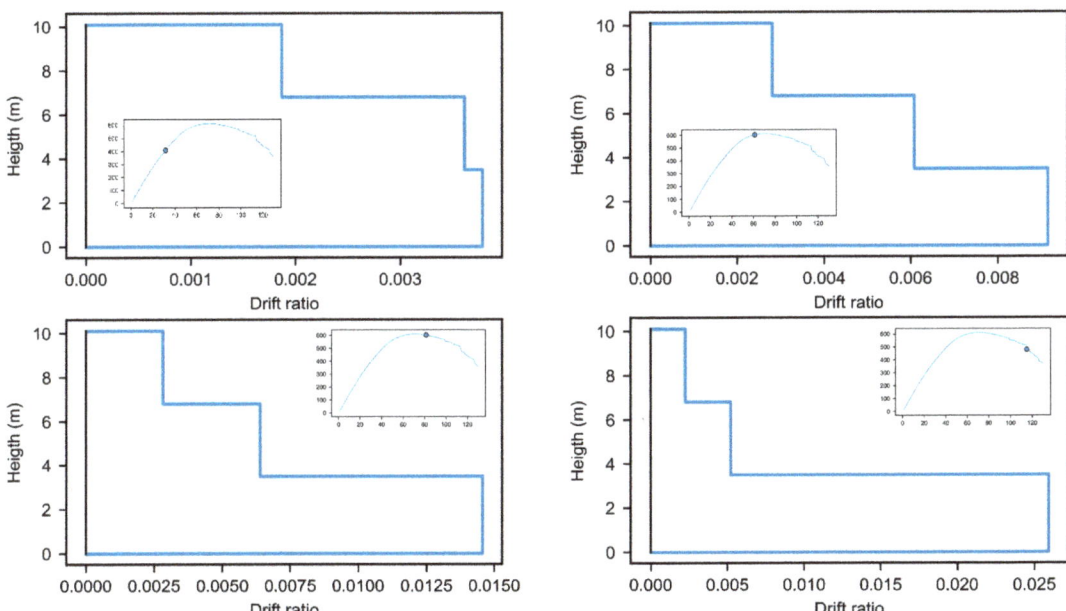

Figure 16. Existing building. Uniform pushover analysis in X direction: evolution of the inter-story drift distributions.

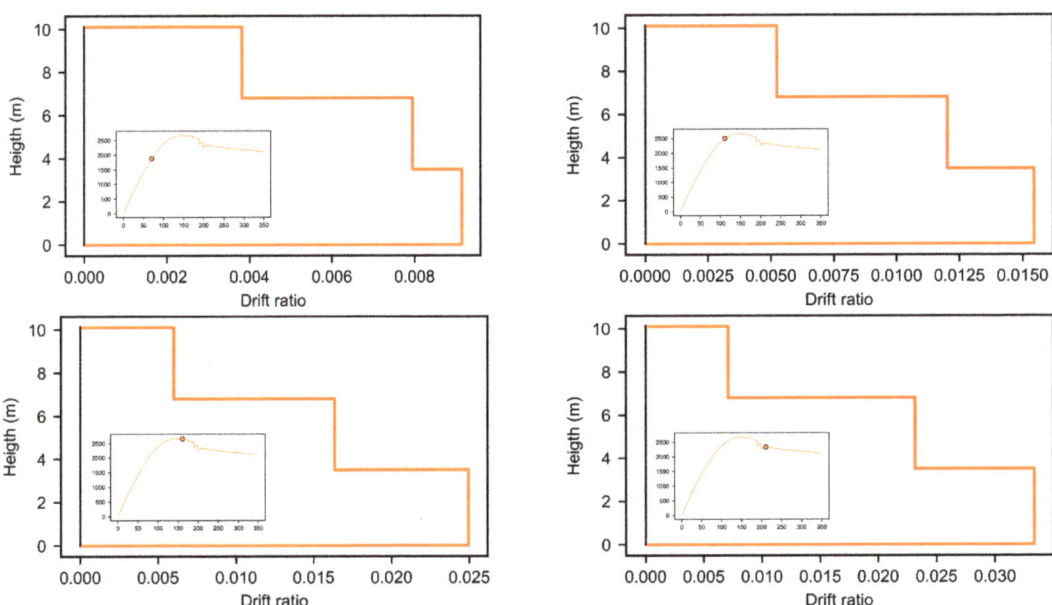

Figure 17. Retrofitted building. Uniform pushover analysis in X direction: evolution of the inter-story drift distributions.

On the contrary, in the weak Y direction both the existing and retrofitted buildings (Figures 18 and 19) exhibit a more uniform distribution of drift.

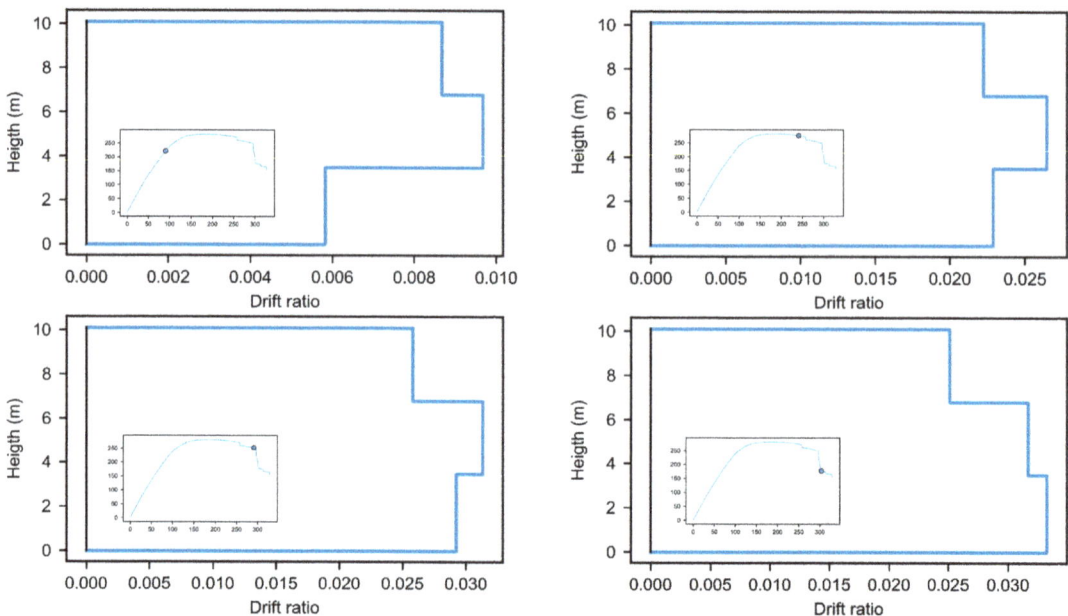

Figure 18. Existing building. Uniform pushover analysis in Y direction: evolution of the inter-story drift distributions.

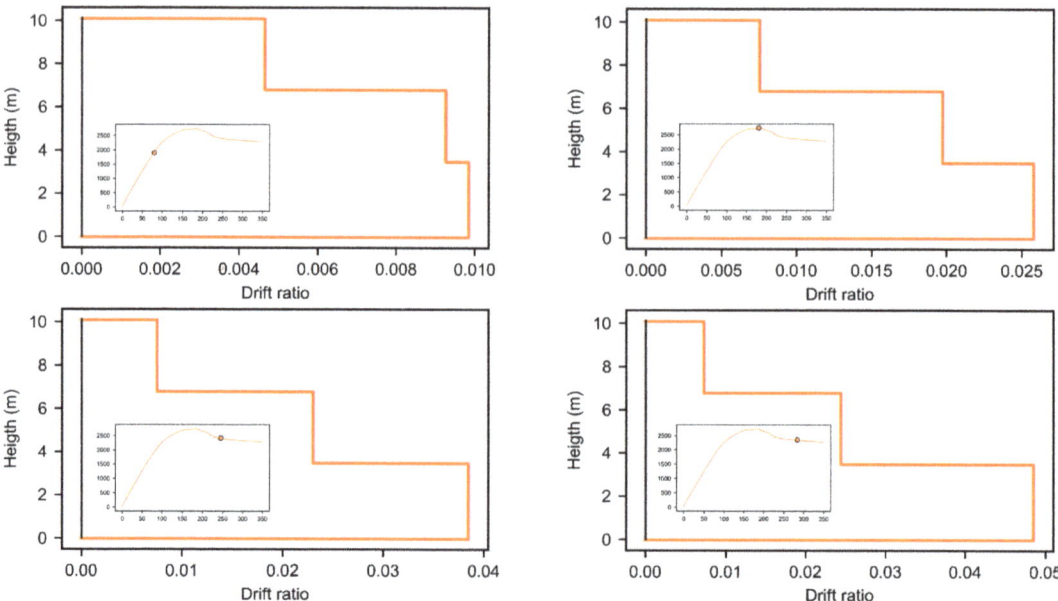

Figure 19. Retrofitted building. Uniform pushover analysis in Y direction: evolution of the inter-story drift distributions.

7. Conclusions

The proposed RC-framed skin consists of cladding the existing building envelope with an innovative system, obtaining a retrofitted structure with a significant improvement of seismic performances, other than of energy efficiency, see also [13]. One of the main

Figure 1. The former tobacco factory in Cervinara: (**a**) outdoor; (**b**) indoor.

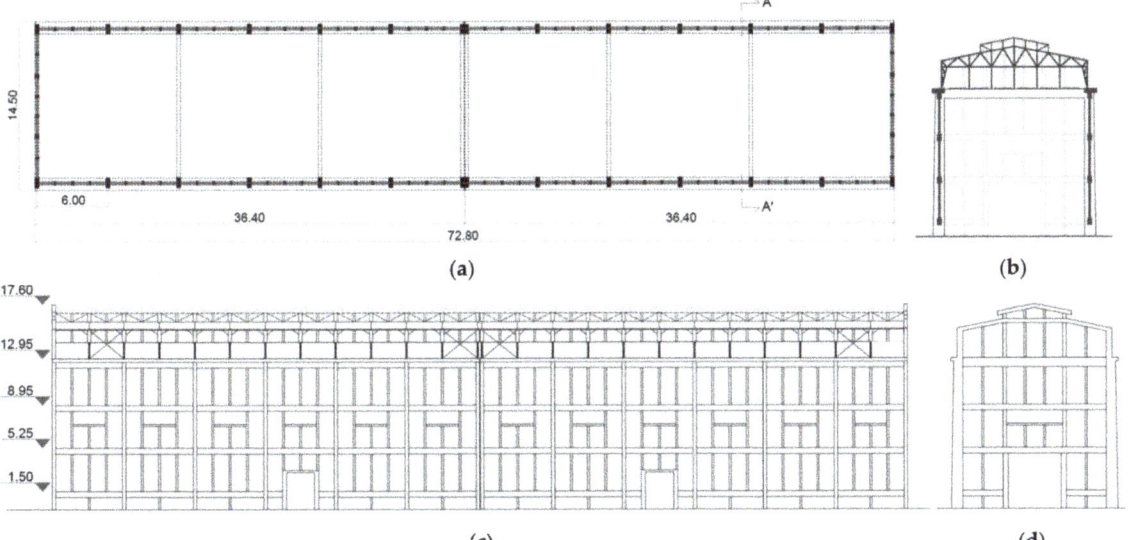

Figure 2. The former tobacco factory in Cervinara: (**a**) fourth-floor plan; (**b**) transverse section A–A'; (**c**) south side; (**d**) east side.

2.2. Characterization of Materials and Soil

As mentioned above, all the structural elements are made of reinforced concrete and all the frames are filled with hollow brick walls. Since it was not possible to perform destructive and non-destructive tests on building materials, the mechanical characterization of materials is carried out thanks to several studies about evolution of mechanical properties of materials over the years and with in situ tests and visual inspections. All these studies are based on the Italian standards in force at that time [14]. Therefore, many experimental test values detected from literature sources on existing RC buildings analogous to the building under study are considered. This means that the mechanical properties found automatically consider the decrease of strength and Young's moduli deriving from environmental actions and loading history applied. Therefore, the range of values suggested from the literature sources is considered and, aiming at performing analysis on the safe side, it was decided to use the lowest value of this range.

Regarding concrete, according to a literature study [15], it is possible to define three generations of materials based on the production age:
1. First generation: from 1930 to 1955;
2. Second generation: from 1955 to 1980;
3. Third generation: from 1980 to today.

The industrial building under investigation belongs to the second generation, so the minimum resistance considered is C12/15, as shown in Figure 3.

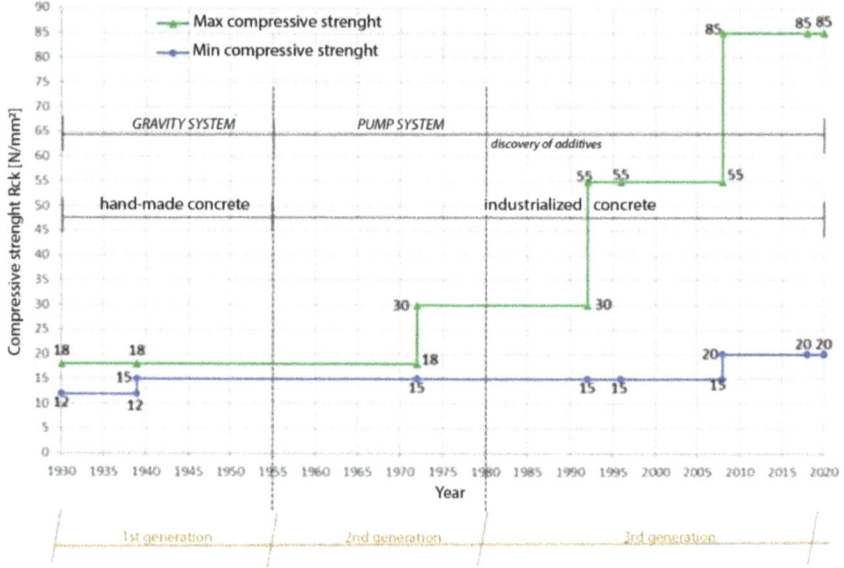

Figure 3. Evolution of the concrete strength from 1930 to today [15].

The mechanical properties of reinforcement steel are defined with the "STIL v1.0" software, which was defined based on extensive research in the field [16–18]. By entering information about production age and steel bars type, the software indicates the average values of yield strength f_y, hardening ratio f_u/f_y, and elongation at failure.

In this case, the considered average value of yield strength is 350 MPa. Visive inspection of degraded elements has allowed to assume the disposition of bars.

With regard to mechanical characterization of the steel trussed roof, a previous study indicated that steel used in 1960 has properties comparable with S235 steel [19].

According to the Italian standards [20,21], the knowledge level of the structure is represented by a confidence factor that reduces the values of the mechanical parameters of the materials. In this case, the confidence factor FC is equal to 1.2, which is the value representative of an intermediate knowledge level.

Regarding subsoil, surveys carried out by the city of Cervinara during the drafting of the 2018 Municipal Urbanistic Plan PUC provided a seismic category B with a topographic category T1, which means a ground slope between 0° and 15°.

2.3. Evaluation of Gravity Loads

The considered building has no intermediate floors and no roof structure, so the only gravity loads are those deriving from both the vertical framed structures and the walls.

Beams, columns, and walls weights are to be calculated according to Italian Code NTC 2018, considering concrete specific weight γ_{RC} equal to 25.0 kN·m^{-3} and hollow brick specific weight γ_{wall} equal to 8.0 kN·m^{-3}. In particular, the SAP2000 software allows to calculate beams and columns weights automatically, while the infill walls weight G_{infill} is considered as distributed load on the beams with a value of 915.78 kN·m^{-1}.

The presence of openings is considered as percentage decrease of walls weight: 10% for the first floor, 20% for the second floor, and 0% for the other floors.

2.4. FEM Modeling and Nonlinear Analysis

The FEM model of the building before intervention (Figure 4) was set up by means of the SAP2000 v20.0 software according to the Italian Code NTC 2018. Beams and columns are defined as one-dimensional elements by entering all the acquired data about materials and geometries of sections. Interaction between soil and building is schematized by assigning full constraints at the bases of columns. The roof structure is schematized with linear elements able to represent a constraint between east and west facades. Beam-column nodes are considered as rigid with their own stiffness. Infill walls are schematized as distributed loads on the frame members, neglecting their contribution in terms of stiffness. This assumption is justified since the spans are filled with weak cladding walls having thickness of 10 cm only, which are also provided with some openings. Therefore, the contribution given by infill walls to the building lateral stiffness is neglected. In the nonlinear analyses, concentrated plastic hinges are assigned to all the structural members.

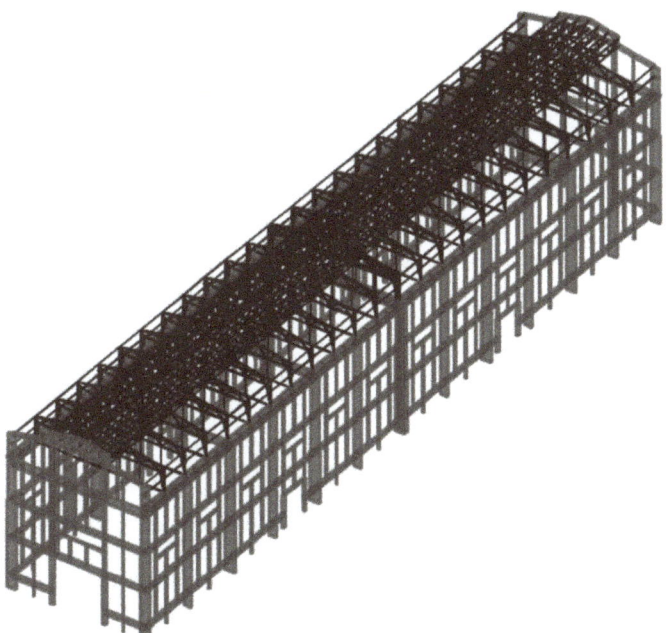

Figure 4. FEM model of the factory.

The seismic vulnerability evaluation of the building is carried out using N2 method [22,23]. This method is based on distribution of horizontal forces acting along the main building directions x and y, parallel to the long and short facades of the building, respectively.

For each direction, two distributions of forces are considered:

4. Horizontal forces equivalent to a uniform distribution of horizontal accelerations, that are proportional to the building seismic mass, and are herein called Accel_x and Accel_y.
5. Horizontal forces proportional to the elastic modal strain of the main vibration mode, Φ_{1x} and Φ_{1y}, that are herein called Modal_x and Modal_y.

The evaluation of dynamic properties of the building is carried out through a modal analysis on the structural model. The analysis results of the structure in terms of the force–displacement curves obtained from the SAP2000 software are represented in Figure 5.

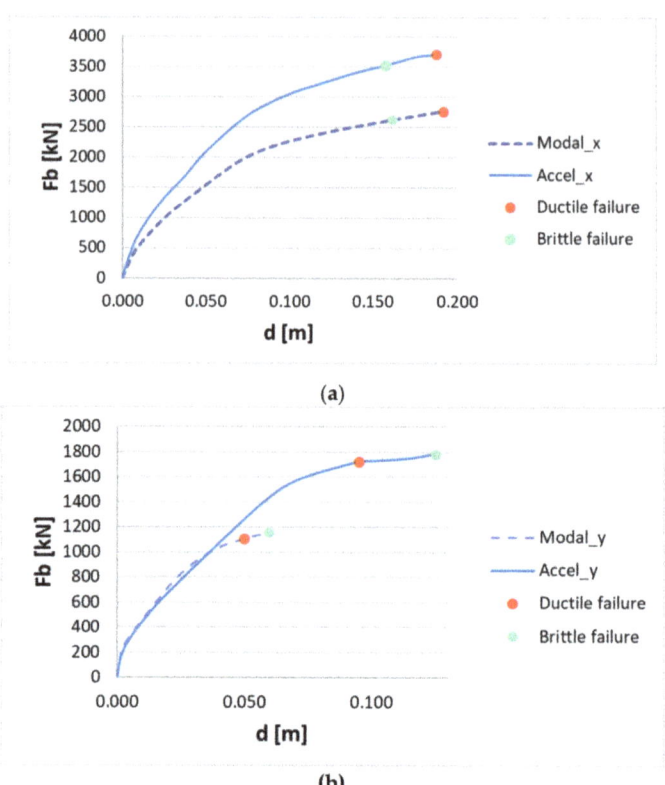

Figure 5. Capacity curves: (**a**) x-direction; (**b**) y-direction.

Ductile failure and brittle failure are represented on the curves of Figure 6 with cyan and red points, respectively.

By representing the capacity curves on the ADRS plane and by comparing them to the elastic response spectrum of the soil, it is possible to obtain the required displacements (Figure 6). The ratio between capacity displacement and demand one is the safety index (IR), whose values referring to the two analysis directions and seismic force distributions are depicted in Table 1. In this table $d_{u,d}*$ and $d_{u,b}*$ are the ductile (bending moment) capacity displacement and the brittle (shear) capacity one, respectively, d_r* is the demand displacement required by the earthquake, and IR,d and IR,b are the safety indices related to the ductile failure and the brittle failure, respectively. As it is observed in Table 1, safety index related to Modal_y case is less than 1, so that the most critical situation is identified precisely along this direction.

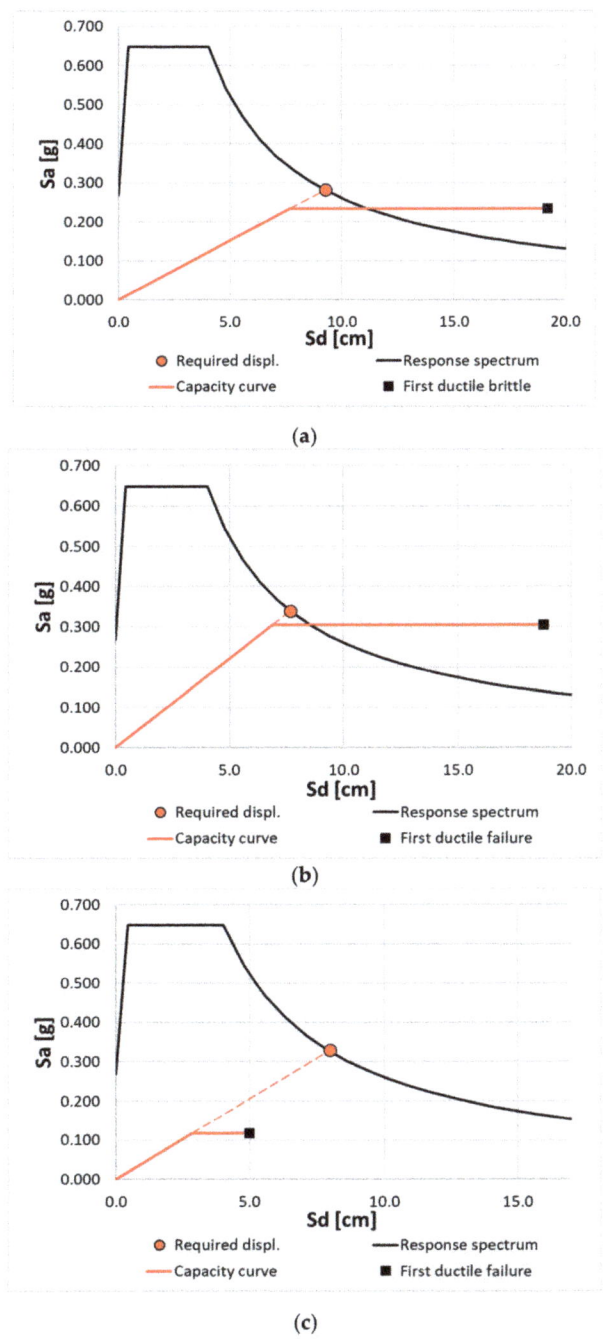

Figure 6. *Cont.*

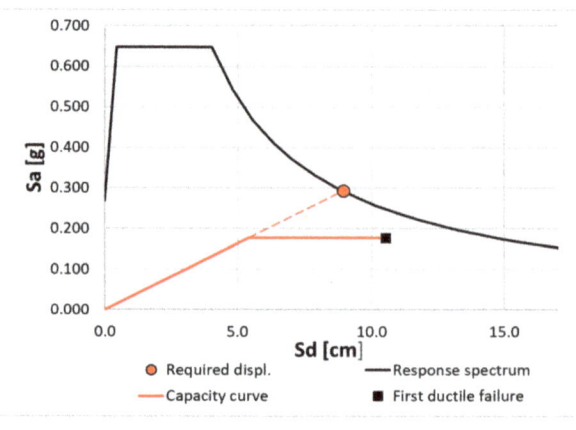

(d)

Figure 6. Seismic checks in the ADRS plane: (**a**) modal_x; (**b**) accel_x; (**c**) modal_y; (**d**) accel_y.

Table 1. Safety indices.

Distribution	du,d* (cm)	du,b* (cm)	dr* (cm)	IR,d	IR,b
Modal_x	19.2	16.2	9.3	2.1	1.7
Accel_x	18.8	15.8	7.7	2.4	2.1
Modal_y	5.0	6.0	8	0.6	0.8
Accel_y	10.5	12.5	8.9	1.2	1.4

2.5. Design of Exoskeletons

To improve the seismic behavior of the factory, both orthogonal and parallel exoskeletons are designed and placed along the building transverse direction (y-direction). From the elastic response spectrum, a target displacement Δ^*_{tar} is defined, so that the retrofitted structure remains in the elastic field. In particular, $\Delta^*_{tar} \leq d^*_y$, d^*_y being the yielding displacement of the structure.

To evaluate the global lateral stiffness of the retrofitted structure K_d, the following equation is used:

$$K_d = \frac{F^*_e}{\Delta^*_{tar}} = \frac{m^* \cdot S_{ADRS}(\Delta^*_{tar})}{\Delta^*_{tar}} \quad (1)$$

where m* is the mass of the structure and S_{ADRS} is the elastic spectral acceleration for the displacement Δ^*_{tar}. The formulation above is valid under the hypothesis that the equivalent mass and the modal participation factor of the existing construction and the retrofitted ones remain the same. Furthermore, it is valid under the hypothesis that the yielding displacement of the exoskeleton corresponds to the yielding displacement of the existing structure. Under the hypothesis of parallel coupling, the global lateral stiffness of the exoskeleton Ke can be evaluated by the following equation:

$$K_e = K_d - K_{EX} \quad (2)$$

where K_{EX} is the global lateral stiffness of the existing structure, that can be evaluated as

$$K_{EX} = \frac{F^*_y}{d^*_y} \quad (3)$$

where F^*_y and d^*_y are yielding force and displacement of the system, respectively [24–27].

With regard to the typological choice, based on the seismic-resistant scheme, and the dimensional choice, related to the first attempt for system design, concentric bracing frames (CBS) and circular hollow sections (CHS) are chosen to comply with the previous provisions related to global lateral stiffness. Columns have diameter of 193.7 mm with thickness from 5 to 10 mm on the longest sides (north and south) and of 219.1 mm with thickness from 8 to 16 mm on the shortest sides (east and west). CHS profiles of 168.3 × 8 mm are used for beams. Bracings along the east and west sides are made of CHS sections with diameter of 114.3 mm and thickness ranging from 3.2 to 8 mm. Instead, bracings of the exoskeletons placed on the building north and south sides have diameter of either 88.9 mm (thickness of 3.2 mm) or 101.6 mm (thickness from 3.2 to 6.3 mm).

In Figure 7, the disposition of exoskeleton is shown. The orthogonal exoskeletons are connected to horizontal St. Andrew's cross bracings at every floor.

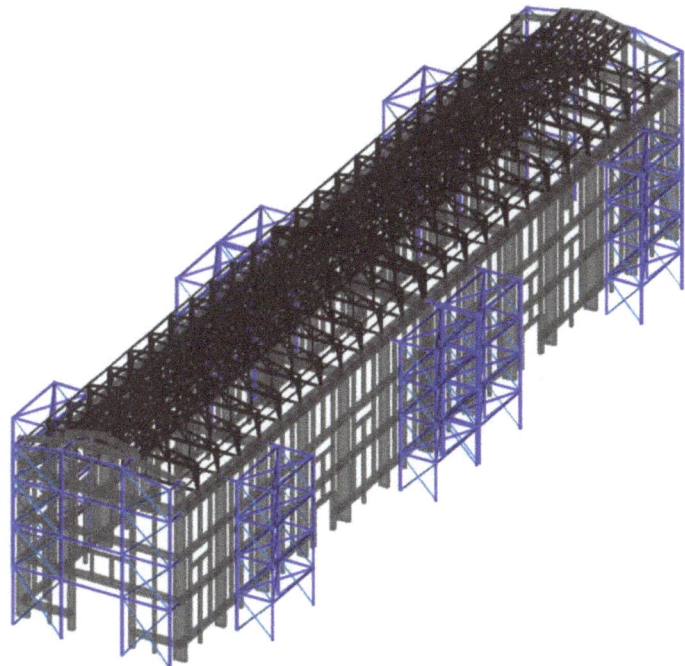

Figure 7. FEM model of the building in the post-intervention phase.

2.6. Photovoltaic Plant

The huge area and the southern exposure of the roof were used to install a photovoltaic plant with high-efficiency panels. According to the Italian Standard [28], the minimum power of the plant is calculated through the following equation:

$$P = \frac{S}{K} \quad (4)$$

where S is the building area and K is an index equal to 50 $\frac{m^2}{kW}$.

In this case, all the six buildings of the area, whose dimensions are shown in Table 2, are considered. The analysis results are shown in Table 3.

Table 2. Geometrical dimensions of the buildings of the examined industrial area.

Building	L (m)	l (m)	S (m^2)
Building 1	52.80	14.60	770.0
Building 2	52.80	14.60	770.0
Building 3	42.00	14.60	615.3
Building 4	36.85	14.60	538.0
Building 5	65.00	14.60	948.0
Building 6	36.85	14.60	538.0

Table 3. Power of the plant.

Building	Pmin (kW)	N. of Panels	Ptot (kW)
Building 1	15.40	57	22.8
Building 2	15.40	57	22.8
Building 3	12.31	45	18.0
Building 4	10.76	39	15.6
Building 5	18.96	69	27.6
Building 6	10.76	39	15.6

The chosen panel is the Sunpower Maxeon 3, illustrated in Figure 8. The main features of this panel are reported in Table 4.

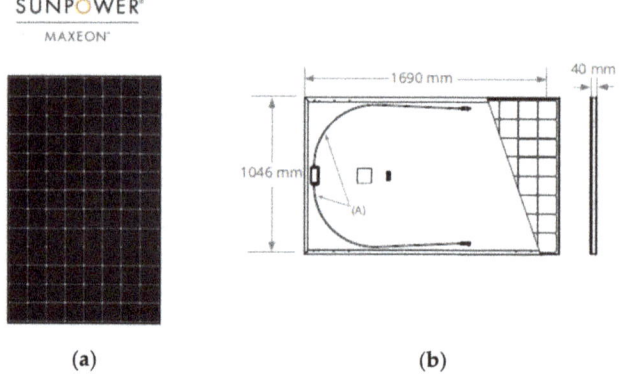

(a)　　　　　　　　　　　(b)

Figure 8. Sunpower Maxeon 3 panel: (**a**) picture of the panel; (**b**) dimensions.

Table 4. Features of the chosen panel.

Pnom (W)	L (mm)	H (mm)
400	1046	1690

Subsequently, to calculate the annual energetic production of the panel, the PVGIS (Photovoltaic Geographical Information System) software is used. PVGIS is a simulator offered by The European Commission's Science and Knowledge Service, and, by entering some data, such as panel features, its inclination and orientation, and the place coordinates, it gives back the monthly average of energy efficiency (kWh), the monthly average of irradiation per square meter, and the annual production of the system.

By considering an angle of inclination equal to 11° and a percentage of system losses equal to 14%, the simulator provided the output data shown in Figure 9.

PVGIS-5 estimates of solar electricity generation:

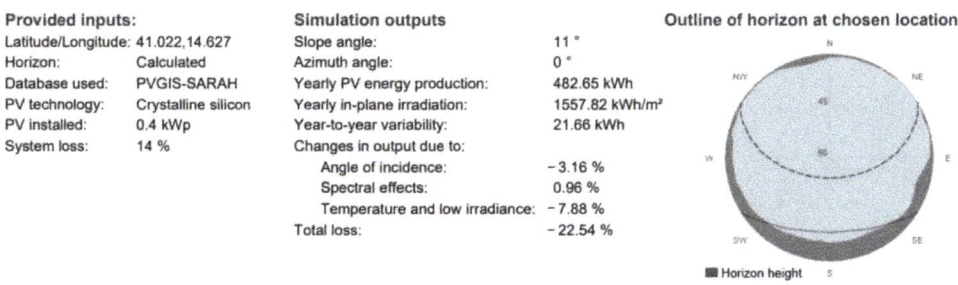

Figure 9. Output data by PVGIS simulator.

3. Results

3.1. Seismic Retrofitting

As in the previous case, FEM model of the building after intervention was set up by means of the SAP2000 v20.0 software. Beams, columns, and diagonals of exoskeletons are modeled using one-dimensional elements by entering all the acquired data on both materials and geometries of sections. In this case, interaction between soil and building is schematized by assigning pinned restraints at the bases of columns. The performed nonlinear static analyses give the capacity curves shown in Figure 10.

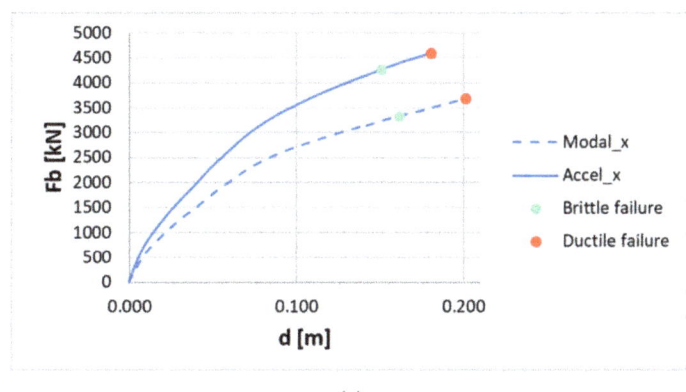

(a)

Figure 10. *Cont.*

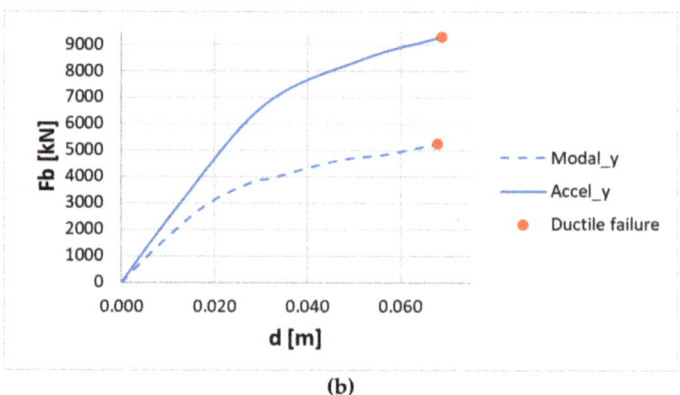

(b)

Figure 10. Capacity curves: (**a**) x-direction; (**b**) y-direction.

By representing the capacity curves in the ADRS plane and by comparing them to the elastic response spectrum of the soil, it is possible to see, as shown in Figure 11, that the seismic retrofitting and, therefore, new safety indices, are attained only in y-direction (Table 5).

Table 5. Safety indices in the post-intervention phase.

Distribution	du,d* (cm)	du,b* (cm)	dr* (cm)	IR,d	IR,b
Modal_x	20.2	16.2	9.2	2.2	1.8
Accel_x	18.1	15.1	8.0	2.3	1.9
Modal_y	6.8	10.0	3.2	2.1	3.1
Accel_y	6.9	35.0	2.1	3.3	16.7

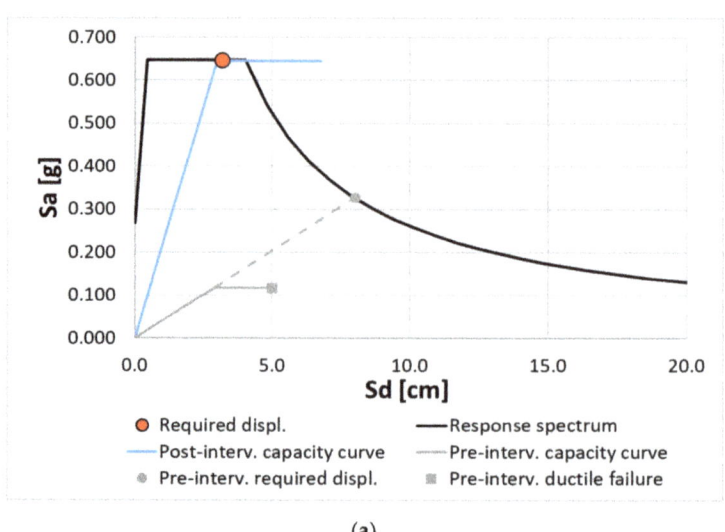

(a)

Figure 11. *Cont.*

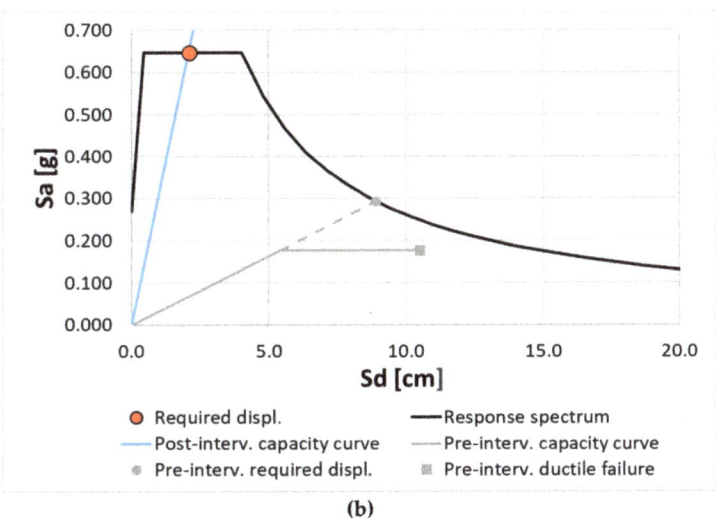

Figure 11. ADRS plane: (**a**) modal_y; (**b**) accel_y.

As shown, exoskeletons brought a significant increase of the safety indices, which attained values greater than 1, so that a marked improvement of the seismic behavior of the building is noticed. Exoskeletons allow to reach an increase of seismic safety of about 30% along the considered directions, by acting only on the building lateral stiffness. This is a satisfactory result that testifies the validity of this intervention.

3.2. Photovoltaic Plant Efficiency

By considering the previous data, it is possible to calculate the annual production of photovoltaic plant per square meter (Table 6).

Table 6. PV yearly energy production.

Building	Annual Prod. (kWh)	Annual Prod./m² (kWh)
Building 1	27,561.2	35.79
Building 2	27,561.2	35.79
Building 3	21,758.9	35.36
Building 4	18,857.7	35.05
Building 5	33,363.6	35.19
Building 6	18,857.7	35.05

These values are compared to the estimated annual energy consumption of several building types and to the benchmark and target data related to schools and offices to assume a hypothetic consumption for the building under investigation, as well as to understand if the plant can provide the entire building's energy requirement.

From the analysis carried out, it appears that the range of indices is very wide, being framed within a minimum value equal to 60 kWh/m² and a maximum value equal to 170 kWh/m² [29]. This means that the percentage of energy requirement of the studied construction may vary between 20% and 60% of that of school and office buildings. Therefore, in the current case, the photovoltaic plant could satisfy the entire energy requirement.

4. Conclusions

In this paper, the seismic and energy rehabilitation of an abandoned RC tobacco factory in Cervinara, in the district of Avellino (Italy), is proposed by designing steel exoskeletons and by installing a photovoltaic plant on the roofs of the building. In situ tests and visual inspections allowed to understand all of the geometrical and mechanical features of the building structural elements used to obtain a FEM model through the SAP2000 software, which allowed to carry out pushover analysis on the building. While the factory showed a good performance in the longitudinal (x) direction with satisfactory safety indices, in the transverse (y) direction, the absence of an appropriate number of seismic-resistant frames reduced the seismic performances, with safety indices less than one. The detected seismic deficiencies led to the design of the optimal seismic intervention to reach an improvement of the seismic features. Steel exoskeletons allowed to improve the global behavior of the building, without neglecting architectural quality, sustainability, and environmental issues. In the weakest building direction (Y-direction), these systems allowed to reach increases in terms of strength and displacements, in order to satisfy the required seismic demand, to attain the complete seismic retrofitting of the structure. In particular, exoskeletons allowed to reach an increase of seismic safety of about 30% by acting only on the building lateral stiffness. Therefore, the significant increase of the safety indices, with values greater than one, decidedly improved the seismic behavior of the building. This satisfactory result testified the validity of this intervention from a seismic viewpoint.

On the other hand, from an energy viewpoint, a photovoltaic plant was designed using high-efficiency panels installed on the wide roof area. The annual production of photovoltaic plant per square meter was compared that of schools and offices, variable in the range from 60 kWh/m^2 to 170 kWh/m^2, to understand if the plant can provide the entire building's energy requirement. The performed analyses showed that the energy consumed by the retrofitted building is between 20% and 60% lower than that of school and office buildings. This means that, using these new energy generation systems, the photovoltaic plant allowed to generate an amount of energy able to satisfy the entire building requirements.

As a conclusion, the proposed study showed the validity of exoskeletons as efficient systems to improve both seismic features of RC buildings and values of Italian industrial heritage, which should be preserved and restored since they represent irreplaceable testaments to the industrial archaeology widespread on the Italian territory.

Author Contributions: Conceptualization, A.F.; methodology, A.F.; software, Y.M.; validation, A.F.; former analysis, Y.M.; investigation, A.F. and Y.M.; resources, A.F. and Y.M.; data curation, A.F. and Y.M.; writing—original draft preparation, Y.M.; writing—review and editing, A.F.; visualization, A.F. and Y.M.; supervision, A.F.; project administration, A.F. All authors have read and agreed to the published version of the manuscript.

Funding: This research received no external funding.

Institutional Review Board Statement: Not applicable.

Informed Consent Statement: Not applicable.

Data Availability Statement: Not applicable.

Acknowledgments: The work was developed in the framework of the WP2 task "Cartis" of the Italian DPC-ReLUIS 2019–2021 research project, which is gratefully acknowledged.

Conflicts of Interest: The authors declare no conflict of interest.

References

1. Mainardi, M. The preservation of industrial heritage in Italy: Traces of history, interpretation, methods. *Stor. Futuro* **2013**. Available online: http://storiaefuturo.eu/la-conservazione-del-patrimonio-industriale-in-italia-tracce-di-storia-interpretazione-metodi/ (accessed on 13 September 2021).
2. Negri, M. To the origins of Italian industrial archaeology. In *Industrial Archaeology in ITALY*; Ciuffetti, A., Parisi, R., Eds.; Franco Angeli: Milano, Italy, 2007.
3. Chiapparino, F. Archaeology, heritage and landscape of industry. The general evolution and the case of the Marches. In *Turismo e Sviluppo Locale*; Novelli, R., Ed.; Cattedrale: Ancona, Italy, 2010; pp. 70–83.
4. Mazzolani, F.M. *Refurbishment by Steelwork*; Arcelor Mittal: Luxembourg, 2007.
5. Comité Européen du Béton-Fédération Internationale du Béton (CEB-FIB). *Seismic Assessment and Retrofit of Reinforced Concrete Buildings, CEB-FIB Bulletin No. 24*; State-of-art Report, Task Group 7.1: 2003; International Federation for Structural Concrete (fib): Lausanne, Switzerland, 2003.
6. Federal Emergency Management Agency (FEMA). *Techniques for the Seismic Rehabilitation of Existing Buildings, FEMA 547/2006*; FEMA: Washington, DC, USA, 2006.
7. Bellini, O.E.; Marini, A.; Passoni, C. Adaptive exoskeleton systems for the resilience of the built environment. *TECHNE-J. Technol. Architect. Environ.* **2018**, *15*, 71–80.
8. Caverzan, A.; Lamperti Tornaghi, M.; Negro, P. Taxonomy of the redevelopment methods for non-listed architecture: From façade refurbishment to the exoskeleton system, JRC, Conference and workshop Reports. In Proceedings of the Safesust Workshop, Ispra, Italy, 26–27 November 2016.
9. Foraboschi, P.; Giani, H. Exoskeletons: Architectural and structural prerogatives (Part I). *Structural* **2017**, *214*, 1–23.
10. Foraboschi, P.; Giani, H. Exoskeletons: Seismic retrofit and architectural regeneration (Part I). *Structural* **2018**, *215*, 1–23.
11. Marini, A.; Passoni, C.; Riva, P.; Negro, P.; Romano, E.; Taucer, F. *Technology Options for Earthquake Resistant, Eco-Efficient Buildings in Europe: Research Needs*; Report EUR 26497 EN. JRC87425; Publications Office of the European Union: Luxemburg, 2014; ISBN 978-92-79-35424-3. [CrossRef]
12. Marini, A.; Belleri, A.; Feroldi, F.; Passoni, C.; Preti, M.; Riva, P.; Giuriani, E.; Plizzari, G. Coupling energy refurbishment with structural strengthening in retrofit interventions. In Proceedings of the Safesust Workshop, Ispra, Italy, 26–27 November 2015.
13. Marini, A.; Passoni, C.; Belleri, A.; Feroldi, F.; Preti, M.; Metelli, G.; Riva, P.; Giuriani, E.; Plizzari, G. Combining seismic retrofit with energy refurbishment for the sustainable renovation of RC buildings: A proof of concept. *Eur. J. Environ. Civ. Eng.* **2017**, 1–20. [CrossRef]
14. Royal Decree, n. *Royal Decree n. 2229 16.11.1939 Rules for the Execution of Simple or Reinforced Concrete Structures*; Official Gazette of the Italian Republic n. 92 of 18/04/1940—Ordinary Supplement n. 92: Rome, Italy, 1939.
15. Petrungaro, F.; Basile, A.; Brandonisio, G. Evolution of the Concrete Strength from 1930 to Today. *Ingenio* **2020**. Available online: https://www.ingenio-web.it/28708-come-sono-evolute-le-resistenze-del-calcestruzzo-dagli-anni-30-ad-oggi (accessed on 20 May 2021).
16. Verderame, G.M.; Ricci, P.; Esposito, M.; Manfredi, G. STIL v1.0. Software Guideline. 2012. Available online: https://www.reluis.it/images/stories/manuale%20STIL%20v1_0.pdf (accessed on 20 May 2021).
17. Verderame, G.M.; Ricci, P.; Esposito, M.; Sansiviero, F.C. Mechanical properties of reinforcement steel for RC structures built from 1950 to 1980. In Proceedings of the XXVI National Conference AICAP "Development Prospects of Concrete Structures in the Third Millennium", Padova, Italy, 19–21 May 2011.
18. Verderame, G.M.; Stella, A.; Cosenza, E. Mechanical properties of reinforcement steel for RC structures built in 60s. In Proceedings of the X National Conference ANIDIS "The Seismic Engineering in Italy", Potenza-Matera, Italy, 9–13 September 2001.
19. Di Lorenzo, G.; Formisano, A.; Terracciano, G.; Landolfo, R. Iron alloys and structural steel from XIX century until today: Evolution of mechanical properties and proposal of a rapid identification method. *Constr. Build. Mater.* **2021**, *302*, 124132. [CrossRef]
20. NTC. *Ministerial Decree, D.M. 20 February 2018 Updating of Technical Standards for Construction*; NTC: Quezon City, Philippines, 2018.
21. Ministry of Infrastructure and Transport. *Instructions for the Application of the New Technical Code for Constructions, N. 35*; Ministry of Infrastructure and Transport: Rome, Italy, 2019.
22. Fajfar, P. Capacity Spectrum Method Based on Inelastic Demand. *Spectra Earthq. Engng. Struct. Dyn.* **1999**, *28*, 979–993. [CrossRef]
23. Fajfar, P. A nonlinear analysis method for performance based seismic design. *Earthq. Spectra* **2000**, *16*, 573–592.
24. Formisano, A.; Di Lorenzo, G.; Colacurcio, E.; Di Filippo, A.; Massimilla, A.; Landolfo, R. Steel Orthogonal Exoskeletons for Seismic Retrofit of Existing Reinforced Concrete and Prestressed Concrete Buildings: Design Criteria and Applications. *Costr. Met.* **2020**, *Nov–Dec*, 40–50. Available online: https://www.torrossa.com/it/resources/an/4548259 (accessed on 13 September 2021).
25. Landolfo, R.; Formisano, A.; Di Lorenzo, G.; Colacurcio, E.; Di Filippo, A. Steel Exoskeletons for RC Buildings Retrofit: Project Methodology and Application to a Case Study. *Ingenio* **2021**. Available online: https://www.ingenio-web.it/31636-esoscheletri-in-acciaio-per-il-retrofit-di-edifici-in-ca-metodologia-progettuale-e-applicazione-ad-un-edificio (accessed on 13 September 2021).
26. Landolfo, R.; Formisano, A.; Di Lorenzo, G.; Colacurcio, E.; Di Filippo, A. Steel Exoskeletons for RC Buildings Retrofit: Structural Concept and Applications. *Ingenio* **2021**. Available online: https://www.ingenio-web.it/31096-esoscheletri-in-acciaio-per-il-retrofit-strutturale-di-edifici-in-ca-concept-strutturale-e-applicazioni (accessed on 13 September 2021).

27. Landolfo, R.; Formisano, A.; Di Lorenzo, G.; Colacurcio, E.; Di Filippo, A. Steel Exoskeletons for RC Buildings Retrofit: State of the Art and Definitions. *Ingenio* **2021**. Available online: https://www.ingenio-web.it/30264-esoscheletri-in-acciaio-per-il-retrofit-strutturale-di-edifici-esistenti-in-ca-stato-dellarte-e-definizioni (accessed on 13 September 2021).
28. President of the Republic. *Legislative Decree n.28 03.03.2011 Implementation of the Directive 2009/28/CE about the Promotion of the Use of Renewable Energy*; Ministry of Economic Development: Rome, Italy, 2011.
29. De Pasquale, A. *Energy Consumption Benchmark of Office Buildings in Italy*; ENEA: Rome, Italy, 2019.

Article

Evaluation of the Seismic Retrofitting of Mainshock-Damaged Reinforced Concrete Frame Structure Using Steel Braces with Soft Steel Dampers

Fujian Yang [1,2], Guoxin Wang [1,2,*] and Mingxin Li [3]

[1] State Key Laboratory of Coastal and Offshore Engineering, Dalian University of Technology, Dalian 116024, China; fjyang@mail.dlut.edu.cn
[2] Institute of Earthquake Engineering, Faculty of Infrastructure Engineering, Dalian University of Technology, Dalian 116024, China
[3] Shandong Provincial Key Laboratory of Civil Engineering Disaster Prevention and Mitigation, Shandong University of Science and Technology, Qingdao 266590, China; jiexia1991@outlook.com
* Correspondence: gxwang@dlut.edu.cn; Tel.: +86-411-8470-7364

Abstract: Most reinforced concrete (RC) frames would exhibit different degrees of damage after mainshock excitations, and these mainshock-damaged RC (MD-RC) frames are highly vulnerable to severe damage or even complete collapse under aftershock excitations. In the present study, the effectiveness of utilizing soft steel damper (SSD) as a passive energy dissipation device for seismic retrofitting of MD-RC frame under aftershock actions was investigated. A common three-story RC frame in the rural area was employed and a numerical evaluation framework of retrofitting analysis of the MD-RC frame was also proposed. Based on proposed evaluation framework, nonlinear dynamic time history analysis of the MD-RC frame with and without retrofitting schemes was conducted to evaluate the retrofit effect of the retrofitting schemes on the MD-RC frame. The results revealed that the retrofitting schemes could effectively improve the natural vibration characteristics of the MD-RC frame, especially the first-order natural frequency with a maximum increase of nearly four times. The retrofit effect of the MD-RC frame under pulse-like aftershocks is better than non-pulse-like aftershocks and the retrofit effect of minor damage MD-RC frame is slightly better than that of severe damage. In addition, only retrofitting the bottom story of MD-RC frame might cause aggravate structural damage.

Keywords: seismic retrofitting; mainshock-damaged RC frame; soft steel damper; seismic performance; mainshock-aftershock seismic sequence

1. Introduction

Post-earthquake disaster surveys have shown that numerous building structures might suffer from different levels of damage after mainshock excitations, and these mainshock-damaged (MD) structures will face the threat of frequent aftershocks again, which could cause serious failures or even complete collapse [1,2]. Especially for the widely used reinforced concrete (RC) frame structures, they easily form a weak-story mechanism during mainshocks. In that case, the concrete covers fall off, steel rebars yield, and RC frames will produce large story drift, which results in the loss of structural bearing capacity. Nevertheless, some seismic disaster investigations have found that most RC frames still have a certain residual capacity to resist collapse after mainshocks [1,3]. For example, about 60% of RC frames could continue to be used and 32% of RC frames needed to be reinforced before use after the 2008 Wenchuan earthquake [3]. In addition, due to the uncertainty of earthquake occurrence, these mainshock-damaged RC (MD-RC) frames might be further damaged by aftershocks. Consequently, these MD-RC frames that still have retrofitting significance and they should be strengthened as soon as possible to improve their seismic capacity and prevent them from further damage during aftershocks, and then ensure the safety of people in the disaster area and reduce the waste of resources caused by the demolition of structures.

In recent years, various seismic retrofitting technologies for RC frames have been proposed in order to ensure structural seismic safety [1,4], such as replacing concrete, enlarging section, bonded steel plate, external encased steel, fiber-reinforced polymer (FRP) [5–8] composites or carbon/glass fiber reinforced polymers (CFRP/GFRP) composites [9–13], and so on. More efforts have also been devoted to the investigation of seismic performance of retrofitted structure while using either model tests or numerical modeling in such researches. The results have shown that the use of these retrofitting technologies can significantly improve the strength, stiffness, and ductility of RC frame or their members. However, there still existed some insufficiencies for these retrofitting measures and evaluation methods. For example, some conventional retrofitting measures, like replacing concrete and enlarging section, would cost much time to reach the target strength of concrete materials, and other retrofitting techniques, like bonded steel plate, external encased steel, and FRPs, are also relatively complex and time-consuming [1]. In addition, due to the time of aftershocks occurrence being usually short and frequent, quickly and effectively strengthen the mainshock-damaged structure is more helpful in reducing the further damage during aftershocks. Recently, several passive energy dissipation devices (PEDD), such as metallic dampers, friction dampers, viscoelastic dampers [4], viscous fluid dampers [14–16], and buckling-restrained braces dampers [17–19], have been widely used as global (i.e., structure-level) modification strategy [20–22]. Among these PEDDs, metallic dampers have attracted increasing attention and have become the favorite damping device for seismic retrofitting due to the following advantages, e.g., inexpensive, easy fabrication, and stable hysteretic behavior, etc. [20,23,24]. In addition, metallic dampers devices could dissipate the most input energy of earthquake through excellent plastic deformation potential under seismic excitation, thereby reducing seismic action on structures. The research on metallic dampers originated from the works of Kelly et al. [25] and Skinner et al. [26]. Subsequently, different types of metallic damper devices have been proposed and tested, such as shear yielding damper (e.g., shear links [21,27–30], steel shear wall damper (SSWD) [31], slit damper (SD) [32–34], dual-function metallic damper (DFMD) [35], etc.), flexural yielding damper (e.g., added damping and added stiffness (ADAS) [36–38], pipe damper (PD) [39,40], U-shaped energy dissipation damper (UEDD) [41–43], etc.), and combined metallic hysteretic dampers [22,25,26,44], etc. The literature reviews show that metallic dampers with steel brace could significantly enhance the lateral stiffness, strength, and deformation capacity of RC frames, which indicates that these metallic dampers could play a positive role in strengthening RC frames [22,23]. However, some researchers [45,46] also found that it is precisely due to the increase in structural overall stiffness that these retrofitting devices might have a potential adverse effect on the structure. To this end, a novel adaptive hysteretic damper has been developed by Gandelli et al. [46], and it effectively improved this unacceptable situation.

Nonetheless, it is worth noting that the aforementioned studies mainly conducted the retrofitting analysis of new-built or existing structures or structural members, and few studies for the post-mainshock damaged structures [20,22]. In other words, most of experimental or numerical studies focused on structural members, and the retrofit effect of the entire structures was rarely investigated, especially for MD-RC frames that might suffer from adverse failure or collapse under aftershocks. It is generally believed that retrofitting analysis of the damaged structure belongs to the category of secondary force, and the stiffness and strength of the structural material will decrease significantly after the first excitation (i.e., mainshock). Because of the randomness of earthquakes (or aftershocks), it is obviously insufficient to only consider the perfect structures, which might overestimate the actual bearing capacity of structure. To this end, some investigations [1,20,22] have aimed at the retrofitting study of earthquake-damaged structures through experiments or numerical simulation techniques in recent years.

The objective of this research aims at studying the retrofitting analysis of the MD-RC frames under aftershock excitations. In order to understand the seismic performance of MD-RC frames with and without retrofitting schemes, a typical low-standard designed RC frame in the rural area is taken as the research object and a numerical evaluation framework of retrofitting analysis of MD-RC frame is also proposed that is based on the "element live and death" technology of ABAQUS program [47]. In addition, a retrofitting device that is based on combined soft steel dampers (CSSD) and three layout schemes is designed for MD-RC frames. Finally, through nonlinear dynamic time history analysis of MD-RC frame with and without retrofitting schemes, the retrofit effects of the CSSD retrofitting devices on MD-RC frames under three layout schemes are evaluated in terms of natural frequency, displacement response, interstory drift response, and shock absorption rate of MD-RC frames.

2. Methodology

Numerical analysis of post-mainshock retrofitting of mainshock-damaged structure can be broadly classified into two commonly used methods, namely the strength reduction (SR) method [23,48] and stepwise (SW) method [20,22]. As the name implies, the SR method is to artificially reduce the strength of structural materials or the performance level of structure in order to promote the structure to reach a target damage state. Although SR method is simple and direct, damaged structure that is obtained by artificially setting the strength reduction is more or less different from the real damage state of the structure after earthquakes excitation. However, the SW method is to make the intact structure reach a specific damage state by performing a single earthquake excitation. Obviously, the SW method is more consistent with the actual earthquake process, and it is more reasonable to perform subsequent aftershock response analysis by adding retrofit members. Therefore, this study will adopt SW method in the post-earthquake retrofitting research of earthquake-damaged structure under aftershocks. In the SW method, the 'element birth and death' technology of numerical software is adopted in order to realize the modelling of retrofitting devices for MD-RC frames. Furthermore, because the interstory drift ratio (IDR) is an appropriate indicator of structural damage levels for RC or steel frame, the maximum IDR is selected as the evaluating indicator of structural damage state in the present study. Table 1 summarizes the structural damage states corresponding to the IDR limit values, according to China seismic code [49].

Table 1. Structural damage states corresponding to the interstory drift ratio (IDR) limit values.

Damage States	Description	IDR [1] (%)
Neglected	No damage or localized minor cracking	<0.4
Minor damage	Slight cracking throughout	0.4~1.0
Moderate damage	Severe cracking, localized spalling	1.0~2.0
Severe damage	Crushing of concrete, reinforcement exposed	2.0~4.0
Collapse	Collapsed	>4.0

[1] IDR: Maximum interstory drift ratio.

Based on previously mentioned method and technology, a post-earthquake retrofitting evaluation framework of MD-RC frame is proposed (Figure 1) and the detailed steps are summarized, as follows: (a) build the intact structural model and perform gravity analysis (i.e., applying gravity acceleration to structural model) to simulate the real force balance state of structures in the first step; (b) determine the target damage states of MD-RC frame that are to be analyzed and perform nonlinear dynamic time history analysis of intact structure under the excitations of mainshock with different intensity to make the intact structure enter target damage states (step 2); (c) design retrofitting schemes for MD-RC frame and retrofit the MD-RC frame by designed retrofitting schemes (step 3); (d) perform a nonlinear dynamic time history analysis of the MD-RC frame with and without retrofitting schemes under aftershock excitations in step 4; and, (e) evaluate the seismic performance of the MD-RC frame under given retrofitting schemes and analyze the retrofit effect of the retrofitting schemes on the MD-RC frames under aftershock excitations (step 5).

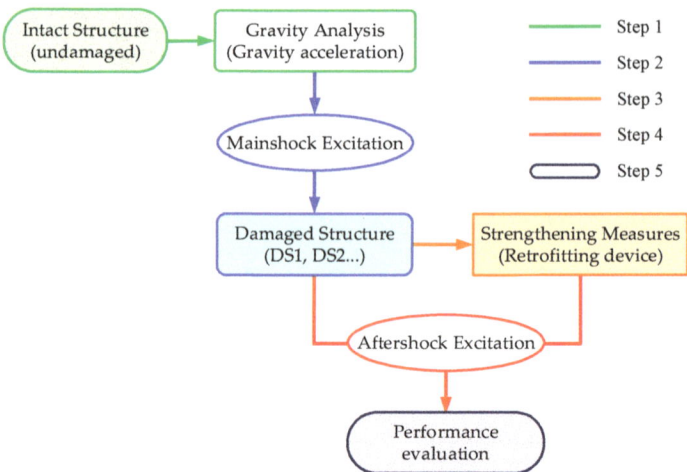

Figure 1. Post-earthquake retrofitting evaluation framework of the mainshock-damaged reinforced concrete (MD-RC) frame.

3. Building System under Investigation

3.1. Intact RC Frame Structure

A typical three-story, two-span RC frame building that is located in a high-seismicity site (Fortification Intensity VIII and Site Class II) of China is considered in this research (see Figure 2). The case-building is representative of the typical low-rise RC frame with low-standard seismic design in the rural area of China [49,50], which has a 12 m × 12 m rectangular plan and a total height of 11.5 m (among that the first-floor height is 4.5 m and the others are 3.5 m). The cross-section size of columns and beams are 0.3 m × 0.3 m and 0.5 m × 0.3 m, respectively. More specifically, the thickness of the slabs and cover concrete is 0.12 m and 0.03 m, respectively. The design dead load (DL) and live load (LL), excluding floor slab self-weight, are considered to be 3.00 kN/m^2 and 0.5 kN/m^2 on the roof, 2.25 kN/m^2 and 2.0 kN/m^2 on typical floors respectively and the representative value of gravity load is combined by 1.0 × DL + 0.5 × LL according to China load design code [51]. All of the concrete materials of the RC frame are the C30 grade, and the longitudinal rebars and stirrup are HRB335 and HPB300 grade. Figure 2 shows the geometries and reinforcement details of this RC frame.

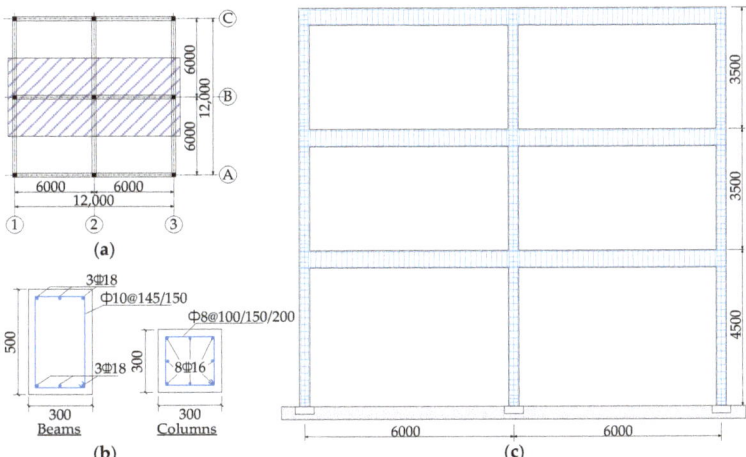

Figure 2. Geometries and reinforcement details for RC frame (all dimensions in mm): (**a**) plan view; (**b**) reinforcement details of the beam and column section; (**c**) elevation view.

3.2. Numerical Model and Material Paramters

In this research, structural modeling is performed while applying the ABAQUS nonlinear finite element software [47]. The middle frame (see Figure 2) is selected for planar modeling due to the symmetry of the analyzed structure. In the numerical, the concrete and rebars are modelled separately, and concrete and rebars of beam, column, and joint are simulated while using the Solid (C3D8R) and Truss (T3D2) elements, respectively. The rebars are embedded in the concrete to simulate the interaction between rebars and concrete, meanwhile the ideal bond is assumed and the influence of bond-slip between rebars and concretes is neglected. In addition, the influence of infill wall and soil-structure interaction are also neglected. It should be mentioned that the representative values of slab weight and gravity load are converted into the model density, in order to simplify the modeling of the RC frame. Figure 3 shows the planar three-dimensional (3D) numerical model of intact RC frame. In this numerical model, the total number of structural elements is 43,863, which includes 37,125 C3D8R elements and 6738 T3D2 elements.

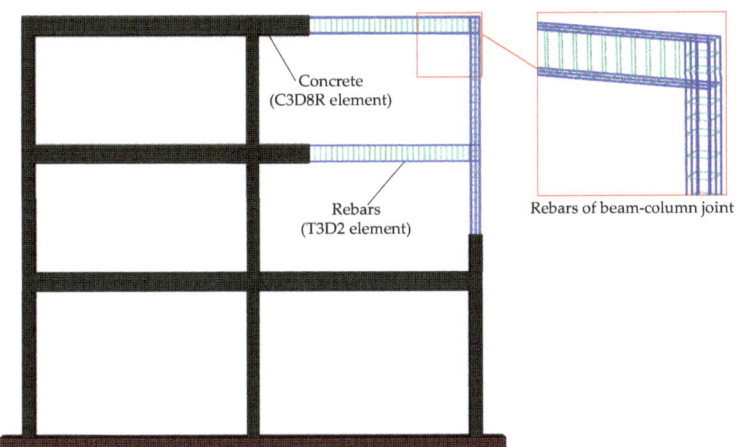

Figure 3. The planar three-dimensional (3D) numerical model of intact RC frame.

In the numerical analysis, the rebars material adopt the dynamic hardening bilinear elastoplastic constitutive model based on the Von–Mises yield criterion. The concrete damage plasticity (CDP) constitutive model in ABAQUS material library is employed to take into account the strain hardening/softening behavior of concrete materials. The CDP model is based on the isotropy assumptions, while using elastic damage combined with tensile and compressive plasticity in order to replace the inelastic behavior of concrete, and considers the degradation of elastic stiffness, due to plastic strain in the process of tension or compression and the stiffness recovery under cyclic loading [52]. Figure 4 shows the stress–strain relations of CDP model under uniaxial cycle loading. Their constitutive relations under uniaxial tension and compressive can be expressed, respectively, by the following Formula [52],

$$\sigma_t = (1-d_t)E_0(\varepsilon_t - \varepsilon_t^{pl})$$
$$\sigma_c = (1-d_c)E_0(\varepsilon_c - \varepsilon_c^{pl})$$
(1)

where, σ_t and σ_c are the tensile and compressive stress of concrete. respectively; E_0 is the initial (undamaged) elastic stiffness; ε_t^{pl} and ε_c^{pl} are equivalent plastic strain tensors in tensile and compressive conditions; d_t and d_c are the tensile and compressive damage factors respectively, where their values range from 0 (undamaged) to 1 (complete damaged). The two damage variables could consider the strength degradation of concrete materials [4] and they can be computed according to China current code [50].

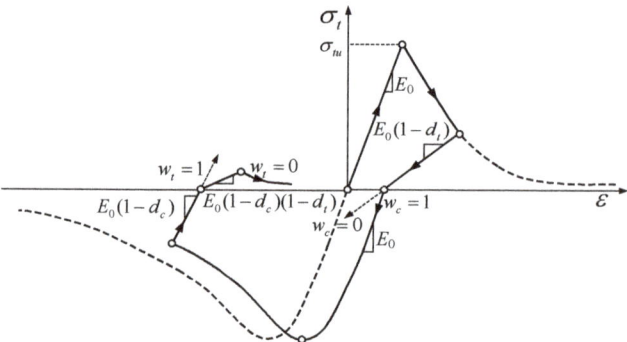

Figure 4. Stress-strain relations under uniaxial cycle loading of concrete damage plasticity (CDP) model.

In addition, in order to consider the complex degradation mechanisms of concrete materials under uniaxial cyclic conditions, the weight factors (stiffness recovery factor) w_t and w_c, which can control the recovery of the tensile and compressive stiffness upon load reversal are defined in the CDP model. The default values for the stiffness recovery factors w_t and w_c are 0 and 1 in ABAQUS program, respectively [52], which means that the tensile stiffness does not recover and the compressive stiffness completely recovers under reverse loading.

Table 2 summarizes the material input parameters of concrete and rebars constitutive models. In addition, the five plasticity parameters of the CDP model [52], such as dilation angle (ψ), eccentricity (ε), strength ratio (f_{b0}/f_{c0}), K, and viscosity parameter, are set to 30.0, 0.1, 2/3, 1.16, and 0.0005, respectively. Moreover, the Rayleigh damping ratio of 5% is employed to account for the energy dissipation in structural system.

Table 2. Material input parameters of concretes and rebars in this study.

Material Types	Constitutive Model	Input Parameters	Values
Concretes (C30)	CDP	Mass density, ρ_c (kg/m^3)	2400
		Elastic modulus, E_c (MPa)	30,000
		Poisson's ratio, ν	0.2
		Compressive strength, f_c (MPa)	20.1
		Tensile strength, f_t (MPa)	2.01
Rebars HRB335 (HPB300)	Bilinear Elastoplastic	Mass density, ρ_s (kg/m^3)	7830
		Elastic modulus, E_s (GPa)	200 (210)
		Poisson's ratio, ν	0.3
		Yield Strength, f_y (MPa)	335 (300)
		Ultimate strength, f_u (MPa)	450 (345)

3.3. Verification of Numerical Model

The natural vibration characteristics of planar and spatial 3D frame structure are compared in order to verify the applicability of the planar 3D numerical model of this study. Table 3 shows the first three frequencies of two frame systems in the x-direction. It is evident that the natural frequencies of the planar 3D frame model are in good agreement with the spatial 3D model, except the third-order frequency, which has a large difference of 5%. However, the seismic response of low-rise buildings is mainly controlled by low-order modals. Therefore, the planar 3D frame structure modeling will be used to carry out the retrofit analysis of the damaged frame structure subjected to the unidirectional earthquake in order to reduce the computational time cost.

Table 3. The first three frequencies of the planar and spatial 3D frame model.

Mode	Natural Frequency (Hz)		
	Planar 3D Frame Model	Spatial 3D Frame Model	Error (%)
1	1.23	1.24	0.80
2	4.05	4.00	1.25
3	7.03	6.71	4.77

4. Modeling of MD-RC Frame

4.1. Input Motions

According to the methodology that is introduced in Section 2, the MD-RC frame models with different damage states (DS) can be obtained through the excitation of mainshocks with different intensities. Subsequently, the seismic performance analysis of the MD-RC frames with and without retrofitting schemes that are subjected to aftershocks is performed. For this purpose, the mainshock-aftershock (MS-AS) seismic sequences should be chosen before the retrofit analysis of the MD-RC frame. More specifically, in order to investigate the influence of earthquake type on retrofit effect, a widely used artificial seismic sequence method (i.e., randomized approach) introduced in literature [53] will be employed in this study. In the randomized approach, irrespective of the source distance (R) and earthquake magnitude (M), the ratios between the peak ground accelerations (PGA) for the two event cases are given by,

$$\frac{PGA_{(2-EVENTS)}}{PGA_{(1-EVENT)}} = \frac{PGA_{(M-0.3010)}}{PGA_{(M)}} = \frac{10^{0.49+0.23(M-6-0.3010)-\log\sqrt{R^2+8^2}-0.0027\sqrt{R^2+8^2}}}{10^{0.49+0.23(M-6)-\log\sqrt{R^2+8^2}-0.0027\sqrt{R^2+8^2}}} = 0.8526 \quad (2)$$

Furthermore, in order to construct near-fault MS-AS seismic sequences, six recorded near-fault strong ground motions (include four pulse-like and two non-pulse-like ground motions) are selected from the Pacific Earthquake Engineering Research Center (PEER) NGA-West2 ground motion database as the input motions of the numerical simulation, according to the reference [54]. Table 4 summarizes the information of these selected recorded near-fault ground motions. In this table, the pulse-like ground motion recorded at SVC station in the 1989 Loma Prieta earthquake and JGB station in the 1994 Northridge earthquake are regarded as the mainshock ground motions (remark MS1 and MS2, respectively). Meanwhile, in order to investigate the impact of aftershock types, two pulse-like and two non-pulse-like motion records (i.e., do not contain strong velocity pulses.) are selected as the aftershock ground motions (remark AS1, AS2, AS3, and AS4, respectively). Furthermore, the peak values (i.e., PGA) of the mainshocks and aftershocks are unified as the same sign in order to avoid the polarity influence of MS-AS seismic sequence on the response and behavior of MD-RC frame. In this study, the peak values of mainshocks and aftershocks are uniformly set to be positive in the constructed artificial near-fault seismic sequence.

Table 4. Information of selected recorded near-fault ground motions in this study.

No.	Earthquake	Year	M_w [1]	Station and Comp. [2]	R_{RUP} [3]	V_{S30} [4]	Pulse
MS1	Loma Prieta	1989	6.9	SVC270	9.3	347.9	yes
MS2	Northridge	1994	6.7	JGB022	5.4	525.8	yes
AS1	Northridge	1994	6.7	RRS228	6.5	282.3	yes
AS2	Imperial Valley	1979	6.5	E06230	1.4	203.2	yes
AS3	Gazli, USSR	1976	6.8	GAZ090	5.5	259.6	no
AS4	Loma Prieta	1989	6.9	BRN090	10.7	462.2	no

[1] Moment magnitude; [2] Ground motion components; [3] Closest distance to rupture plane, unit: km; [4] Average shear velocity of top 30 m, unit: m/s.

4.2. Numerical Modeling of MD-RC Frame

In order to obtain the numerical modeling of MD-RC frame, the initial damage state (DS) of MD-RC frame after mainshock should be first defined, as was mentioned previously. Severe damaged structures might be demolished and lose possibility of repair, according to the definition of structural damage state classification (see in Table 1). In addition, disaster investigation found that most of the earthquake-damaged structures are in a moderately damaged state or below [3]. To this end, two damage states (i.e., minor and moderate damage) of MD-RC frames are considered to investigate the damping effect of the retrofitting schemes on MD-RC frame under different damage levels in the present study. In the first damage state (remark DS1), the maximum IDR of intact structure should be in the range of 1/250 to 1/100. In the second damage state (remark DS2), the maximum IDR of intact structure should be in the range of 1/100 to 1/50. By the trial calculation, the intact structure will enter the above damage state (i.e., DS1 and DS2) when the PGA of the MS1 and MS2 mainshock were both 0.15 g and 0.30 g, respectively. Figure 5 presents the IDR distribution and damage limitation value under MS1 and MS2 mainshock excitation. In Figure 5, it is evident that the maximum IDR values of MD-RC frame fall within the range of minor and moderate damage under MS1 and MS2 mainshock excitation. In addition, all of the maximum IDR occur on the structural bottom floors, so the bottom floors must be retrofitted. Finally, four numerical models of MD-RC frame containing two damage states (i.e., DS1 and DS2) are obtained.

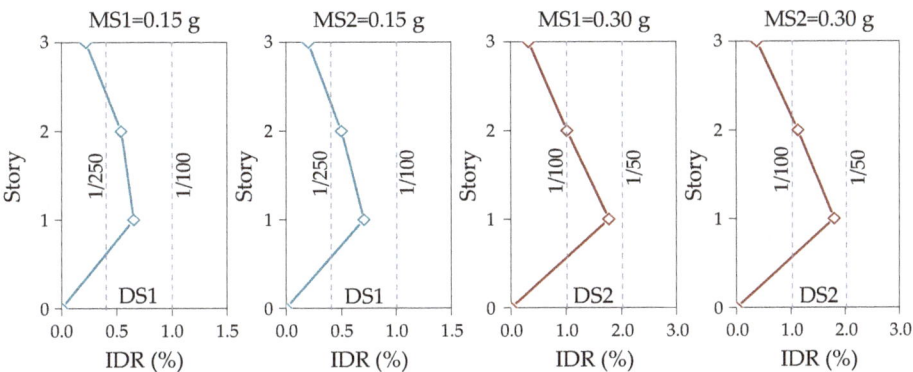

Figure 5. IDR distribution and damage limitation value under MS1 and MS2 mainshock excitation.

Table 5 lists the natural frequencies of the intact RC frame and four MD-RC frames. It is clear that the natural frequencies of the intact frame are significantly reduced, which indicates that the stiffness and strength of the structural materials have decreased to varying degrees.

Table 5. Natural frequencies of the intact RC frame and MD-RC frame (Hz).

Mode	Intact Structure	After MS1		After MS2	
		DS1	DS2	DS1	DS2
1	1.23	0.99	0.76	1.11	0.93
2	4.05	3.62	3.13	3.80	3.46
3	7.04	6.39	5.86	6.65	6.22

5. Strengthening for MD-RC Frame Structure

5.1. Metallic Energy Dissipator

Soft steel damper (SSD) has high plastic deformation ability and low cycle fatigue resistance, which can enter the plastic energy dissipation stage earlier than structural members during an earthquake. Therefore, SSD could reduce the structural damage by dissipating part of the seismic input energy, and it has been widely used as the metallic energy dissipator (MED) for structural retrofitting [20,55]. Generally, typical SSD is designed to shear yielding mode (e.g., shear link or slit damper), because shear yielding dampers could provide higher initial stiffness under small earthquakes and significant energy dissipation potential during large earthquakes by inelastic deformation [28,30,55]. In addition, some studies have adopted opening-hole (or window) shaped SSD [30,35,55–57] or combined shear-and-flexural [44] SSD in order to limit stress concentration and out-of-plane buckling of SSD. Accordingly, a combined strip-shaped shear-and-flexural SSD (CSSD) MED has been designed and improved in this research by drawing on the above MEDs. This CSSD device consists of a strip-shaped SSD [35,55,57] (shear) and two flange plates (flexure) on the end of the SSD. A displacement-based design procedure has been used in order to design the capacity of the CSSD. The similar design process of these properties and capacity of the CSSD can be referred to following literature works [35,55,57]. The same capacity of the CSSD (i.e., designed according to maximum story yielding force) is employed for all stories in order to simplify the calculation. Figure 6 illustrates the configuration information of the CSSD device used in this research. As shown in Figure 6, all SSDs with an overall size of 400×400 mm^2 and 12 mm thickness (t_s) has been used in this study. Two flange plates of 12 mm thickness (t_f) are welded to both ends of the SSD. Table 6 shows the material properties used in the CSSD device.

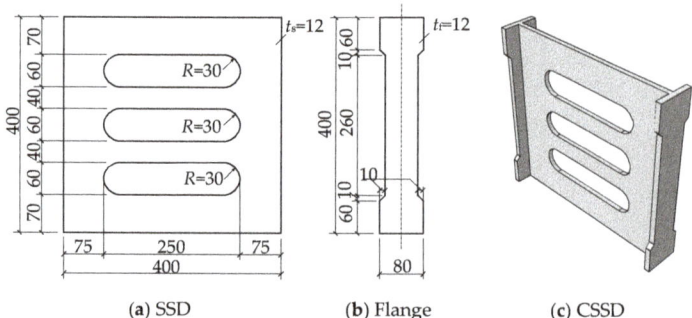

Figure 6. Configuration of the combined soft steel dampers (CSSD) device (all dimensions in mm): (**a**) soft steel damper (SSD); (**b**) flange; (**c**) CSSD device.

Table 6. Material properties of the CSSD device in this study.

Material Types	Elastic Modulus E_s (GPa)	Yield Strength f_y (MPa)	Ultimate Strength f_u (MPa)
SSD	210	100	350
Flange	210	235	441

Because the CSSD device has two kinds of steel materials and complex geometric shapes, it is difficult to obtain a simplified calculation formula for computing the yield displacement, yield, and ultimate force of the CSSD device [55]. To this end, elastoplastic pseudo-static numerical analysis of the CSSD device under cyclic loadings is conducted while using the ABAQUS program to obtain the capacity of the CSSD device. Figure 7 shows the hysteretic curves of the CSSD numerical model under cyclic loadings. From this figure, it is clearly that CSSD device has larger initial stiffness and the hysteresis loop has a plump shape, which means that the CSSD has a higher ductility and a better energy dissipation performance after yielding.

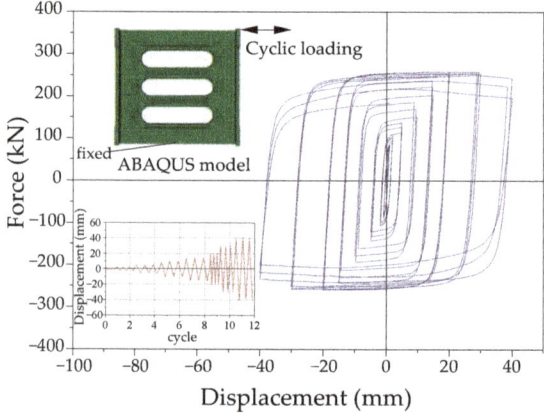

Figure 7. Hysteresis loop of the CSSD numerical model under cyclic loadings.

5.2. Layout Scheme of CSSD Retrofitting Device

CSSD is usually installed between the braces and the beam, forming the CSSD brace (CSSDB) system [55]. In this study, the CSSDB system is composed of the CSSD device and chevron steel brace. During an earthquake, the CSSD device could provide significant energy dissipation potential through inelastic deformation and the steel braces could provide the lateral stiffness in order to prevent large deformation of the structural system. However, when subjected to strong ground motion, the brace buckling might lead to the degradation of lateral strength and stiffness of the structural system [58,59]. Thus, the braces should be designed to remain elastic for an axial force that is greater than corresponding to the failure strength of the CSSD device [60]. Hence, a simple H-shaped Beam (H-Beam, H150 × 150 × 10 × 7 mm) of steel bracing system is employed based on the design-criterion available in literature [35,55,57] in this study. In addition, three layout schemes (LS) of CSSDB systems are considered in this study in order to evaluate different CSSDB retrofitting schemes in terms of structural seismic performance. It should be noted that the same CSSD and steel braces are employed in these CSSDB systems. The specific LSs are introduced, as follows:

- LS-1: Arranged CSSDB system on the bottom floor of MD-RC frame (Figure 8a);
- LS-2: Arranged CSSDB system on the first and second floors of MD-RC frame (Figure 8b); and,
- LS-3: Arranged CSSDB system on all floors of MD-RC frame (Figure 8c).

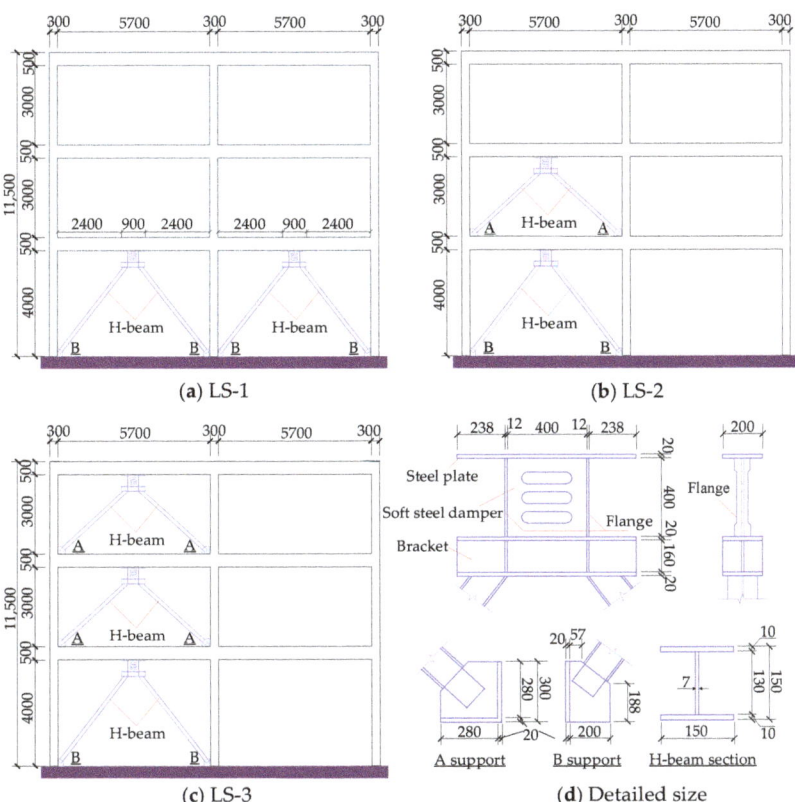

Figure 8. Layout schemes (**a**–**c**) and detailed size (**d**) of CSSDB systems for MD-RC frame (all dimensions in mm).

Figure 8 shows the layout information and detailed size of CSSDB system. In all of the CSSDB systems, the CSSD device is attached to the steel plates at the top and bottom using the welded connections. The steel plates and H-beams are made of Q345 steel with a Young's modulus of 206 GPa, yielding strength of 345 MPa, and Poisson's ratio of 0.3. In these retrofitting systems, the connections of steel plates and beam-column are welding.

Rigid connection between CSSDB systems and concrete beam-column is assumed, and the influence of slippage and peeling between steel plates and concretes are neglected.

6. Analysis of Retrofit Effect of SSDB Systems

In this section, nonlinear dynamic time history analysis of MD-RC frames with and without retrofitting schemes are performed under four aftershocks for the purpose of evaluating the retrofit effect of the CSSDB systems on MD-RC frames under different layout schemes. Four MD-RC frames that contain two kinds damage states (DS1 and DS2) according to Section 4 are considered (i.e., corresponding MS amplitude is 0.15 g and 0.30 g, respectively). The AS amplitudes are scaled using the artificial seismic sequence method. In addition, the natural vibration characteristics, story displacement response, maximum IDR, and shock absorption rate are chosen as the indicators of structural seismic performance.

6.1. Natural Vibration Characteristic

The natural frequency of structure is an important parameter that reflects the dynamic characteristics of structures, and it is also related to the structural mass and stiffness. In general, during strong motions, concrete cracking and rebars yield might induce the stiffness of structure decrease and make the structure lose its bearing capacity, further leading to the change of natural vibration characteristics of structure. Therefore, the effects of CSSDB systems on the dynamic characteristics of MD-RC frames under different layout schemes are investigated in this section. For this purpose, frequency ratio (FR) factor is defined as $FR = f_r/f_{ur}$, where f_{ur} is the frequency of MD-RC frame without retrofitting (i.e., un-retrofitted) and f_r is the frequency of retrofitted MD-RC frame under three layout schemes. Figure 9 shows the natural frequencies ratio (FR) of retrofitted MD-RC frame and un-retrofitted MD-RC frame under DS1 and DS2 damage states. All FR values exceed 1.0, and the maximum value of FR is about 4.0 and the minimum value of FR is about 1.5, indicating that retrofitting systems have effectively improved the overall stiffness of MD-RC frames, as shown in Figure 9, intuitively. In addition, no matter which mainshock (i.e., MS1 and MS2) induced MD-RC frame, the FR value increase with the order of layout schemes (i.e., LS-1, LS-2, and LS-3). Especially for the first-order frequency, the amplitude of FR significantly increases. For the second- and third-order frequency, the changes in FR is not evident from LS-1 to LS-2, while both retrofitting schemes are different from LS-3. The results indicate that, with the increase of the number of retrofitted stories of MD-RC frame, the structural stiffness increases more obviously. Furthermore, for MD-RC frame with different damage states (i.e., DS1 and DS2), the natural frequency of moderate damaged structures (DS2) has significantly increased than that of minor damage (DS1), and with the decrease of modal order, the difference between DS1 and DS2 becomes increasingly obvious.

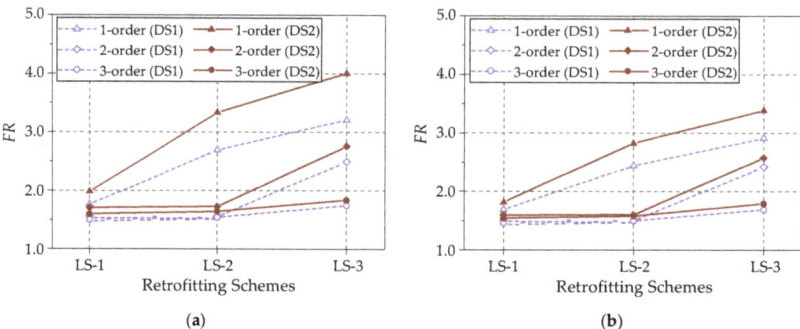

Figure 9. Frequency ratios (FR) of DS1 and DS2 damage states MD-RC frames with and without retrofitted by three layout schemes: (**a**) MS1 mainshock; and, (**b**) MS2 mainshock.

6.2. Story Displacement Response

In this section, story displacement response analysis of MD-RC frames with and without retrofitting schemes is carried out for the purpose of evaluating the retrofit effect of CSSDB systems under three layout schemes. As was mentioned previously, two damage states (i.e., DS1 and DS2) of the MD-RC frame after MS1 and MS2 excitations are considered. Meanwhile, four aftershocks, including near-fault pulse-like earthquakes (AS1 and AS2) and non-pulse-like earthquakes (AS3 and AS4), are selected as secondary excitation. As an example, Figures 10 and 11 show the story displacement responses of retrofitted and un-retrofitted MD-RC frames subjected to AS1 and AS3 aftershock. In Figures 10 and 11, the first column shows the displacement time histories of the intact frame under the mainshock excitations, the second and third columns show the relative displacement during aftershocks with respect to the residual displacement after mainshock excitations. From the overall view of Figures 10 and 11, the peak story displacement of MD-RC frame after retrofitting significantly drops when compared with un-retrofitted MD-RC frame, which indicates that CSSDB systems can effectively control the peak displacement response of MD-RC frame. In addition, the peak story displacement of retrofitted MD-RC frame under AS1 aftershock decrease more significantly than that of AS3 aftershock. This result indicates CSSDB systems are more effective in controlling the dynamic response under pulse-like aftershock.

Similar to the above conclusion, the residual displacement of MD-RC frames under pulse-like aftershocks also decreases significantly. As the number of retrofitted floors increases (i.e., from LS-1 to LS-3), the decrease of residual displacement becomes more obvious. However, for non-pulse-like aftershocks, the residual displacement of MD-RC frame slightly and irregularly changes. Especially for moderate damaged (DS2) structures (see Figures 10b and 11b), the retrofitting devices caused a greater permanent displacement for MD-RC frames. The reason may be that the retrofit of MD-RC frame will cause the redistribution of internal force in the structure, which results in the change of structural natural period and mode shape. When non-pulse-like aftershocks (containing rich high-frequency components) act, the presence of high-frequency components will excite high-order mode shapes of the MD-RC frame, resulting in a complex displacement response and permanent displacement. Especially when the mainshock damage state of MD-RC frame is more serious, this phenomenon is more obvious. Therefore, in the retrofit design of seismic-damaged structures, it is necessary to ensure that the retrofitted structure has a reasonable internal force distribution form. In addition, the above results also indicate that both of retrofitting schemes and aftershock types have a significant influence on displacement response of MD-RC frame. Furthermore, when comparing the mainshock damage state of MD-RC frame, the retrofit effect of CSSDB systems on the MD-RC frame with minor damage (DS1) is slightly better than that of moderate damage (DS2) no matter that aftershock types. A similar conclusion could be observed from the results of other aftershocks listed in Table 4.

Moreover, in order to quantitative assess the retrofit effect of CSSDB systems under three layout schemes on MD-RC frame, the peak displacement degradation percentage (PDDP) is defined by,

$$PDDP = \frac{d_{ur}^i - d_r^i}{d_{ur}^i} \times 100\% \qquad (3)$$

where, d_{ur}^i and d_r^i are the maximum displacement at ith story of un-retrofitted and retrofitted MD-RC frame, respectively. Table 7 presents PDDP values (%) of MD-RC frame under different retrofit schemes for each story. Intuitively, the mean values of PDDPs under pulse-like aftershocks (i.e., AS1 and AS2) are generally greater than that of non-pulse-like aftershocks (i.e., AS3 and AS4) no matter which retrofitting schemes. It indicates that CSSDB retrofitting systems have a better damping effect for pulse-like aftershocks than non-pulse-like aftershocks. In addition, when comparing the three layout schemes, LS-1 can significantly control the maximum bottom story displacement of MD-RC frame, and the maximum PDDP is as high as 94.8%. However, the effect of LS-1 on the other floors of

MD-RC frame is obviously not as good as other retrofitting schemes. Especially for the top displacement response of MD-RC frame induced by MS2 mainshock decreased by −2.3% and −24% under AS3 and AS4 aftershock, respectively. This observation indicates that only retrofitting the bottom story (i.e., LS-1) of MD-RC frame might have an adverse effect on damaged frame, especially with non-pulse-like aftershocks excitation. This can be explained in that the structural beam-column joints have been various damaged after the mainshock excitation, especially the beam-column joints damage of structural bottom story will be more serious. In this case, only reinforcing the bottom story of MD-RC frame will cause the stiffness of the structural bottom story to be significantly increased when compared to the upper stories, which results in an obvious weak story mechanism, causing structural damage to move to the upper stories.

In addition, when comparing the results of LS-2 and LS-3, it can be found that most of the mean PDDPs under LS-3 are greater than LS-2, and the difference between them under pulse-like aftershocks is not higher than 17%, while the difference under the non-pulse-like aftershocks is up to 34%. For the LS-3 retrofitting scheme, the story displacements of the MD-RC frame have been effectively controlled and the peak displacement degradation range from 40~80%. In general, the greater numbers of reinforced structural stories, the more obvious retrofit effect, and the story number with the maximum PDDP is equal to the number of strengthened stories (i.e., bold font in Table 7). However, as the number of reinforced stories increases, both construction and economy will worsen, so it is necessary to seek a balance between economy and retrofit effect.

Table 7. Peak displacement degradation percentage (PDDP) (%) of MD-RC frames under different retrofitting schemes.

MS DS	Story	AS1			AS2			AS3			AS4		
		LS-1	LS-2	LS-3	LS-1	LS-2	LS-3	LS-1	LS-2	LS-3	LS-1	LS-2	LS-3
DS1(MS1)	1	**93.8**	72.5	76.5	**93.9**	64.0	76.6	**89.9**	45.2	59.2	**91.0**	49.1	42.1
	2	66.7	**76.4**	79.1	60.7	**70.7**	78.6	27.3	**49.2**	64.7	41.5	**56.4**	51.8
	3	55.7	72.6	**79.9**	50.5	60.6	**79.4**	3.4	34.7	**65.2**	23.6	44.3	**55.2**
	Mean	72.1	73.8	78.5	68.4	65.1	78.2	40.2	43.0	63.0	52.0	49.9	49.7
DS1(MS2)	1	**94.8**	72.2	75.2	**94.3**	69.4	79.7	**87.2**	44.7	60.3	**90.3**	45.6	52.8
	2	67.5	**76.3**	78.3	74.1	**75.6**	81.6	20.5	**49.3**	64.7	45.8	**51.9**	63.4
	3	56.2	71.8	**79.2**	65.1	67.1	**82.1**	−2.3	33.0	**65.6**	30.8	35.1	**66.8**
	Mean	72.8	73.4	77.6	77.8	70.7	81.1	35.1	42.3	63.5	55.6	44.2	61.0
DS2(MS1)	1	**75.2**	44.1	41.1	**93.6**	82.5	81.8	**93.8**	27.0	25.9	**88.3**	38.5	31.2
	2	41.0	**45.3**	44.5	76.8	**83.1**	82.2	42.0	**32.2**	33.1	39.9	**46.0**	43.1
	3	26.0	40.1	**46.7**	68.9	80.9	**82.5**	21.6	25.5	**35.8**	18.7	42.4	**47.7**
	Mean	47.4	43.2	44.1	79.8	82.2	82.2	52.5	28.2	31.6	49.0	42.3	40.7
DS2(MS2)	1	**79.7**	50.8	52.1	**92.1**	77.9	77.2	**90.6**	34.6	49.4	**74.7**	26.6	18.6
	2	47.8	**51.7**	54.5	73.3	**79.0**	78.0	33.5	**40.7**	54.5	6.7	**34.8**	33.8
	3	33.7	46.2	**56.5**	65.1	76.5	**78.5**	10.5	34.1	**56.7**	−24.0	22.8	**38.2**
	Mean	53.7	49.6	54.4	76.8	77.8	77.9	44.9	36.5	53.5	19.1	28.1	30.2

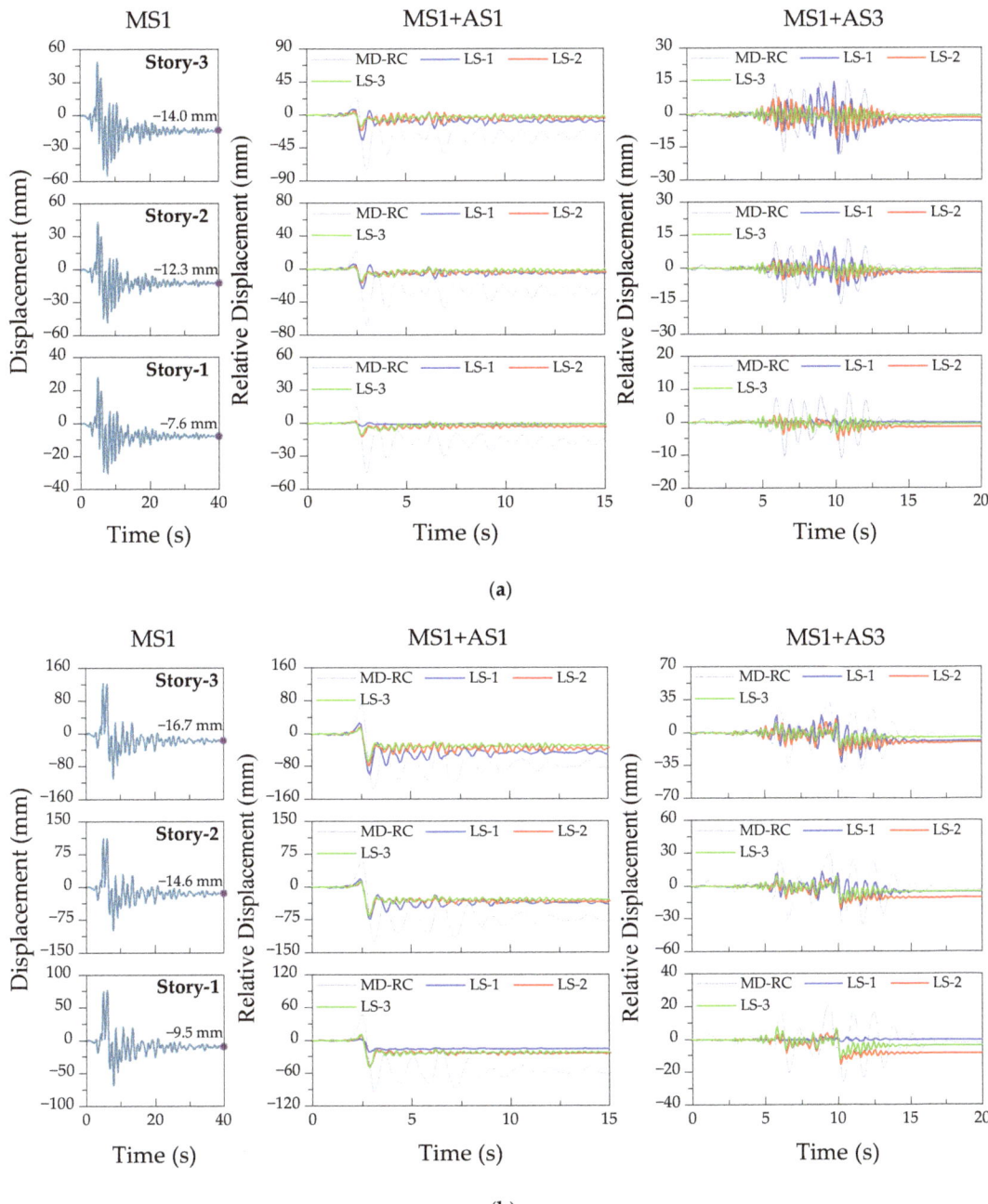

Figure 10. Displacement time histories of DS1 (**a**) and DS2 (**b**) damage states MD-RC frame due to MS1 mainshock before and after retrofitting subjected to AS1 and AS3 aftershocks.

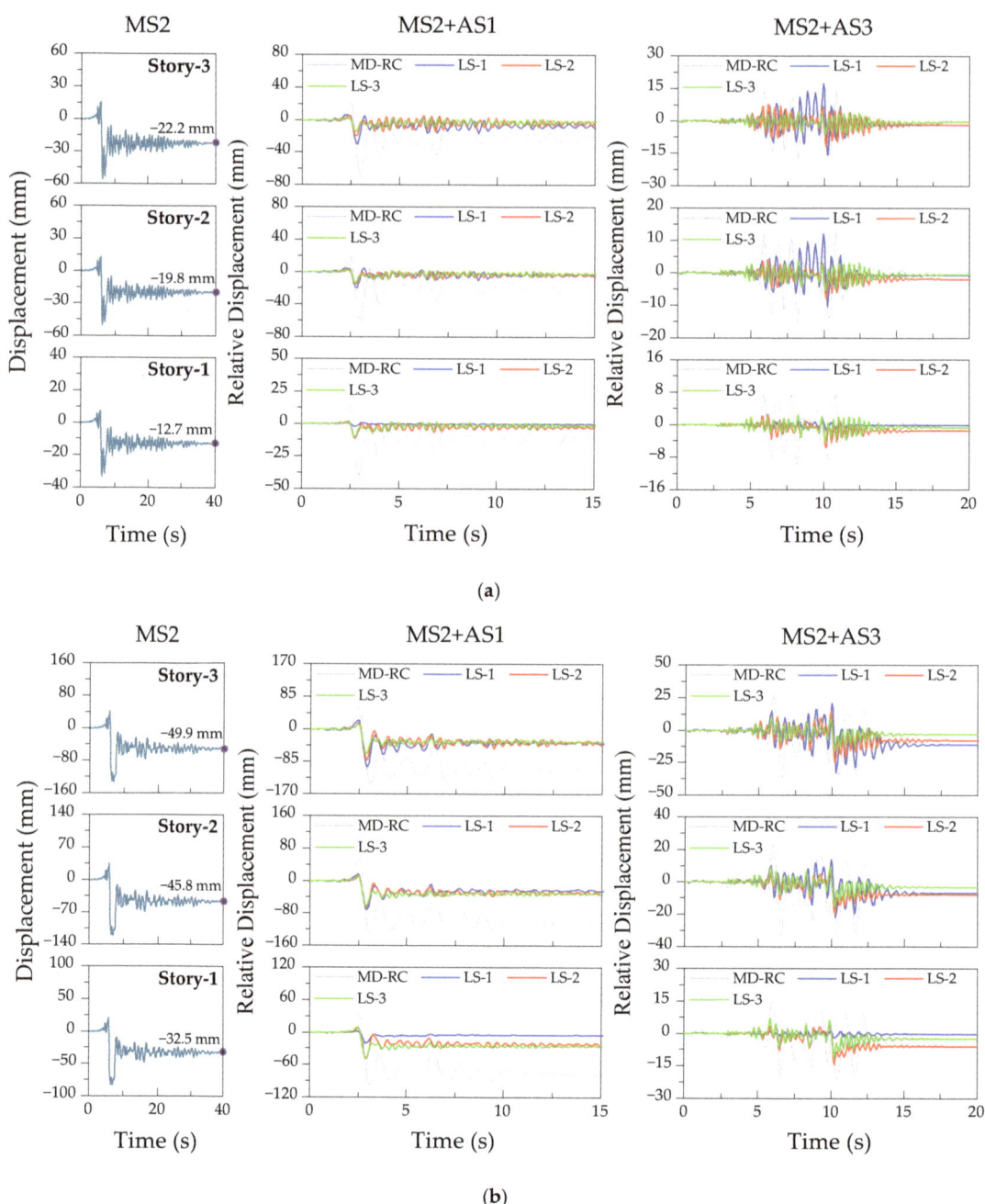

Figure 11. Displacement time histories of DS1 (**a**) and DS2 (**b**) damage states MD-RC frame due to MS2 mainshock before and after retrofitting subjected to AS1 and AS3 aftershocks.

6.3. Interstory Drift Ratio (IDR)

IDR is one of the macroscopic parameters reflecting the structural story deformation and stiffness, and it is also one of the important indicators for judging the structural damage levels in engineering, as was mentioned previously. To this end, Figures 12 and 13 shows the IDR distribution along with the height of MS1 and MS2 MD-RC frame with minor (DS1) and moderate (DS2) damage state before and after retrofitting under aftershocks excitations, respectively. From these figures, it can be initially observed that, in the case of LS-1, the maximum IDR reduction rate is up to 97.7% (first story) when compared with un-retrofitted MD-RC frame. LS-1 has a better performance in reducing the IDR of the first story of MD-RC frame structure as compared with other retrofitting schemes. However, the IDR reduction rate of second and third story of MD-RC frame is lower than other schemes. Especially for the non-pulse-like aftershocks scenarios, in some MD-RC frames retrofitted while using LS-1, the IDRs of second and third story even exceed those of un-retrofitted structure. The maximum IDR is 276.6% (third story) as large as the un-retrofitted case. It further illustrates that the LS-1 retrofitting scheme is not suitable for the retrofitting of MD-RC frame in this study. For the LS-2 retrofitting scheme, it is evident that the IDRs are lower than un-retrofitted MD-RC frame, except for the top story. Especially for the LS-3 retrofitting scheme, the IDR of each story of MD-RC frame are significantly reduced no matter the damage state. More specifically, the lowest reduction rate of IDR is 18.6%, and the highest reduction rate of IDR is up to 87.9%, which indicates that LS-3 plays a more positive effect on the seismic performance of MD-RC frame as compared with other retrofitting schemes. Moreover, when comparing the results of LS-2 and LS-3, although LS-2 does not reinforce the top story of MD-RC frame, the maximum IDR of the first and second story are relatively close under LS-2 and LS-3 schemes. Although the IDR of the top story under the LS-2 retrofitting scheme exceeds other stories, the maximum IDRs are significantly lower than the un-retrofitted case. Therefore, with the comprehensive consideration of economy and installation portability, the engineering applicability of LS-2 scheme is better than LS-3 scheme.

In addition, to quantitatively analyze the retrofit effect of these CSSDB systems on MD-RC frame, in this study the shock absorption rate (SAR) index is used and it is defined as,

$$SAR = \frac{MIDR_{ur} - MIDR_r}{MIDR_{ur}} \times 100\% \quad (4)$$

where, $MIDR_{ur}$ and $MIDR_r$ are the maximum IDR (MIDR) of un-retrofitted and retrofitted MD-RC frame, respectively.

Table 8 summarizes the MIDR and SAR values of MD-RC frame before and after retrofitting under different LSs. From this table, it is evident that the MIDR of retrofitted MD-RC frame has dropped significantly for LS-2 and LS-3 schemes, the maximum drop is nearly 5.6 times. However, for the LS-1 scheme, similar to the previous observations, the MIDRs of some retrofitted MD-RC frames even exceed that of un-retrofitted structures, especially under non-pulse-like aftershocks. More specifically, for the minor damage MD-RC frame, the MIDR under LS-1 scheme is usually 2.0~4.0 times as large as that of LS-2 and LS-3 schemes; for the moderate damage MD-RC frame, the MIDR under LS-1 scheme is usually 1.4~2.2 times as large as that of LS-2 and LS-3 schemes. In terms of SAR; it is evident that the SAR under LS-1 is usually the lowest and even negative under non-pulse-like aftershocks, with a maximum SAR of −74.1%. Obviously, it is not sufficient to carry out only retrofit on structural bottom story for MD-RC frame due to the randomness and uncertainty of ground motion. Furthermore, no matter the mainshock excitation or mainshock damage state, the SAR under LS-2 and LS-3 schemes ranges from 47.1% to 83.3% during pulse-like aftershocks, and ranges from 18.5% to 68.2% during non-pulse-like aftershocks, which further proves that the retrofit effect of CSSDB systems under pulse-like aftershocks is significantly better than that of the non-pulse-like. Overall, there is no noticeable difference in the SAR of MD-RC frames that were retrofitted by LS-2 and LS-3 schemes. However, the LS-2 scheme is more suitable for the retrofit of MD-RC frame in this research, due to the fact that LS-2 scheme retrofits less stories when compared with LS-3 scheme.

Table 8. Maximum IDR and shock absorption rate (SAR) of MD-RC frame before and after retrofitting.

Aftershocks	Schemes	DS1 (MS1)		DS1 (MS2)		DS2 (MS1)		DS2 (MS2)	
		MIDR	SAR (%)	MIDR	SAR (%)	MIDR	SAR (%)	MIDR	SAR (%)
AS1 (Pulse-like)	UR	0.0101		0.0089		0.0206		0.0219	
	LS-1	0.0058	42.4	0.0052	41.8	0.0147	28.6	0.0150	31.4
	LS-2	0.0027	73.5	0.0024	73.0	0.0106	48.8	0.0100	54.4
	LS-3	0.0023	77.3	0.0021	75.8	0.0109	47.2	0.0105	52.1
AS2 (Pulse-like)	UR	0.0037		0.0040		0.0246		0.0188	
	LS-1	0.0027	27.0	0.0019	52.1	0.0081	67.1	0.0073	61.3
	LS-2	0.0013	64.9	0.0012	71.0	0.0041	83.3	0.0040	78.5
	LS-3	0.0008	78.4	0.0008	80.7	0.0044	82.1	0.0043	77.2
AS3 (Non-pulse-like)	UR	0.0024		0.0022		0.0057		0.0048	
	LS-1	0.0032	−33.3	0.0031	−41.1	0.0057	0.0	0.0058	−19.9
	LS-2	0.0012	50.0	0.0011	42.9	0.0034	40.4	0.0031	34.6
	LS-3	0.0008	66.7	0.0007	68.9	0.0027	52.6	0.0024	49.4
AS4 (Non-pulse-like)	UR	0.0034		0.0029		0.0084		0.0054	
	LS-1	0.0036	−8.8	0.0031	-3.8	0.0089	−6.1	0.0094	−73.2
	LS-2	0.0019	44.1	0.0019	34.5	0.0041	51.2	0.0040	26.6
	LS-3	0.0015	55.9	0.0014	51.7	0.0043	48.8	0.0044	18.6

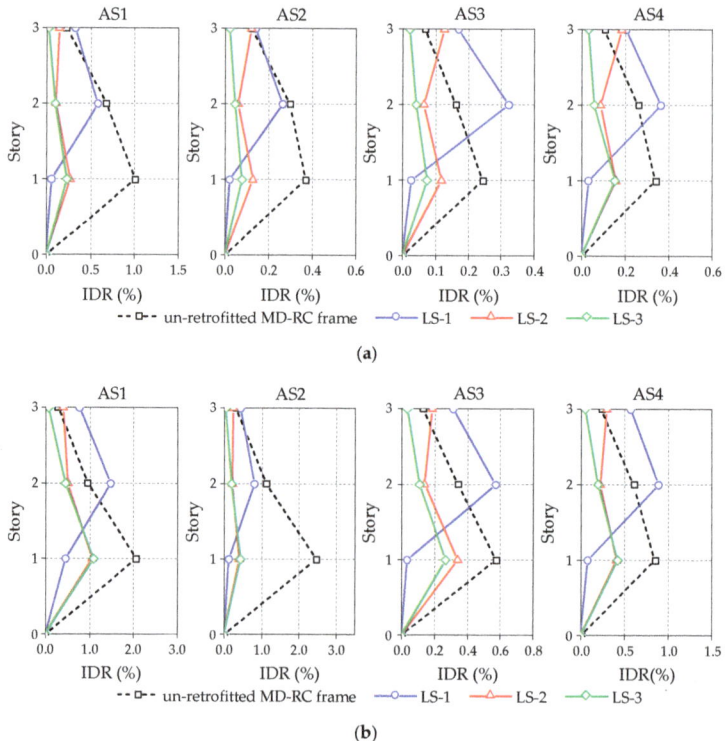

Figure 12. IDR distribution of MS1 MD-RC frame with DS1 (**a**) and DS2 (**b**) damage state before and after retrofitting.

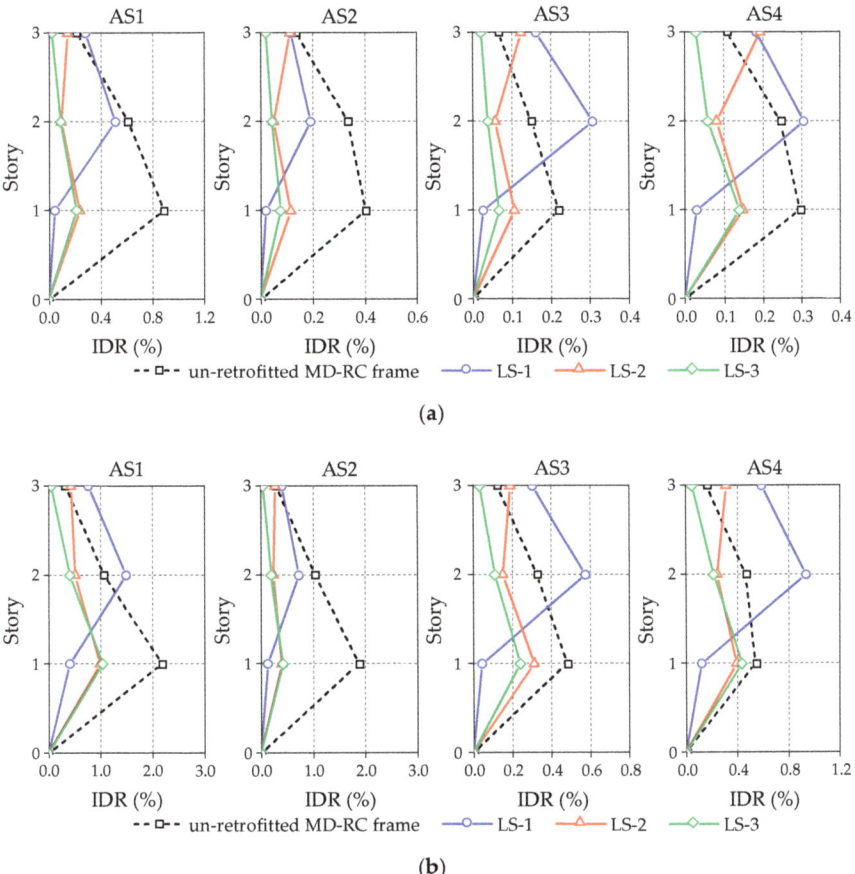

Figure 13. IDR distribution of MS2 MD-RC frame with DS1 (**a**) and DS2 (**b**) damage state before and after retrofitting.

7. Conclusions

This paper mainly conducted the post-earthquake retrofitting analysis of the mainshock-damaged RC (MD-RC) frame while using soft steel damper as the passive energy dissipation device under near-fault aftershocks excitation. Firstly, a numerical evaluation framework for post-earthquake retrofitting of the MD-RC frame was proposed. Subsequently, three retrofitting schemes based on soft steel dampers and steel brace were designed for the MD-RC frame structure. Finally, through the nonlinear dynamic time history analysis of the MD-RC frame structure with and without retrofitting schemes, the retrofit effect of the three retrofitting schemes on the MD-RC frame was evaluated, and some conclusions were obtained, as follows:

(1) The CSSDB retrofitting systems have effectively improved the natural frequency (stiffness) of the MD-RC frame. The first-order natural frequency has a largest increase, with a maximum increase of nearly four times and the second and third-order natural frequencies have a maximum increase of three times. In addition, with the increase of the number of retrofitted stories of the MD-RC frame, the structural stiffness increases more obviously. The natural frequency of moderate damaged structures has increased significantly than that of minor damage. Especially for the first-order mode, the frequency increase in moderately damaged structures is more significant.

(2) The reduction of the maximum story displacement and residual displacement of the retrofitted MD-RC frame under pulse-like aftershocks is more obvious than that under non-pulse-like aftershocks, which indicates that aftershock type has an important influence on the retrofit effect of CSSDB systems for the MD-RC frame. In addition, the retrofit effect of CSSDB systems on the MD-RC frame with minor damage is slightly better than that of the structure with severe damage, indicating that the damage state of the MD-RC frame should also be considered when carrying out the seismic retrofitting design of earthquake-damaged structures.

(3) In the case of LS-1 scheme, the maximum IDR reduction rate is up to 97.7% (first story) when compared with un-retrofitted MD-RC frame. LS-1 scheme has better performance in reducing the IDR of the first story of MD-RC frame structure as compared with the other retrofitting schemes. However, the IDR reduction rate of the second and third story of MD-RC frame is lower than other schemes. Especially for the non-pulse-like aftershocks scenarios, in some MD-RC frames that were retrofitted using LS-1 scheme, and the IDRs of second and third story even exceed those of un-retrofitted MD-RC frame. The maximum IDR is 276.6% (third story) as large as the un-retrofitted case. For the LS-2 scheme, it is evident that the IDRs are lower than un-retrofitted MD-RC frame, except for the top story. Especially for LS-3, the IDR of each story of the MD-RC frame are significantly reduced no matter which damage state. Therefore, it can be seen that LS-3 plays a more positive effect on the seismic performance of the MD-RC frame.

(4) The shock absorption rate (SAR) under only retrofitting bottom story (i.e., LS-1) of the MD-RC frame is lower than that of retrofitting more stories, and the difference is approximately 1.2~2.0 times lower than other retrofitting schemes for pulse-like aftershocks. However, under non-pulse-like aftershocks, the SAR of only retrofitting the structural bottom story is negative with a maximum amplitude of -74.1%. It indicates that only retrofitting the bottom story of the structure is not usually sufficient, especially for the non-pulse-like aftershocks., In addition, there is no noticeable difference in the SAR of MD-RC frames between retrofitting two stories (LS-2) and three stories (LS-3). In contrast, LS-2 might have a better economy and installation portability in engineering applications in this research.

It should be noted that the above conclusions are based on the results of a three-story building and, when only one SSD retrofitting configuration is used, the influence of the geometric sizes of SSD device on the retrofit effect are ignored in this study. In addition, a limited set of earthquakes inputs is considered in the numerical investigation (i.e., two mainshocks and four aftershocks earthquakes). However, due to the randomness of a future earthquake, the main conclusions of the present study are only valid for the examined earthquake scenarios. Therefore, the generalization of these conclusions for other buildings and more seismic inputs still needs greater investigations. Moreover, a more thorough life-cycle cost analysis needs to be conducted in order to select the most optimal retrofitting scheme for an earthquake-damaged building.

Author Contributions: Conceptualization, F.Y. and G.W.; methodology, F.Y. and G.W.; software, F.Y. and M.L.; validation, F.Y. and M.L.; investigation, F.Y. and G.W.; writing—original draft preparation, F.Y.; writing—review and editing, F.Y. and G.W. All authors have read and agreed to the published version of the manuscript.

Funding: This research was funded by the Ministry of Science and Technology of China (National Key R&D Program), grant number 2018YFD1100405 and the National Natural Science Foundation of China, grant number 51578113.

Institutional Review Board Statement: Not applicable.

Informed Consent Statement: Not applicable.

Data Availability Statement: The ground-motion records used in this study were retrieved from the PEER NGA-West2 database (http://ngawest2.berkeley.edu, last accessed December 2018). The results presented in this study are available on request from the corresponding author. The data are not publicly available due to the reason that the authors are conducting further analysis on the same structural model.

Acknowledgments: The authors would like to thank Wang Rui and Ding Yang for their suggestions on numerical analysis and paper writing. The authors also would like to thank the editor-in-chief and four anonymous reviewers for their constructive comments that have helped improve this paper substantially.

Conflicts of Interest: The authors declare no conflict of interest.

References

1. Liu, R.Y.; Yang, Y. Experimental study on seismic performance of seismic-damaged RC frame retrofitted by prestressed steel strips. *Bull. Earthq. Eng.* **2020**, *18*, 6475–6486. [CrossRef]
2. Di Sarno, L. Effects of multiple earthquakes on inelastic structural response. *Eng. Struct.* **2013**, *56*, 673–681. [CrossRef]
3. Tsinghua University; Southwest Jiaotong University; Beijing Jiaotong University. Analysis of Building Damage in Wenchuan Earthquake. *J. Build. Struct.* **2008**, *29*, 1–9. (In Chinese)
4. Dong, Y.R.; Xu, Z.D.; Li, Q.Q.; Xu, Y.S.; Chen, Z.H. Seismic behavior and damage evolution for retrofitted RC frames using haunch viscoelastic damping braces. *Eng. Struct.* **2019**, *199*, 109583. [CrossRef]
5. Di Ludovico, M.; Prota, A.; Manfredi, G.; Cosenza, E. Seismic strengthening of an under-designed RC structure with FRP. *Earthq. Eng. Struct. Dyn.* **2008**, *37*, 141–162. [CrossRef]
6. Sasmal, S.; Novák, B.; Ramanjaneyulu, K. Numerical analysis of fiber composite-steel plate upgraded beam-column sub-assemblages under cyclic loading. *Compos. Struct.* **2011**, *93*, 599–610. [CrossRef]
7. Sasmal, S.; Khatri, C.P.; Karusala, R. Numerical simulation of performance of near-surface mounted FRP-upgraded beam–column sub-assemblages under cyclic loading. *Struct. Infrastruct. Eng.* **2015**, *11*, 1012–1027. [CrossRef]
8. Mostofinejad, D.; Hosseini, S.M.; Tehrani, B.N.; Eftekhar, M.R.; Dyari, M. Innovative warp and woof strap (WWS) method to anchor the FRP sheets in strengthened concrete beams. *Construct. Build. Mater.* **2019**, *218*, 351–364. [CrossRef]
9. Le-Trung, K.; Lee, K.; Lee, J.; Lee, D.H.; Woo, S. Experimental study of RC beam-column joints strengthened using CFRP composites. *Compos. B Eng.* **2010**, *41*, 76–85. [CrossRef]
10. Singh, V.; Bansal, P.P.; Kumar, M.; Kaushik, S.K. Experimental studies on strength and ductility of CFRP jacketed reinforced concrete beam-column joints. *Construct. Build. Mater.* **2014**, *55*, 194–201. [CrossRef]
11. Esmaeeli, E.; Barros, J.A.; Sena-Cruz, J.; Fasan, L.; Prizzi, F.R.L.; Melo, J.; Varum, H. Retrofitting of interior RC beam-column joints using CFRP strengthened SHCC: Cast-in-Place Solution. *Compos. Struct.* **2015**, *122*, 456–467. [CrossRef]
12. Hsieh, C.T.; Lin, Y. Detecting debonding flaws at the epoxy-concrete interfaces in near-surface mounted CFRP strengthening beams using the impact-echo method. *NDT E Int.* **2016**, *83*, 1–13. [CrossRef]
13. Prado, D.M.; Araujo, I.D.G.; Haach, V.G.; Carrazedo, R. Assessment of shear damaged and NSM CFRP retrofitted reinforced concrete beams based on modal analysis. *Eng. Struct.* **2016**, *129*, 54–66. [CrossRef]
14. Reinhorn, A.M.; Li, C.; Constantinou, M.C. *Experimental and Analytical Investigation of Seismic Retrofit of Structures with Supplemental Damping: Part 1-Fluid Viscous Damping Devices*; Report No. NCEER-95-0001; National Center for Earthquake Engineering Research: Buffalo, NY, USA, 1995.
15. Lin, W.H.; Anil, K.C. Earthquake response of elastic SDF systems with non-linear fluid viscous dampers. *Earthq. Eng. Struct. Dyn.* **2002**, *31*, 1623–1642. [CrossRef]
16. Goel, R.K. Effects of supplemental viscous damping on seismic response of asymmetric-plan systems. *Earthq. Eng. Struct. Dyn.* **1998**, *27*, 125–141. [CrossRef]
17. Di Sarno, L.; Manfredi, G. Experimental tests on full-scale RC unretrofitted frame and retrofitted with buckling-restrained braces. *Earthq. Eng. Struct. Dyn.* **2012**, *41*, 315–333. [CrossRef]
18. Di Sarno, L.; Manfredi, G. Seismic retrofitting with buckling restrained braces: Application to an existing non-ductile RC framed building. *Soil Dyn. Earthq. Eng.* **2010**, *30*, 1279–1297. [CrossRef]
19. Takeuchi, T.; Nakamura, H.; Kimura, I.; Hasegawa, H.; Saeki, E.; Watanabe, A. Buckling Restrained Braces and Damping Steel Structures. U.S. Patent US20050055968A1, 17 March 2005.
20. Vafaei, M.; Sheikh, A.M.O.; Alih, S.C. Experimental study on the efficiency of tapered strip dampers for the seismic retrofitting of damaged non-ductile RC frames. *Eng. Struct.* **2019**, *199*, 109601. [CrossRef]
21. Sahoo, D.R.; Rai, D.C. Design and evaluation of seismic strengthening techniques for reinforced concrete frames with soft ground story. *Eng. Struct.* **2013**, *56*, 1933–1944. [CrossRef]
22. Oinam, R.M.; Sahoo, D.R. Seismic rehabilitation of damaged reinforced concrete frames using combined metallic yielding passive devices. *Struct. Infrastruct. Eng.* **2017**, *13*, 816–830. [CrossRef]
23. Lee, C.H.; Ryu, J.; Kim, D.H.; Ju, Y.K. Improving seismic performance of non-ductile reinforced concrete frames through the combined behavior of friction and metallic dampers. *Eng. Struct.* **2018**, *172*, 304–320. [CrossRef]
24. Morelli, F.; Piscini, A.; Salvatore, W. Seismic behavior of an industrial steel structure retrofitted with self-centering hysteretic dampers. *J. Construct. Steel Res.* **2017**, *139*, 157–175. [CrossRef]

25. Kelly, J.M.; Skinner, R.I.; Heine, A.J. Mechanisms of energy absorption in special devices for use in earthquake resistant structures. *Bull. N. Z. Soc. Earthq. Eng.* **1972**, *5*, 63–88.
26. Skinner, R.I.; Kelly, J.M.; Heine, A.J. Hysteretic dampers for earthquake-resistant structures. *Earthq. Eng. Struct. Dyn.* **1974**, *3*, 287–296. [CrossRef]
27. Durucan, C.; Dicleli, M. Analytical study on seismic retrofitting of reinforced concrete buildings using steel braces with shear link. *Eng. Struct.* **2010**, *32*, 2995–3010. [CrossRef]
28. Rai, D.C.; Annam, P.K.; Pradhan, T. Seismic testing of steel braced frames with aluminum shear yielding dampers. *Eng. Struct.* **2013**, *46*, 737–747. [CrossRef]
29. Dusicka, P.; Itani, A.M.; Buckle, I.G. Cyclic behavior of shear links of various grades of plate steel. *J. Struct. Eng.* **2010**, *136*, 370–378. [CrossRef]
30. Nuzzo, I.; Losanno, D.; Caterino, N.; Serino, G.; Rotondo, L.M.B. Experimental and analytical characterization of steel shear links for seismic energy dissipation. *Eng. Struct.* **2018**, *172*, 405–418. [CrossRef]
31. Hitaka, T.; Matsui, C. Experimental study on steel shear wall with slits. *J. Struct. Eng.* **2003**, *129*, 586–595. [CrossRef]
32. Chan, R.W.; Albermani, F. Experimental study of steel slit damper for passive energy dissipation. *Eng. Struct.* **2008**, *30*, 1058–1066. [CrossRef]
33. Tagawa, H.; Yamanishi, T.; Takaki, A.; Chan, R.W. Cyclic behavior of seesaw energy dissipation system with steel slit dampers. *J. Construct. Steel Res.* **2016**, *117*, 24–34. [CrossRef]
34. Saffari, H.; Hedayat, A.A.; Nejad, M.P. Post-Northridge connections with slit dampers to enhance strength and ductility. *J. Construct. Steel Res.* **2013**, *80*, 138–152. [CrossRef]
35. Li, H.; Li, G. Experimental study of structure with "dual function" metallic dampers. *Eng. Struct.* **2007**, *29*, 1917–1928. [CrossRef]
36. Martinez-Romero, E. Experiences on the use of supplementary energy dissipators on building structures. *Earthq. Spectra* **1993**, *9*, 581–625. [CrossRef]
37. Perry, C.L.; Fierro, E.A.; Sedarat, H.; Scholl, R.E. Seismic upgrade in San Francisco using energy dissipation devices. *Earthq. Spectra* **1993**, *9*, 559–579. [CrossRef]
38. Tsai, K.C.; Chen, H.W.; Hong, C.P.; Su, Y.F. Design of steel triangular plate energy absorbers for seismic-resistant construction. *Earthq. Spectra* **1993**, *9*, 505–528. [CrossRef]
39. Mahjoubi, S.; Maleki, S. Seismic performance evaluation and design of steel structures equipped with dual-pipe dampers. *J. Construct. Steel Res.* **2016**, *122*, 25–39. [CrossRef]
40. Maleki, S.; Bagheri, S. Pipe damper, Part II: Application to bridges. *J. Construct. Steel Res.* **2010**, *66*, 1096–1106. [CrossRef]
41. Tena-Colunga, A.; Pérez-Moreno, D. Seismic upgrading of a nine-story building at Mexico City's lake-bed zone using U-Shaped energy dissipation devices. In Proceedings of the 9th International Seminar on Earthquake Prognostics, San Jose, Costa Rica, 19–23 September 1994; Volume 10, pp. 1991–9684.
42. Tena-Colunga, A.; Del Valle, E.; Pe'rez-Moreno, D. Issues on the seismic retrofit of a building near resonant response and structural pounding. *Earthq. Spectra* **1996**, *12*, 567–597. [CrossRef]
43. Rahnavard, R.; Rebelo, C.; Craveiro, H.D.; Napolitano, R. Numerical investigation of the cyclic performance of reinforced concrete frames equipped with a combination of a rubber core and a U-shaped metallic damper. *Eng. Struct.* **2020**, *225*, 111307. [CrossRef]
44. Sahoo, D.R.; Singhal, T.; Taraithia, S.S.; Saini, A. Cyclic behavior of shear-and-flexural yielding metallic dampers. *J. Construct. Steel Res.* **2015**, *114*, 247–257. [CrossRef]
45. Gandelli, E.; Taras, A.; Distl, J.; Quaglini, V. Seismic retrofit of hospitals by means of hysteretic braces: Influence on acceleration-sensitive non-structural components. *Front. Built Environ.* **2019**, *5*, 100. [CrossRef]
46. Gandelli, E.; Chernyshov, S.; Distl, J.; Dubini, P.; Weber, F.; Taras, A. Novel adaptive hysteretic damper for enhanced seismic protection of braced buildings. *Soil Dyn. Earthq. Eng.* **2020**, *141*, 106522. [CrossRef]
47. ABAQUS. *Version ABAQUS. 6.14 Document*; ABAQUS Inc.: Johnston, RI, USA, 2014.
48. Imjai, T.; Setkit, M.; Garcia, R.; Figueiredo, F.P. Strengthening of damaged low strength concrete beams using PTMS or NSM techniques. *Case Stud. Construct. Mater.* **2020**, *13*, e00403. [CrossRef]
49. Ministry of Housing and Urban-Rural Development of the People's Republic of China. *Code for Seismic Design of Buildings*; GB50011-2010; Ministry of Housing and Urban-Rural Development of the People's Republic of China: Beijing, China, 2010. (In Chinese)
50. Ministry of Housing and Urban-Rural Development of the People's Republic of China. *Code for Design of Concrete Structures*; GB50010-2010; Ministry of Housing and Urban-Rural Development of the People's Republic of China: Beijing, China, 2010. (In Chinese)
51. Ministry of Housing and Urban-Rural Development of the People's Republic of China. *Load Code for the Design of Building Structures*; GB50009-2010; Ministry of Housing and Urban-Rural Development of the People's Republic of China: Beijing, China, 2012. (In Chinese)
52. ABAQUS. *ABAQUS Theory Manual, "Version 6.13"*; ABAQUS Inc.: Johnston, RI, USA, 2014.
53. Hatzigeorgiou, G.D.; Beskos, D.E. Inelastic displacement ratios for SDOF structures subjected to repeated earthquakes. *Eng. Struct.* **2009**, *31*, 2744–2755. [CrossRef]
54. Yang, F.; Wang, G.; Ding, Y. Damage demands evaluation of reinforced concrete frame structure subjected to near-fault seismic sequences. *Nat. Hazards* **2019**, *97*, 841–860. [CrossRef]

55. Li, H.N.; Li, G.; Li, Z.; Xing, F. Earthquake-Resistant design of reinforced concrete frame with metallic dampers of "dual function". *J. Build. Struct.* **2007**, *28*, 36–43. (In Chinese)
56. Nuzzo, I.; Losanno, D.; Serino, G.; Bozzo, L. A Seismic-resistant Precast r.c. system equipped with shear link dissipators for residential buildings. In Proceedings of the Second International Conference on Advances in Civil, Structural and Environmental Engineering-ACSEE 2014, Zurich, Switzerland, 25–26 October 2014; Volume 2.
57. Li, G. Theoretical and Experimental Research on the Structure with New Type of Metallic Dampers. Ph.D. Thesis, Dalian University of Technology, Dalian, China, 2006. (In Chinese).
58. Dicleli, M.; Mehta, A. Seismic retrofitting of chevron-braced steel frames based on preventing buckling instability of braces. *Int. J. Struct. Stab. Dyn.* **2009**, *9*, 333–356. [CrossRef]
59. Li, G.; Dong, Z.Q.; Li, H.N. Simplified collapse-prevention evaluation for the reserve system of low-ductility steel concentrically braced frames. *J. Struct. Eng.* **2018**, *144*, 04018071. [CrossRef]
60. Sahoo, D.R.; Rai, D.C. Seismic strengthening of non-ductile reinforced concrete frames using aluminum shear links as energy-dissipation devices. *Eng. Struct.* **2010**, *32*, 3548–3557. [CrossRef]

Article

Modelling Strategies for the Numerical Simulation of the Behaviour of Corroded RC Columns under Cyclic Loads

Filippo Molaioni *, Fabio Di Carlo and Zila Rinaldi

Department of Civil Engineering and Computer Science Engineering, University of Rome Tor Vergata, via del Politecnico 1, 00133 Rome, Italy; di.carlo@ing.uniroma2.it (F.D.C.); rinaldi@ing.uniroma2.it (Z.R.)
* Correspondence: filippo.molaioni@students.uniroma2.eu

Abstract: Rebars corrosion phenomena can modify the structural behaviour of reinforced concrete (RC) members and consequently the seismic performance of RC structures. Since many existing RC structures are affected by this phenomenon, the influence of the reinforcement corrosion on the seismic performance is still under examination, especially when the corrosive attack is localized in the dissipative areas of the plastic hinges. In this work, the effect of localized corrosion is numerically investigated, through the adoption of a suitable finite element model, object of validation with the outcomes of an experimental campaign carried out in the Laboratory of the University of Rome "Tor Vergata", on un-corroded and corroded RC columns subjected to axial load and cyclic horizontal actions. Particular attention has been paid to the definition of the three-dimensional model and to the modelling of the corroded rebars and their corrosion morphology. Indeed, different modelling strategies are proposed with the aim to properly simulate the cyclic behaviour of the corroded columns. The main results show how more refined strategies taking into account the morphological aspects of the corrosion phenomenon produce a better fit with the experimental results for both Damage Control and Life Safety limit states performance.

Keywords: RC corroded columns; localised corrosion; numerical analyses; modelling strategies; cyclic actions

Citation: Molaioni, F.; Di Carlo, F.; Rinaldi, Z. Modelling Strategies for the Numerical Simulation of the Behaviour of Corroded RC Columns under Cyclic Loads. *Appl. Sci.* **2021**, *11*, 9761. https://doi.org/10.3390/app11209761

Academic Editor: Jorge de Brito

Received: 24 September 2021
Accepted: 15 October 2021
Published: 19 October 2021

Publisher's Note: MDPI stays neutral with regard to jurisdictional claims in published maps and institutional affiliations.

Copyright: © 2021 by the authors. Licensee MDPI, Basel, Switzerland. This article is an open access article distributed under the terms and conditions of the Creative Commons Attribution (CC BY) license (https://creativecommons.org/licenses/by/4.0/).

1. Introduction

Reinforced concrete (RC) columns play a fundamental role on the seismic performance of the concrete structure, as they represent its main force-transmitting element. One of the major degradation causes of RC structures consists in the steel reinforcement corrosion, as witnessed by recent cases worldwide, especially when the corrosive attack is localized in the dissipative areas of the plastic hinge and pitting phenomena occur. Steel bars in concrete are naturally exposed to a high pH environment, which allows the formation of a protective passivating film. As a consequence of a decrease of the pH value or of a high concentration of chloride ions, this protective film can undergo to a disruption, causing the trigger of the corrosion process of the steel reinforcement. In presence of localized concentrations of chlorides, pitting corrosion of the steel bars occur, with a localized reduction of the bar section through pits, affecting its internal layers. These phenomena represent a cause of concern for several RC buildings, particularly when low strength concrete is used. Besides the obvious reduction of the resistant section, a reinforcement ductility reduction can take place [1,2]. The formation of expansive corrosion products induces concrete cracking and high stresses, affecting the bond behaviour between concrete and steel bars [3–6].

These phenomena can cause severe damages leading to structural unexpected crisis and to failure mechanisms of corroded structures very different from the ones of new or sound constructions.

Therefore, it is fundamental to examine the effect of the corrosion damage, for a proper assessment of the seismic capacity of RC columns. Since several RC structures have nowadays an age close to, or higher than, their design life, this aspect is becoming a worthy issue, with the need to be clearly underlined and codified, starting from the analysis of actual structural cases, experimental and theoretical research [7–13].

The interest in the evaluation of the influence of corrosion on the cyclic behaviour of RC column is witnessed by the many experimental studies available in literature, carried out for example on columns characterized by square [14–18] and circular [19] cross sections, and RC moment-resisting frame [20]. In general, the experimental studies show a reduction in ductility and load-bearing capacity of RC columns, with increasing corrosion of the steel rebars.

Consequently, in recent years, several modelling approaches for the numerical assessment of the cyclic behaviour of RC columns subjected to corrosion of steel reinforcement have been proposed, in order to progressively investigate the numerous involved parameters. In [21] fiber-based models carried out with Opensees are used to numerically predict the cyclic behaviour of RC hollow bridge piers with corroded rebars, showing a reduction of strength, secant stiffness and energy loss due to corrosion. In [22] a multi-layered shell Finite Element (FE) model based on the fixed crack approach is used to consider the corrosion effects on RC elements subjected to cyclic loadings, in order to calibrate correction coefficient for the ultimate rotational capacity prediction, also in presence of buckling. In [23] some of the authors proposed a three-dimensional (3D) FE model accounting for steel corrosion and interface decay, highlighting the onset of peculiar mechanism related to the buckling of the corroded rebars. A 3D non-linear FE approach is also used in [24], in which the behaviour of corroded RC columns under seismic loading is studied through a parametric numerical investigation on 240 RC columns, to assess the influence of the several involved parameters in the lateral load resistance and ultimate drift capacity. In [25] a FE model is developed and validated using the results obtained from three different sets of experimental tests available in literature and used to investigate the effect of localised and uneven distribution of corrosion on the cyclic response of RC columns, showing the pitting influence on the plastic hinge length.

However, according to recent guidelines [26], the prediction of the bearing capacity of RC elements affected by reinforcement corrosion can be made with simplified approaches, based on the reduction of the steel area. It is worth to highlight that these methodologies could be suitable for predicting the load bearing capacity of the RC element but could be misleading in the evaluation of the local and global ductility, significantly affected by corrosion.

Aim of the paper is the development of a numerical model for the evaluation of the influence of localized corrosion on the cyclic behaviour and failure modes of RC columns. The paper follows and completes previous research developed by some of the authors in experimental [15] and numerical [21] way, on the influence of the corrosion of the longitudinal rebars on the cyclic response of column specimens.

Indeed, a numerical model is developed, able to catch and highlight all the phenomena occurring in elements characterized by localised corrosion, under cyclic loads, paying much care in the definition of the 3D model and in the simulation of the rebar corrosion. To this aim, different modelling strategies (MS) are adopted, respectively based on: reduction of constitutive law by uniform or pitting corrosion; bar discretization and section reduction; bar discretization and morphology based constitutive law reduction. The described models are validated through a comparison with the results of experimental tests developed at the Laboratory of the University of Rome "Tor Vergata" on four full-scale square RC columns subjected to cyclic loads and characterised by localised rebar corrosion [27].

The reference experimental survey on the four full-scale RC columns, cast and tested under cyclic load is presented in Section 2. The column geometry, the details of the artificial corrosion process of the steel reinforcement, and the cyclic loading system are here presented. The main outcomes of the performed tests are outlined, in terms of effective corrosion amount and morphology evaluated after the tests and in terms of load-horizontal displacements graphs. Section 3 is devoted to the description of the developed numerical models, with particular reference to the definition of the geometry of the specimen, boundary conditions, material properties (concrete and steel reinforcement) and load patterns. As previously described, different strategies are proposed and compared for the modelling of the corroded reinforcement. The section ends with a description of the modelling of the bond between reinforcement and concrete and of the expectations about results, on the basis of the visual inspection and morphological study of the reinforcement. In Section 4 the results of the performed numerical analysis are shown, with reference to un-corroded and corroded elements, to outline the pros and cons of the different modelling strategies to study the effect of non-uniform steel rebars corrosion on the cyclic behaviour of RC columns. Finally, in Section 5 the main findings are summarized, underlining the sharp effect of localised corrosion on the structural behaviour and failure mode of RC columns and the importance to perform cyclic analysis for existing structures subjected to decay phenomena.

2. Reference Experimental Survey

The reference experimental survey was performed at the Laboratory of the University of Rome "Tor Vergata", where four RC columns were cast and tested [27]. The specimens, with a height of 1800 mm and a 300 mm × 300 mm square section, were reinforced with four Φ16 mm longitudinal bars (Figure 1a). Two spacing values, equal to 250 mm and 300 mm, of the closed Φ8 mm stirrups were considered. The concrete cover was equal to 30 mm. The elements were cast on a 1500 mm × 750 mm × 500 mm foundation (Figure 1a).

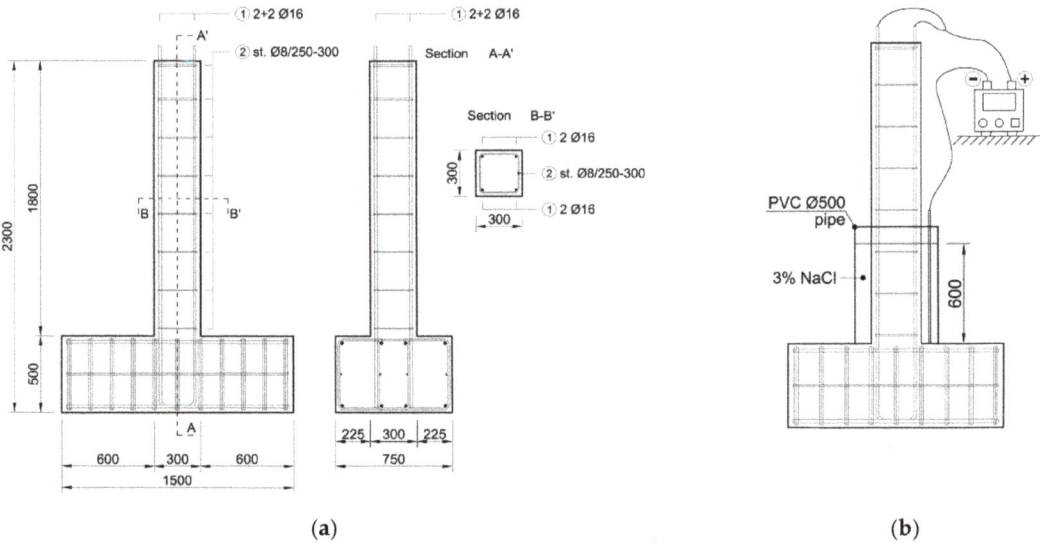

Figure 1. (a) Column geometry; (b) artificial corrosion process.

For each of the two considered stirrups spacing, one column was kept un-corroded for reference (A), while a second one was subjected to a process of artificial corrosion of the steel reinforcement (B) at the column base. A 3% saline solution was contained in a PVC Ø500 mm pipe, placed around the column and fixed to the foundation, up to a height of about 600 mm from the foundation extrados (Figure 1b). The longitudinal rebars were connected to the positive pole of the power supply (anode), while a Ø10 diameter steel bar, placed inside the PVC pipe, acted as the cathode. A current intensity equal to 0.05 A for each bar was adopted. Table 1 shows the layout of the experimental program.

Table 1. Layout of experimental program.

Specimen	Stirrups Spacing	Corrosion
A25	250	-
B25		■
A30	300	-
B30		■

The effective corrosion amount and morphology were evaluated after the tests, by extracting the steel rebars from the specimens B25 and B30, characterised by stirrup spacing of 250 mm and 300 mm, respectively. Both the longitudinal and transversal reinforcements were corroded, and a sharp localization of the corrosive attack took place. In particular, even if the mass loss was about 24% in both the specimens, deep pits were observed in the longitudinal bars up to a diameter reduction of more than 30% in the specimen B25, and up to 50% in the column B30. The corrosion morphology played a fundamental role in defining the failure mode, as highlighted in the following.

The test set-up is shown in Figure 2. An axial load equal to 300 kN was applied to the top column with a self-balanced system and with two high strength rebars, each one connected to a hydraulic jack. After the axial load application, a cyclic horizontal displacement history, with increasing amplitude, was imposed at a height h from the foundation extrados equal to 1.5 m, up to the failure. To this aim, an electro-mechanical jack was fixed to the load frame of the laboratory and linked to the column employing a hinged bar system, in which a load cell was placed. The loading history consisted of three complete cycles, for different values of the column drift, defined as the ratio between the horizontal displacement δ at the load application point and the height h.

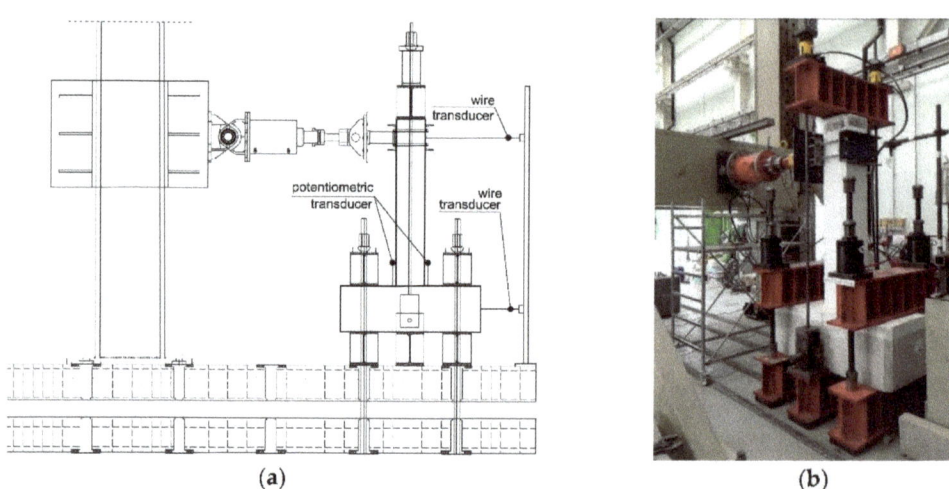

Figure 2. (a) Scheme of Test set-up; (b) test set-up.

The experimental results in terms of load-horizontal displacements are summarised in Figure 3a, with reference to the columns A25 and B25, and in Figure 3b for specimens A30 and B30. In both cases, a decrease of the maximum load of about 30% was measured. A great influence of the corrosion can be observed in the shape of the cycle and then in the ductile behaviour of the elements. Mainly with reference to specimen B30 (sharply affected by very localised corrosion), the pinching effect can be clearly appreciated, due to the onset of brittle mechanisms. Further details can be found in [27].

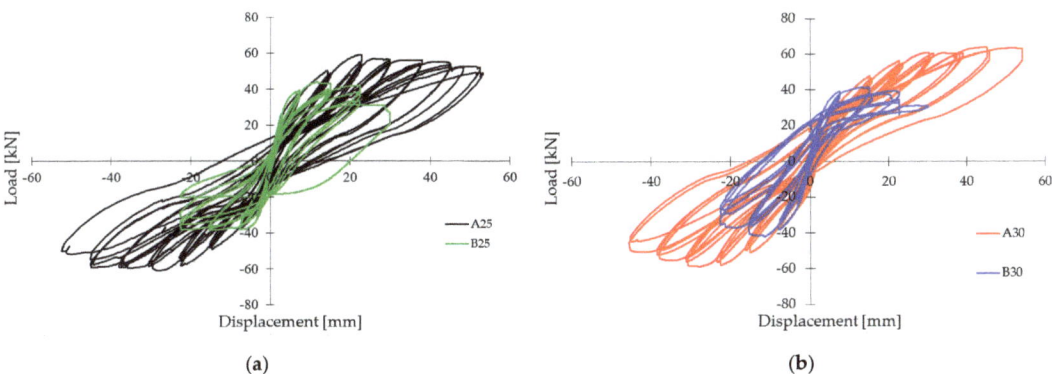

Figure 3. Experimental test results; (**a**) un-corroded A25-corroded B25 specimens; (**b**) un-corroded A30-corroded B30 specimens.

3. FEM Model

Non-linear numerical analyses are performed with the FEM code DIANA 10.5 (2021), to obtain a reliable predictive model for the cyclic behaviour of corroded RC columns. In this section, the geometry of the specimen, the boundary conditions, the material properties and the load patterns, all assumed in agreement with the experimental tests, are presented.

3.1. Geometry, Element Assumption and Analysis

The numerical model is shown in Figure 4a. The considered reference system is superimposed in the same figure. Concrete elements are modeled with the eight-node structural solid element. Through a mesh generator, a desired element size of 50 mm for the column concrete core and the concrete cover is considered, while a size of 100 mm is set for the foundation concrete. The longitudinal reinforcement is modeled as truss bond-slip reinforcement, while the columns transverse reinforcement and the foundation reinforcement are modeled as embedded reinforcement. The "element by element" discretization method is used for reinforcement. In Figure 4b the casting position of the corroded reinforcements, together with the load direction x, is shown.

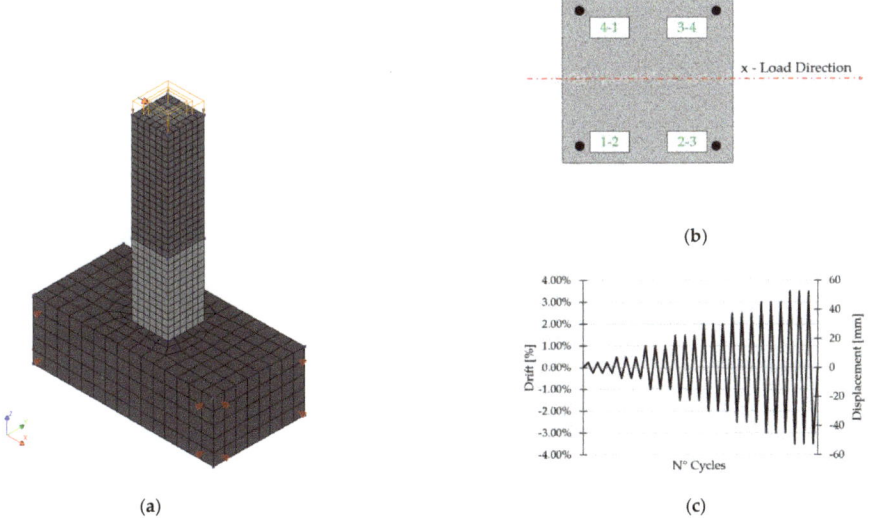

Figure 4. (**a**) Numerical model, (**b**) Column base cross-section and corroded bars casting position, (**c**) Displacement history adopted in the non-linear analyses.

Regarding the boundary conditions, translational restraints in the orthogonal direction to the base and side faces of the foundation are applied, to simulate the foundation clamping provided during the experimental test. A translational restraint at the column top and along the load direction is modeled to perform a displacement control non-linear cyclic analysis. Therefore, a tying constraint for the nodes of the cross-section at the load height is applied to impose the equality of the x-component of the displacement.

The non-linear structural analyses are performed by applying in a first phase the axial load of 300 kN at the top of the column, then the displacement histories shown in Figure 4c is applied in accordance with experimental tests (up to 3.5% Drift for un-corroded specimens and up to 1.5% drift for corroded specimens). The structural non-linear analyses are performed using a regular Newton-Raphson iterative method for the solution of the non-linear equation set (imposing 20 maximum number of iterations for step) with a convergence norm based on displacement and force (convergence tolerance is assumed equal to 0.01).

3.2. Concrete Modelling

The concrete elements are modeled through the Total Strain Crack Model (TSCM) theory [28,29], considering a rotating cracks orientation. The compressive behaviour of concrete is modeled according to the parabolic curve, based on the fracture energy [30,31]. The compressive strength was assumed equal to 20 MPa, in agreement with the experimental test results. The concrete Elastic Modulus and the compressive fracture energy are derived from the compressive strength as indicated in [32]. The concrete stress confinement was considered by adopting the model proposed in [29], through a pre-strain concept in which the lateral expansion effects are accounted for with an additional external loading on the elements. The concrete tensile behaviour is modeled with the exponential softening law. The tensile strength and the Mode-I fracture energy are calculated as a function of the compressive strength according to [32]. The crack bandwidth in agreement with [33] is evaluated as the cube root of the element's volume. The cracking and spalling of the concrete cover due to corrosion products are considered by reducing the concrete cover compressive strength in accordance with [10], in which the model proposed by [28], based on the average tensile strain in the transverse direction, is used. As regards the cyclic behaviour, it is to remark that the TSCM is characterized by a secant unloading in both compression and tension [30], and then the contribution to the energy dissipation of the concrete could be slightly underestimated. Nevertheless, this model was chosen and adopted in the numerical simulation for its stability and reliability in non-linear analyses.

3.3. Steel Reinforcement Modelling

The steel reinforcement has been simulated with the isotropic non-linear constitutive model of Menegotto and Pinto [34]. The constitutive-model parameters were calibrated on the results of tensile tests on the reinforcements used in previous experimental tests made by some of the authors [15]. The model parameters for the un-corroded reinforcement are shown in Table 2. With reference to the corroded reinforcement, one of the goals of this work is to propose and compare different modelling strategies, presented below.

Table 2. Un-corroded steel reinforcement Menegotto-Pinto parameters.

Steel	E_{steel}	Density	$f_{yielding}$	b_0	R_0	A_1	A_2	A_3	A_4
	MPa	kN/m^3	MPa	-	-	-	-	-	-
B450C	210,000	78	520	0.0062	20	18.5	0.15	0.01	7

3.3.1. Corroded Rebars of the Experimental Test

To better understand the effect of both corrosion localization and pitting phenomenon, particular attention was paid to the morphology of the corroded reinforcements. Following the experimental tests, the pieces of the bars subjected to artificial corrosion (i.e., those in the plastic hinge regions) were extracted from the concrete and cleaned of the excess rust to evaluate their mass loss and their corrosion morphology (Figures 5 and 6 for the corroded specimens B25 and B30, respectively). The steel rebars are named according to Figure 4b.

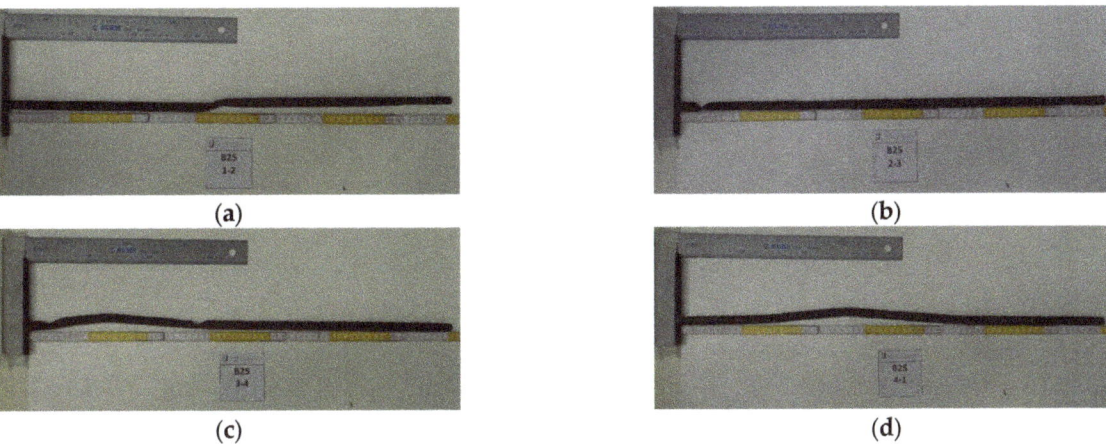

Figure 5. Corroded Rebars, Column B25. (**a**) Bar 1-2, (**b**) Bar 2-3, (**c**) Bar 3-4, (**d**) Bar 4-1.

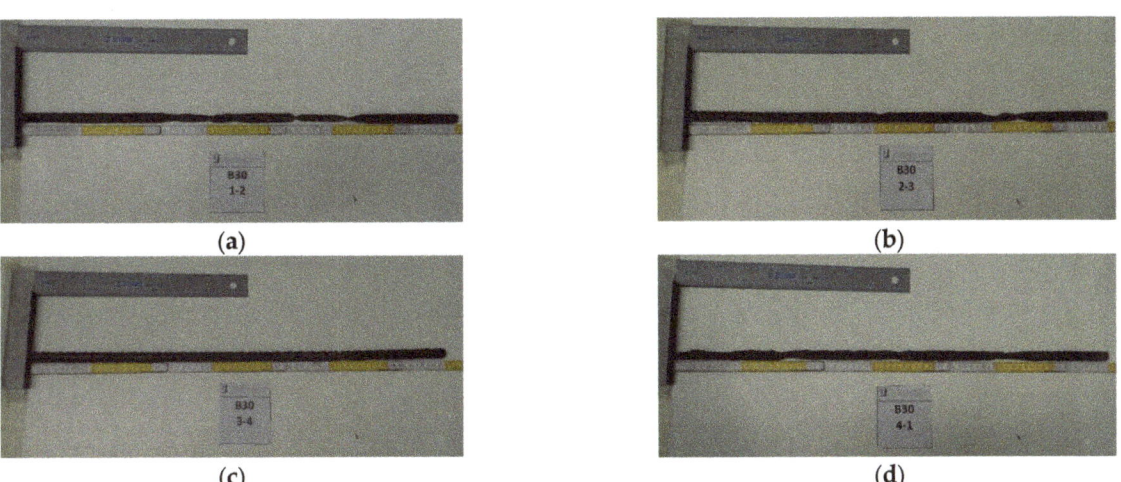

Figure 6. Corroded Rebars, Column B30. (**a**) Bar 1-2, (**b**) Bar 2-3, (**c**) Bar 3-4, (**d**) Bar 4-1.

The average loss mass of each bar was calculated by weighing the specimens, as summarized in Table 3. It is worth noting that the asymmetrical cyclic behaviour of the column B30 (Figure 3b) is due to the difference in the mass loss values of the bars, strongly corroded on a side and moderately corroded on the other (Table 3 and Figure 4).

Measurements of the bars' cross-section minimum diameter, carried out repeatedly along the length of the bars, were performed. These measures are useful for the morphology evaluations, characterizing the modelling strategies 3 and 4 presented in this section.

Table 3. Weight and loss mass value of the corroded steel rebars.

Column	Bar	Lenght	Un-Corroded Weight	Corroded Weight	M_{loss}%
		[mm]	[g]	[g]	-
B-25	1-2	705	1114	840	25%
	2-3	698	1103	810	27%
	3-4	710	1122	835	26%
	4-1	710	1122	920	18%
B-30	1-2	700	1106	750	32%
	2-3	696	1100	900	18%
	3-4	700	1106	915	17%
	4-1	696	1100	760	31%

3.3.2. Modelling Strategy 1: Reduction of Constitutive Law by Uniform Corrosion

The steel constitutive law is broken down according to [35], in which the degradation relationships for uniform corroded steel rebars are proposed. The reduction of the constitutive parameters occurs for each bar, as a function of the actual mass loss (Table 3). In this way, only the portions of the longitudinal reinforcements subjected to corrosion, i.e., the plastic hinge regions, are modeled with a reduced constitutive laws. Table 4 shows the mechanical parameters reduction for modelling strategy 1.

Table 4. Mechanical parameter reduction for modelling strategies 1 and 2.

Column	Bar	Mloss%	$f_{y,uniform}$	$\varepsilon_{y,uniform}$	$f_{y,pitting}$	$\varepsilon_{y,pitting}$
		-	MPa	-	MPa	-
B-25	1-2	25%	334	0.16%	261	0.12%
	2-3	27%	319	0.15%	240	0.11%
	3-4	26%	326	0.16%	250	0.12%
	4-1	18%	386	0.18%	333	0.16%
B-30	1-2	32%	281	0.13%	188	0.09%
	2-3	18%	386	0.18%	333	0.16%
	3-4	17%	393	0.19%	344	0.16%
	4-1	31%	289	0.14%	198	0.09%

3.3.3. Modelling Strategy 2: Reduction of Constitutive Law by Pitting Corrosion

The second modelling strategy, in analogy with the first, provides for the reduction of the steel constitutive law, but in this case considering the phenomenon of pitting corrosion. Pitting causes localized section reductions in the bar through pits, affecting internal layers of the steel bar and further reducing the strength and ductility of the reinforcement. Again in [35], the degradation relationships for pitting corrosion from literature review data are presented. Table 4 shows the assumed mechanical properties according to modelling strategy 2 for both corroded specimens.

3.3.4. Modelling Strategy 3: Bar Discretization and Section Reduction

The "macro-morphology" of the reinforcing bars is considered in the third modelling strategy. In this case, the measurements of the minimum diameter along each corroded bar developed during the experimental tests, are exploited to discretize the bar in pieces over its corroded length. Therefore, an equivalent section is considered for each piece identified. It should be noted that the diameter measurement of bars deformed by test loads, due to yielding or distortion, is omitted from the discretization process. The longitudinal bar is then modeled through connected in series truss bond-slip reinforcement elements, to which a reduced section diameter is assigned. It is remarked that through this strategy no direct reductions are applied on the constitutive laws of the reinforcements.

3.3.5. Modelling Strategy 4: Bar Discretization and Morphology-Based Constitutive Law Reduction

Strategy 4 accounts for both the "macro-morphological" aspects, i.e., the variation of corrosion along with the reinforcement, and the "micro-morphological" aspects, i.e., pitting on the cross-section of the bar. The morphology of the bars' cross-section is considered using the model proposed by Val and Melchers [36], summarized in Figure 7, in which the maximum penetration is linked to the net sectional area of the corroded bar, considering a hemispherical form of pits.

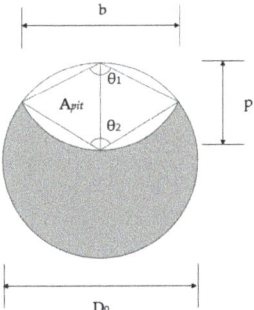

Figure 7. Pit configuration, model by Val and Melchers [36].

$$\begin{cases} A_{pit} = A_1 + A_2 & \text{if } p \leq \frac{D_0}{\sqrt{2}} \\ A_{pit} = A_0 - A_1 + A_2 & \text{if } \frac{D_0}{\sqrt{2}} \leq p \leq D_0 \\ A_{pit} = A_0 & \text{if } p \geq D_0 \end{cases} \quad (1)$$

$$A_1 = 0.5 * \left[\theta_1 \left(\frac{D_0}{2}\right)^2 - b \left| \frac{D_0}{2} - \frac{p^2}{D_0} \right| \right], \quad A_2 = 0.5 * \left[\theta_2 p^2 - b \frac{p^2}{D_0} \right], \quad A_0 = \frac{\pi * D_0^2}{4}, \quad (2)$$

$$b = 2 * p * \sqrt{1 - \left(\frac{p}{D_0}\right)^2}, \quad \theta_1 = 2 * \arcsin\left(\frac{b}{D_0}\right), \quad \theta_2 = 2 * \arcsin\left(\frac{b}{2 * p}\right), \quad (3)$$

Then an inverse method was used to iteratively search the "pitting factor" R (i.e., maximum vs. average penetration ratio) of each bar (Figure 8). Firstly an initial guess value for the "pitting factor" is assumed and, considering the measured maximum penetration of the corroded sections of the bar, the mass loss value for each piece of the corroded bar is evaluated through the Val and Melchers model [36] with Equations (1)–(3). The "calculated mass loss" of each bar is evaluated as the weighted average on the length of each bar piece. If this parameter is equal to the experimental mass loss measured after the tests, the "pitting factor" is then determined. Otherwise, a new "pitting factor" guess value is set in the iterative procedure, until the satisfaction of the check test.

Following this procedure, the influence of the pitting phenomenon has been evaluated for each bar and then the loss mass value assigned to each section of the bar has been optimized according to the pitting factor. In order to significantly compare the numerical results, the same discretization along the bars adopted in the strategy 3 was also applied in this case. Tables 5 and 6 summarize the results obtained with the methodology previously described for all corroded bars, relating to specimens B25 and B30 respectively.

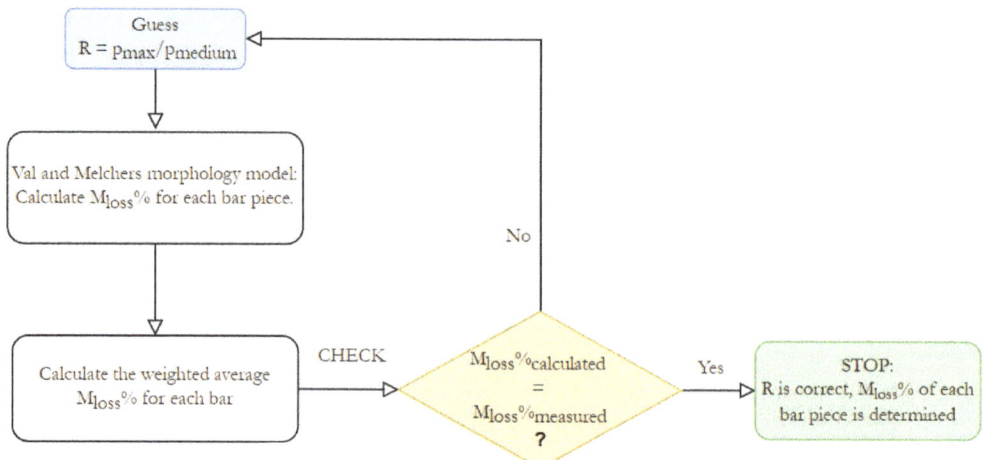

Figure 8. Flow chart of the inverse method for morphology-based mass loss evaluation.

Table 5. Pitting factor and lost mass assumed for strategy 4, column B25.

Column	Bar	Measured M_{loss}%	Pitting Factor	Piece Lenght [mm]	Piece M_{loss}%
B-25	1-2	25%	1	115	27%
				200	14%
				30	40%
				285	16%
				70	40%
	2-3	27%	1	22.5	12%
				52.5	40%
				300	29%
				200	15%
				115	35%
	3-4	26%	1	20	12%
				250	40%
				430	14%
	4-1	18%	1	275	13%
				300	5%
				125	22%

Figures 9 and 10 resume the modelling assumptions made for the "macro-morphology" and "micro-morphology", respectively. Indeed, the bars' discretization is shown through the comparison between measured diameter and assumed diameter. Furthermore, the results of the morphology-based mass loss evaluation are reported through a comparison between the measured mass loss and the assumed mass loss. In these figures, the 0 cm progressive distances of the bar correspond to the column-foundation interface, while the progressive 70 cm corresponds to the top of the corroded bar piece. The longitudinal reinforcement is then modeled through connected in series truss bond-slip reinforcement elements, to which reduced constitutive laws are also set, according to [35] in the case of uniform corrosion, as a function of the calculated loss mass of each bar piece.

Table 6. Pitting factor and lost mass assumed for strategy 4, column B30.

Column	Bar	Measured Mloss%	Pitting Factor	Piece Lenght [mm]	Piece Mloss%
B-30	1-2	32%	2.6	155	25%
				200	40%
				80	28%
				135	40%
				120	20%
	2-3	18%	6.5	275	11%
				65	24%
				150	11%
				135	40%
				65	15%
	3-4	17%	5.9	210	11%
				215	21%
				85	15%
				65	31%
				115	12%
	4-1	31%	4.8	575	35%
				115	15%

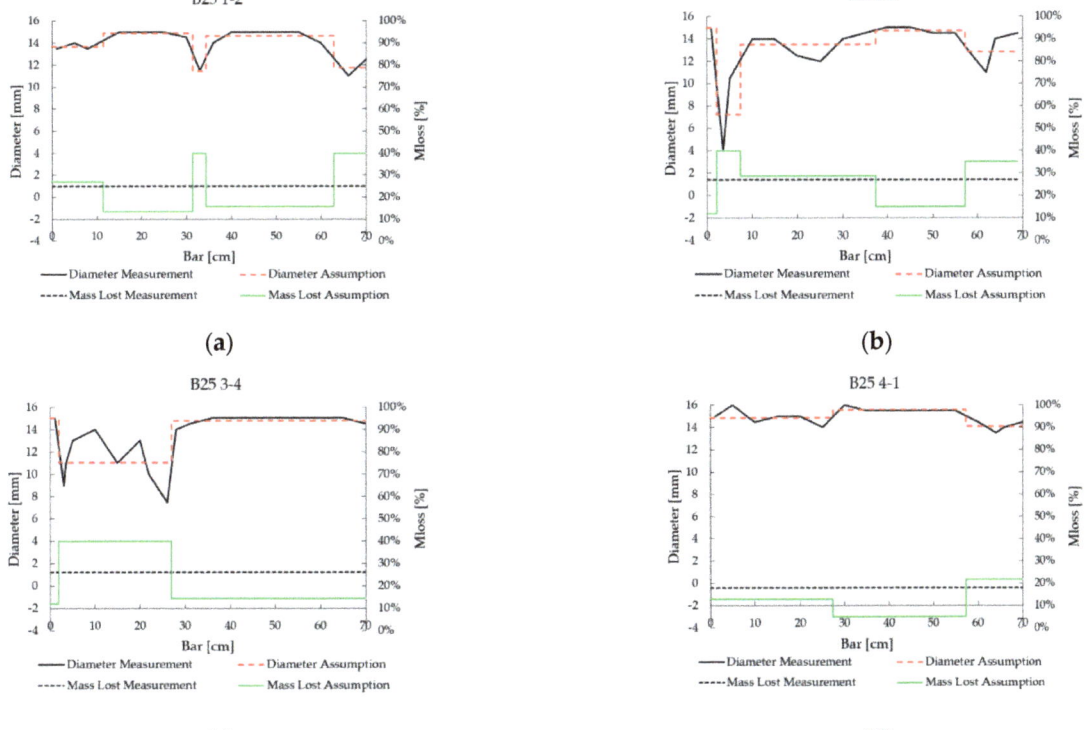

Figure 9. Assumptions for corroded bars (modelling strategies 3 and 4) compared with original measurement, Specimen B25. (a) Bar 1-2, (b) Bar 2-3, (c) Bar 3-4, (d) Bar 4-1.

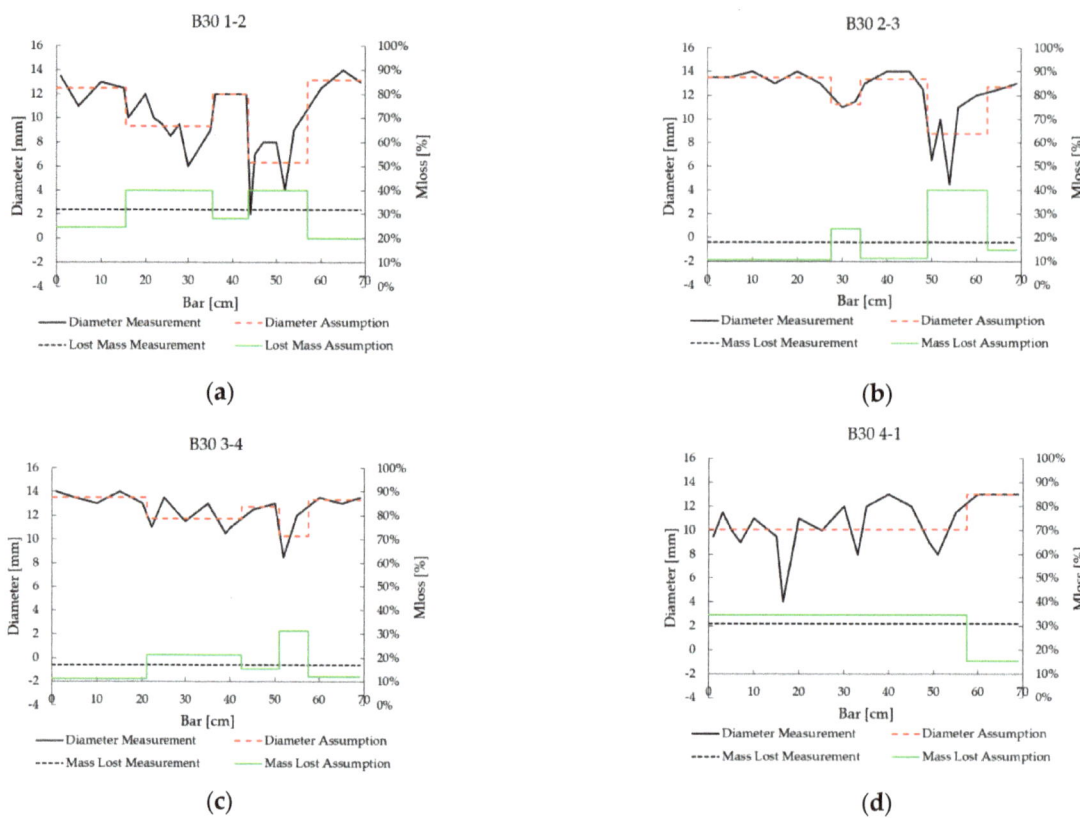

Figure 10. Assumptions for corroded bars (modelling strategies 3 and 4) compared with original measurement, Specimen B30. (**a**) Bar 1-2, (**b**) Bar 2-3, (**c**) Bar 3-4, (**d**) Bar 4-1.

3.4. Bond Modelling

The bond between reinforcement and concrete is modeled according to [32], considering splitting failure and good bond conditions. The differences in the bond laws between the two un-corroded specimens are considered by the different confinement effects given by the stirrups. The expansion of corrosion oxides and the cracking of the concrete cover have a deterioration effect on the bond. Several experimental studies and analytical formulations are available in the literature for the evaluation of the bond between concrete and corroded bars [4,37–39]. In this paper, the law proposed by [40], based on regression analyses of experimental data is adopted for the bond degradation as a function of the corrosion penetration. The model accounts for the amount of transverse reinforcement and predicts the residual bond strength, in relationship to either corrosion penetration or surface crack width. These analyses refer to medium-low corrosion levels, i.e., corrosion penetration up to 0.5 mm. In fact, when the corrosion increases, the bond is reduced until the collaboration between steel and concrete is almost negligible. For this reason, beyond a penetration value equal to 1 mm, a perfect plastic frictional bond behaviour with a bond strength value equal to 0.1 MPa [41,42] is considered. In Figure 11 an example of the bond law reduction assumed for the bars of corroded specimens is presented. As regards the unloading/reloading behaviour, the model unloads with a linear stiffness, that is equal to the initial stiffness, until the opposite residual stress value is reached. This stress value is kept until the back-bone curve in the opposite direction can be picked up [30].

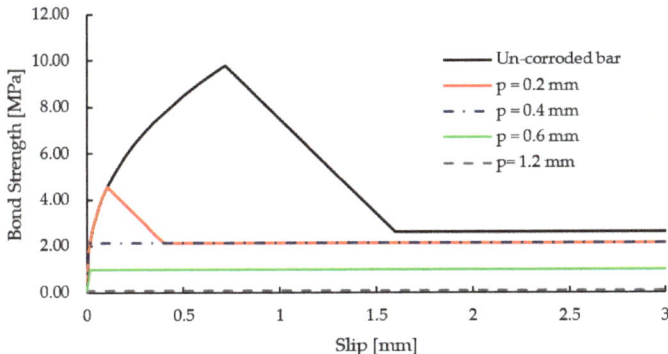

Figure 11. Example of bond law reduction.

3.5. Expectations about Results

Based on the visual inspection of the reinforcement (Figures 5 and 6) and of the pitting factors R identified through the morphological study described in the previous section (Tables 5 and 6), it is highlighted that the pitting phenomenon is more pronounced for specimen B30. Vice versa specimen B25 shows a rather uniform corrosion morphology. Based on this evidence, the following authors' expectations for the B25 specimen are made:

- Modelling strategy 1 could be suitable for capturing the ultimate behaviour of the specimen, while for the pre-yield behaviour there could be some differences with the real response, as an average loss mass value is considered and the influence of the corrosion localization is neglected.
- Modelling strategy 2 should underestimate the capacity of the specimen since the degradation relationships for pitting are too severe in the case of uniform corrosion, as for specimen B25.
- Modelling strategy 3 should improve the result of strategy 1 since the section reduction that is modeled is intrinsically a uniform reduction. Furthermore, considering the geometry of the bar along its length and the relative bond degradation, the stresses in the reinforcement and at the concrete interface should be better understood.
- Modelling strategy 4, which is assumed to be the most complete, should provide the best results, in fact, even if the morphological study did not show strong pitting phenomena, the reduction of the constitutive law is still more refined than the section reduction, as the microscopic effects of corrosion on the bars are considered.

Otherwise, for the B30:

- Modelling strategy 1 should overestimate the ultimate behaviour of the experimental test since medium-high pitting factors were detected for the specimen. Furthermore, it is believed that the results related to the post-cracking behaviour could not be able to provide reliable predictions, since average values of loss mass are used.
- Modelling strategy 2 should better fit the experimental response, with particular reference to the ultimate behaviour, since specific regressions for pitting corrosion are used.
- Modelling strategy 3 should not be suitable for pitting. In particular, the numerical solution could diverge from the real one as the pitting factor could be not representative of the actual condition. Indeed, since the assumption on the section reduction is intrinsically uniform and based on the minimum diameter, generally, the rebar area introduced in the numerical model could be lower than the actual one.
- Modelling strategy 4 should provide better results also in the case of pitting corrosion since the evaluation of the loss mass of each section of the bar is calibrated with the morphological model.

4. Results

4.1. Un-Corroded Specimens

The results of the non-linear numerical analyses, relating to the un-corroded specimens, are shown in Figure 12.

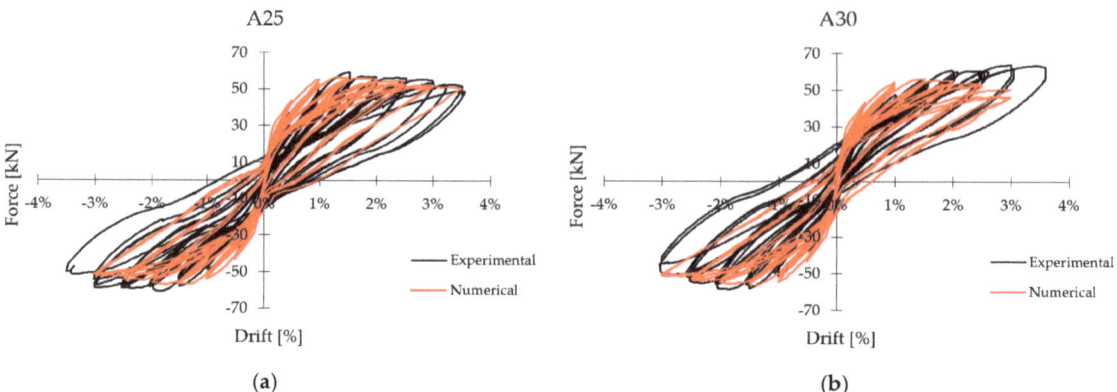

Figure 12. Comparison between experimental and numerical response, un-corroded columns. (**a**) specimen B25, (**b**) specimen B30.

As regards column A25, the main results in terms of load capacity and drift at the maximum capacity are in perfect agreement with the experimental results, as shown in Table 7. For column A30, even if it was not possible to catch the slight asymmetry of the cyclic behaviour, the results of the numerical analyses, reported in Table 7, can be considered satisfactory. Overall, the force-drift cyclic curves obtained through the numerical model show a good fit with the experimental results. The goodness of the results in terms of capacity, displacement, and energy dissipation means that the numerical model is validated by the experimental test and that it represents a reliable base for evaluations regarding the modelling of corroded bars and their effect on the cyclic behaviour of the columns.

Table 7. Main results of un-corroded specimens, experimental vs. numerical.

Specimen		Max. Load (+) kN	Drift at Max. Load (+)	Δload (+)	Max. Load (−) kN	Drift at Max. Load (−)	Δload (−)
A25	Exp.	59.2	1.51%	−4.10%	−60.7	−1.83%	−6.18%
	Num.	56.8	1.30%		−56.9	−1.73%	
A30	Exp	63.7	2.94%	−11.97%	−58.7	−1.85%	−3.75%
	Num	56.0	1.43%		−56.5	−1.43%	

4.2. Corroded Specimen

In this section, the numerical results of corroded specimens, with reference to each of the considered modelling strategies, are reported and compared with the reference experimental tests (Section 2). In Tables 8 and 9 the main results for the different modelling strategies are summarized respectively for columns B25 and B30, in terms of maximum load, drift at the maximum load, and percentage difference of the numerical load capacity with respect to the experimental one.

Table 8. Main results of corroded specimens B25, experimental vs. numerical.

	Specimen	Max. Load (+)	Drift at Max. Load (+)	Δload (+)	Max. Load (−)	Drift at Max. Load (−)	Δload (−)
		kN			kN		
B25	Experimental	43.96	0.78%		−38.42	−1.00%	
	MS-1	40.85	1.17%	−7.07%	−40.47	−1.10%	5.33%
	MS-2	38.58	1.10%	−12.25%	−36.20	−1.03%	−5.76%
	MS-3	42.99	1.00%	−2.20%	−37.39	−0.011	−2.68%
	MS-4	42.77	0.80%	−2.70%	−37.53	−1.30%	−2.32%

Table 9. Main results of corroded specimens B30, experimental vs. numerical.

	Specimen	Max. Load (+)	Drift at Max. Load (+)	ΔLoad (+)	Max. Load (−)	Drift at Max. Load (−)	ΔLoad (−)
		kN			kN		
B30	Experimental	41.65	0.88%		−42.00	−0.75%	
	MS-1	40.76	0.93%	−2.12%	−46.15	−1.23%	9.88%
	MS-2	36.33	0.67%	−12.76%	−43.82	−1.03%	4.32%
	MS-3	31.51	1.00%	−24.34%	−35.76	−1.10%	−14.87%
	MS-4	38.64	0.87%	−7.22%	−42.11	−1.17%	0.25%

4.2.1. Modelling Strategy 1

It is worth reminding that in this strategy the constitutive laws of steel are reduced as a function of the loss mass, that was measured directly on the bars, following the experimental tests. The numerical results, compared with the experimental ones, are reported in Figure 13.

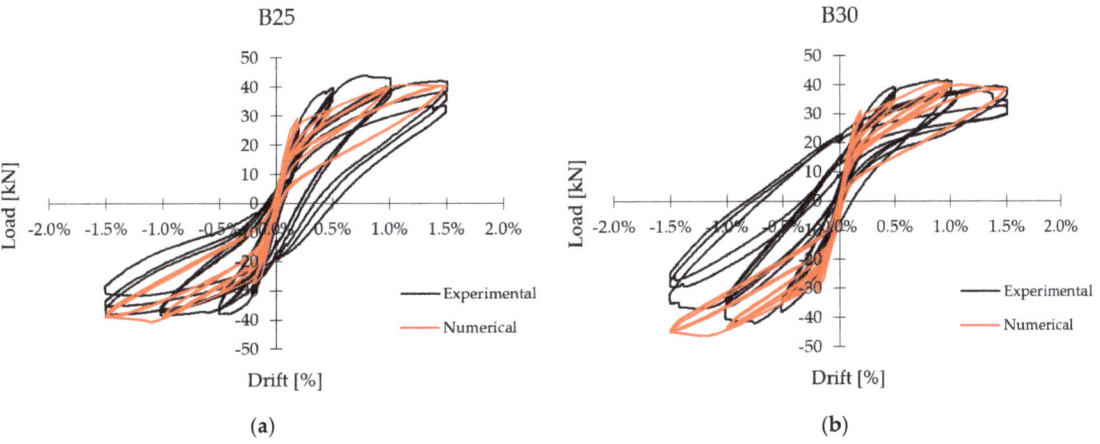

Figure 13. Comparison between experimental and numerical response, corroded specimens, MS 1. (a) column B25, (b) column B30.

The cyclic behaviour, characterized by pinching, is well-caught thanks to the bond reduction, which in the loading and unloading phase reduces the collaboration between steel and concrete. The latter aspect, however, in a modelling strategy that does not consider the variation of the section along the bar, leads to excessive slip of the bars with the consequent over-estimation of the yield drift. The crack patterns obtained through numerical simulations are in good agreement with the experimental observations, with the formation of base cracks that open and close during the cycles as shown below. This crack is precisely due to the bond losses and reinforcement sliding, which also reduces the energy dissipation.

Similar comments can be drawn for column B30: a good approximation of the ultimate load, but a poor prediction of the pre-yield behaviour are found. Furthermore, given the reduction of the mechanical properties due to uniform corrosion, an overestimation of the capacity was expected for the column with pitting phenomenon, as B30. As a matter of fact, this outcome is obtained in the negative load direction only. The main results for the numerical cyclic behaviour of modelling strategy 1 are reported in Tables 8 and 9.

4.2.2. Modelling Strategy 2

In this case, the constitutive law of the reinforcement rebars was reduced with the pitting corrosion degradation relationship as a function of the mass lost by the reinforcement. The numerical cyclic responses for specimens B25 and B30 are reported in Figure 14a, b respectively. Results for B25, reported in Table 8, showed an incorrect prediction of the cyclic behaviour. Neither the ultimate behaviour nor the one preceding the reinforcement yielding are accurately captured and the cyclic dissipation is underestimated. All these results were expected since the reduction of the bond by pitting is too severe for a column that has shown signs of roughly uniform corrosion. The analyses for specimen B30 are not entirely satisfactory. In fact, for this column, which is most affected by pitting, a good fit with the ultimate behaviour was expected. Instead, the numerical result slightly underestimates the experimental one in the positive load direction, while in the negative direction a good fit is obtained. This result may be acceptable since the degradation relationships for pitting used in [35] are made by data regression on several experimental campaigns, therefore a slightly lower reliability of the results is attended when the bars are heavily corroded (i.e., positive load direction). As regards the cyclic behaviour, the same considerations made for specimen B25 apply. This outcome demonstrates/confirms that in case of localization of corrosion in the bar, assuming an average value of mass loss for the whole bar can be misleading and can lead to results different from the real behaviour in the pre-yielding stage.

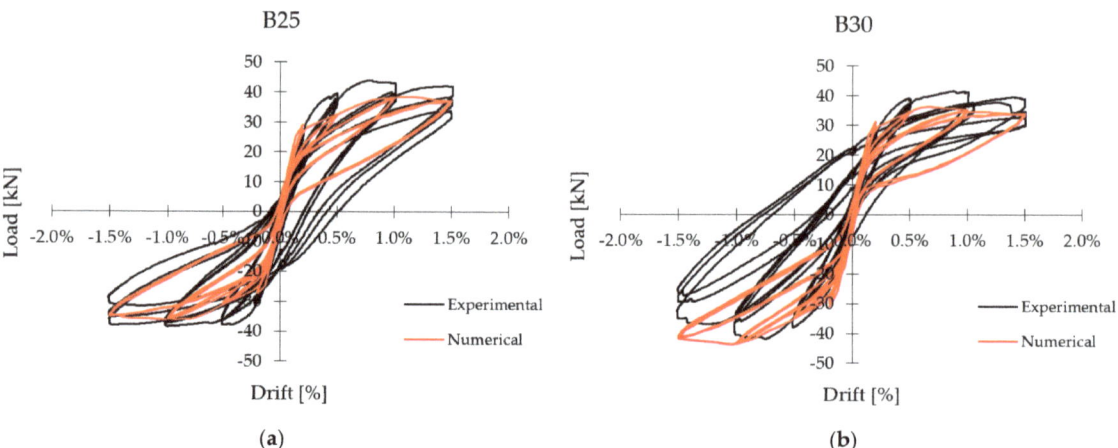

Figure 14. Comparison between experimental and numerical response, corroded specimens, MS 2. (**a**) Column B25, (**b**) Column B30.

4.2.3. Modelling Strategy 3

In this strategy, every single corroded bar was modeled with connected in series truss bond-slip elements, having attributed to each one an equivalent section based on the measurements performed directly on the bar. The numerically predicted behaviours for specimens B25 and B30 are reported in Figure 15. The main results are listed in Tables 8 and 9 respectively.

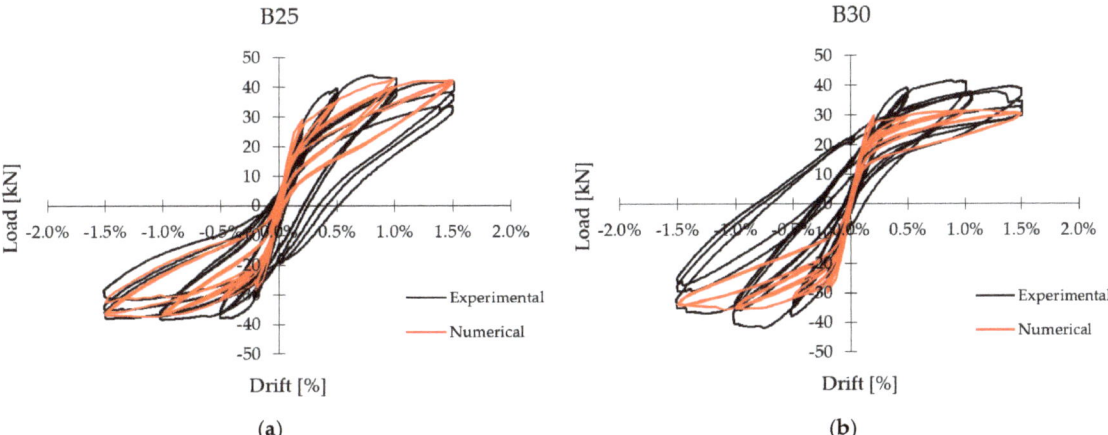

Figure 15. Comparison between experimental and numerical response, corroded columns, MS 3. (**a**) Column B25, (**b**) Column B30.

It can be noted that for specimen B25 the results of the numerical analyses improve if compared to the two previous modelling strategies. A better fit of the post cracking behaviour is identified. Indeed, if the bar is discretized in pieces and constitutive law and bond law are assigned to each part, the stresses in the bar and at the interface with the concrete are better predicted. Furthermore also cyclic dissipation and crack patterns are better described than in previous strategies.

As expected, the result of specimen B30 excessively underestimates the observed experimental response. This outcome was predictable since the section reduction was based on the minimum measured diameter, neglecting the actual pitting morphology (Figure 7). In any case, the results of the numerical analysis, even if they are not able to validate this model for the pitting, confirm that column B30 is affected by the pitting phenomenon and what emerged from the morphological analysis reported in Section 3.

4.2.4. Modelling Strategy 4

The aim of strategy 4 is the modelling of the phenomenon as detailed as possible, considering both the corrosion localization along the bar (macro-morphology) and the corrosion morphology in the cross-section of the bar (micro-morphology). The benefits in terms of fitting the numerical curves emerge from the comparison with the experimental results (Figure 16). About the B25 column, it can be noted that the behaviour in the post-cracking phase has been improved. In fact, both the capacity and the drift for which the first crack and the yielding of the reinforcements occur are captured. Likewise, the ultimate behaviour of the column has been correctly identified. The cyclic dissipation is in agreement with the experimental results, as well as the cracking pattern as shown below. The latter is, also in this case, governed by the formation of the main crack at the base of the column, in which the reinforcements, given the reduction of the bond, slide exchanging minimum stresses with the concrete. This aspect is found in the pinching that characterizes the response curves. In conclusion, the results of the analyses confirm that the numerical model is very suitable for the prediction of the cyclic behaviour of reinforced concrete columns subjected to roughly uniform corrosion. The considerations made concerning column B25 are also valid for column B30. The model provides satisfactory results and is therefore validated by the experimental test. Compared to the other modelling strategies, the results considerably improve, confirming the importance of the morphological study explained in Section 3. Still, a slight underestimation of the capacity in the positive load direction is observed for column B30, but it is considered acceptable.

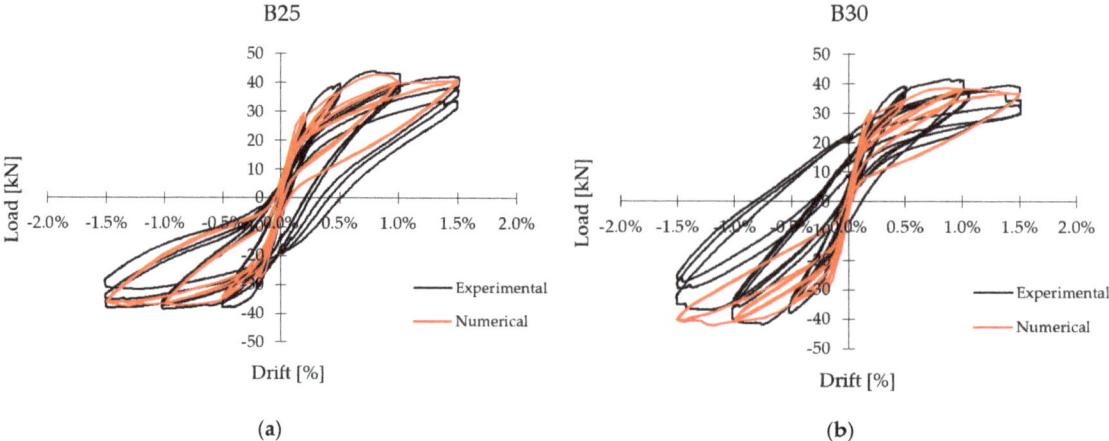

Figure 16. Comparison between experimental and numerical response, corroded specimens, MS 4. (**a**) column B25; (**b**) column B30.

5. Discussion

In the previous section, the main results relating to the numerical models for both specimens B25 and B30 and the proposed modelling strategies have been reported. In general, the authors' expectations regarding the model responses were largely met. From the comparison with the experimental results, it emerged that the modelling strategies based on the constitutive law and bond law reduction as a function of the average loss mass of the corroded bars (i.e., modelling strategies 1 and 2) are suitable for the prediction of the ultimate capacity of the structure. Furthermore, it is noted how, attributing distinct values of the bars' mass loss (instead of a single value for all the bars of the column), allows to catch the cyclic asymmetrical behaviour due to corrosion, as in the case of the column B30. As regards the post-cracking behaviour, it emerged that these strategies (MS 1 and 2) may underestimate the capacity and overestimate the yielding displacement. The authors believe that this deficiency is linked to the fact that the variation of corrosion along the bar is neglected and therefore neither the possible localized bar section reductions nor good bond areas are considered. These strategies risk being not accurate for the prediction of the Damage Control Limit State in the seismic performance assessment, especially as the corrosion degree increases, since the formulations available in the literature are affected by high results' dispersion. The authors also noted that the results for pitting corrosion could be more inaccurate than those for uniform corrosion and therefore believe that, given the danger of the phenomenon, new experimental campaigns must be conducted for the pitting degradation relationship of reinforcements, focusing on the localization of the corroded areas and the influence of the pits. The results relating to modelling strategies 3 and 4 showed that considering detailed micro and macro morphological aspects allows improving the predictivity of the model in main aspects related to the cyclic behaviour of the corroded columns. While strategy 3 proved to be suitable only for columns characterized by roughly uniform corrosion, strategy 4 proved to be very suitable for both specimens and therefore for both types of corrosion. It is also underlined that in addition to the force-drift curves, the numerical models are in agreement with the experimental results with regard to the cracking pattern. This, for corroded columns, is characterized by the opening of a main crack close to the base of the column and by the cyclic slipping of the reinforcement, with a reduction of the collaboration between steel and concrete and of the energy dissipation, as it is emphasized by the pinching in the force-drift curves. Figure 17 shows an example of comparison between the cracking pattern numerically and experimentally obtained.

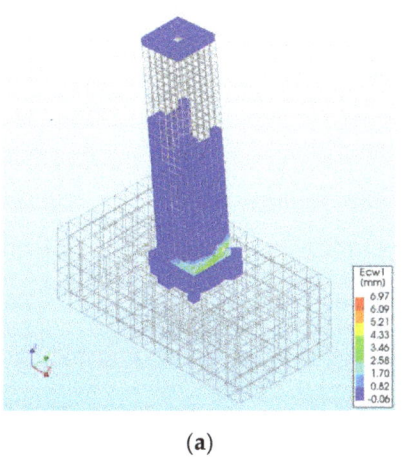
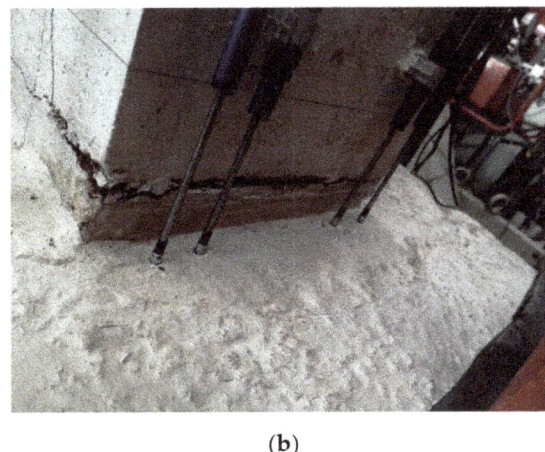

Figure 17. Comparison between cracking pattern, numerical vs. experimental, column B25, drift = −1.5%. (**a**) Numerical, (**b**) Experimental.

The results of the four strategies of numerical analyses and the experimental response have been superimposed in terms of monotonic load-displacement curves, obtained as envelopes of the cyclic ones in Figure 18a,b relatively for B25 and B30.

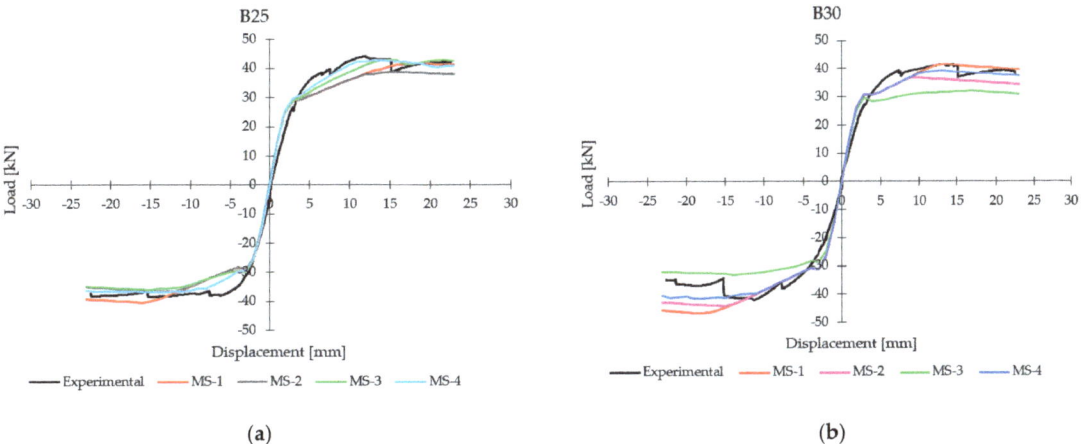

Figure 18. Comparison between experimental and numerical response, monotonic curves. (**a**) Column B25; (**b**) column B30.

From the comparison, the coherence between the models emerges. In particular, strategy 4 stands out for being close to 1 in the case of uniform corrosion and instead tends to migrate towards strategy 2 in the case of pitting corrosion. It should be noted that numerical models tend to underestimate the energy dissipated by the specimens. As mentioned in Section 3.2, this can be due to the model adopted for the concrete (TSCM), that tends to underestimate the energy dissipation of the concrete, and also due to some asymmetric sliding shear phenomena verified during the tests, not fully captured by the numerical model. Finally, it is also noted that almost no model is able to capture the cyclic degradation of strength by a shear mechanism, which can be seen in the last cycles of the experimental tests of corroded columns. These results can be justified by the numerical assumption of considering an average corrosion level in the stirrups, since no detailed measurements are available for the transversal reinforcement.

6. Conclusions

In this work, the cyclic behaviour of RC columns subjected to localized corrosion in the plastic hinge regions was investigated through FE modelling and non-linear analyses. The validation of the numerical models was performed through the comparison with the results of an experimental campaign carried out by some of the authors. Particular attention was paid to the modelling of the corroded rebars and their corrosion morphology. The comparison between the numerical results showed how the most common modelling strategies, which operate according to the section or constitutive law reduction as a function of the average corrosion of the bars, are suitable for the evaluation of the ultimate capacity. Nevertheless, given the uncertainties that characterize the phenomenon from the morphological point of view, they may present deficiencies in the seismic performance assessment (i.e., at the Damage Control performance stage). The strategies based on the morphological study of the corroded bars show, instead, a very good fit for all the cyclic behaviours. The results, therefore, demonstrated how the behaviour of the RC corroded columns depends both on the nature of corrosion (uniform vs. pitting) and on the localization of damage along with the structural element, both on aspects of a random nature that affect the structural response. However, it must be considered that the most refined modelling strategies are difficult to implement for the assessment of real existing structures, as they make use of detailed data and measurements that cannot be obtained with the same precision through on-site surveys. The authors, therefore, believe that to maintain the easy application approach based on the reduction of the constitutive law as a function of the mean mass loss, a new experimental campaign for the mechanical behaviour of corroded bars, which focuses on pitting and localization, is necessary. Furthermore, the assessment of corroded existing RC structures should take into account the probabilistic aspects of the morphology of the corrosion phenomenon.

Author Contributions: Conceptualization, F.M., F.D.C. and Z.R.; methodology, F.M., F.D.C. and Z.R.; formal analysis, F.M.; writing—original draft preparation, F.M.; writing—review and editing, F.M., F.D.C. and Z.R. All authors have read and agreed to the published version of the manuscript.

Funding: This research received no external funding.

Institutional Review Board Statement: Not applicable.

Informed Consent Statement: Not applicable.

Data Availability Statement: Some or all data that support the findings of this study are available from the corresponding author upon reasonable request.

Conflicts of Interest: The authors declare no conflict of interest.

References

1. Cairns, J.; Plizzari, G.A.; Du, Y.; Law, D.W.; Franzoni, C. Mechanical properties of corrosion-damaged reinforcement. *ACI Mater. J.* **2005**, *102*, 256–264.
2. Imperatore, S.; Rinaldi, Z. Mechanical behaviour of corroded rebars and influence on the structural response of R/C elements. In Proceedings of the 2nd International Conference on Concrete Repair, Rehabilitation and Retrofitting, Cape Town, South Africa, 24–26 November 2008; CRC Press Balkema: Boca Raton, FL, USA, 2008.
3. Almusallam, A.A.; Al-Gahtani, A.S.; Aziz, A.R. Effect of reinforcement corrosion on bond strength. *Constr. Build. Mater.* **1996**, *10*, 123–129. [CrossRef]
4. Coronelli, D. Corrosion cracking and bond strength modelling for corroded bars in reinforced concrete. *ACI Struct. J.* **2002**, *99*, 267–276.
5. Prieto, M.; Tanner, P.; Andrade, C. Bond response in structural concrete with corroded steel bars. Experimental results. In *Modelling of Corroding Concrete Structures*; RILEM Book Series; Springer: Dordrecht, The Netherlands, 2011; Volume 5, pp. 231–241.
6. Coccia, S.; Imperatore, S.; Rinaldi, Z. Influence of corrosion on the bond strength of steel rebars in concrete. *Mater. Struct.* **2016**, *49*, 537–551. [CrossRef]

7. Rodriguez, J.; Ortega, L.; Casal, J.; Diez, J.M. Assessing structural conditions of concrete structures with corroded reinforcement. In Proceedings of the International Conference on Concrete in the Service of Mankind, Dundee, UK, 24–26 June 1996.
8. Cairns, J. Assessment of effects of reinforcement corrosion on residual strength of deteriorating concrete structures. In Proceedings of the First International Conference on Behaviour of Damaged Structures, Rio de Janeiro, Brazil, May 1998; Federal University of Fluminense: Niteroi, Brazil, 1998.
9. Castel, A.; Francois, R.; Arligue, G. Mechanical behaviour of corroded reinforced concrete beams—Part 1: Experimental study of corroded beams. *Mater. Struct.* **2000**, *33*, 539–544. [CrossRef]
10. Coronelli, D.; Gambarova, P.G. Structural assessment of corroded reinforced concrete beams: Modelling guidelines. *ASCE J. Struct. Eng.* **2004**, *130*, 1214–1224. [CrossRef]
11. Rinaldi, Z.; Valente, C.; Pardi, L. A simplified methodology for the evaluation of the residual life of corroded elements. *Struct. Infrastruct. Eng. Main Manag. Life-Cycle Des. Perform.* **2008**, *4*, 139–152. [CrossRef]
12. Rinaldi, Z.; Imperatore, S.; Valente, C. Experimental evaluation of the flexural behaviour of corroded P/C beams. *Constr. Build. Mater.* **2010**, *24*, 2267–2278. [CrossRef]
13. Zhang, W.; Liu, X.; Gu, X. Fatigue behaviour of corroded prestressed concrete beams. *Constr. Build. Mater.* **2016**, *106*, 198–208. [CrossRef]
14. Lee, H.S.; Kage, T.; Noguchi, T.; Tomosawa, F. An experimental study on the retrofitting effects of reinforced concrete columns damaged by rebar corrosion strengthened with carbon fiber sheets. *Cem. Concr. Res.* **2003**, *33*, 563–570. [CrossRef]
15. Meda, A.; Mostosi, S.; Rinaldi, Z.; Riva, P. Experimental evaluation of the corrosion influence on the cyclic behaviour of RC columns. *Eng. Struct.* **2014**, *76*, 112–123. [CrossRef]
16. Yang, S.Y.; Song, X.B.; Jia, H.X.; Chen, X.; Liu, X.L. Experimental research on hysteretic behaviours of corroded reinforced concrete columns with different maximum amounts of corrosion of rebar. *Constr. Build. Mater.* **2016**, *121*, 319–327. [CrossRef]
17. Ma, G.; Li, H.; Hwang, H.J. Seismic behaviour of low-corroded reinforced concrete short columns in an over 20-year building structure. *Soil Dyn. Earthq. Eng.* **2018**, *106*, 90–100. [CrossRef]
18. Rajput, A.S.; Sharm, U.K. Corroded reinforced concrete columns under simulated seismic loading. *Eng. Struct.* **2018**, *171*, 453–463. [CrossRef]
19. Ma, Y.; Che, Y.; Gong, J. Behaviour of corrosion damaged circular reinforced concrete columns under cyclic loading. *Constr. Build. Mater.* **2012**, *29*, 548–556. [CrossRef]
20. Liu, X.; Jiang, H.; He, L. Experimental investigation on seismic performance of corroded reinforced concrete moment-resisting frames. *Eng. Struct.* **2017**, *153*, 639–652. [CrossRef]
21. Cardone, D.; Perrone, G.; Sofia, S. Experimental and numerical studies on the cyclic behaviour of R/C hollow bridge piers with corroded rebars. *Earthq. Struct.* **2013**, *4*, 41–62. [CrossRef]
22. Vecchi, F.; Belletti, B. Capacity Assessment of Existing RC Columns. *Buildings* **2021**, *11*, 161. [CrossRef]
23. Di Carlo, F.; Meda, A.; Rinaldi, Z. Numerical evaluation of the corrosion influence on the cyclic behaviour of RC columns. *Eng. Struct.* **2017**, *153*, 264–278. [CrossRef]
24. Vu, N.S.; Yu, B.; Li, B. Prediction of strength and drift capacity of corroded reinforced concrete columns. *Constr. Build. Mater.* **2016**, *115*, 304–318. [CrossRef]
25. Biswas, R.K.; Iwanami, M.; Chijiwa, N.; Nakayama, K. Structural assessment of the coupled influence of corrosion damage and seismic force on the cyclic behaviour of RC columns. *Constr. Build. Mater.* **2021**, *304*, 124706. [CrossRef]
26. fib Bulletin No. 34. *Model Code for Service Life Design*; fib: Lausanne, Switzerland, 2006. [CrossRef]
27. Di Carlo, F.; Isabella, P.; Rinaldi, Z.; Spagnuolo, S. Influence of localized corrosion on the experimental response of R.C. columns under horizontal actions. In Proceedings of the Italian Concrete Days 2020, Naples, Italy, 14–17 April 2021.
28. Vecchio, F.J.; Collins, M.P. Modified Compression-Field Theory for Reinforced Concrete Elements Subjected To Shear. *J. Am. Concr. Inst.* **1986**, *83*, 219–231. [CrossRef]
29. Selby, R.G.; Vecchio, F.J. *Three-Dimensional Constitutive Relations for Reinforced Concrete*; Technical Report 93-02; University of Toronto, Dept. of Civil Engineering: Toronto, ON, Canada, 1993.
30. User's Manual–Release 10.5. Available online: Dianafea.com/manuals/d105/Diana.html (accessed on 23 September 2021).
31. Feenstra, P.H. *Computational Aspects of Biaxial Stress in Plain and Reinforced Concrete*; Delft University of Technology: Delft, The Netherlands, 1993; 159p.
32. fib Model Code for Concrete Structures 2010. Available online: http://hdl.handle.net/1854/LU-4255771 (accessed on 23 September 2021). [CrossRef]
33. Rots, J.G. Computational Modelling of Concrete Fracture. Ph.D. Thesis, Delft University of Technology, Delft, The Netherlands, 1988.
34. Menegotto, M.; Pinto, P.E. Method of Analysis for Cyclically Loaded RC Plane Frames Including Changes in Geometry and Non-Elastic Behaviour of Elements under Combined Normal Force and Bending. In Proceedings of the IABSE Symposium on Resistance and Ultimate Deformability of Structures Acted on by Well Defined Loads, Lisboa, Portugal, 1973; pp. 15–22.
35. Imperatore, S.; Rinaldi, Z.; Drago, C. Degradation relationships for the mechanical properties of corroded steel rebars. *Constr. Build. Mater.* **2017**, *148*, 219–230. [CrossRef]
36. Val, D.V.; Melchers, R.E. Reliability of Deteriorating RC Slab Bridges. *J. Struct. Eng.* **1997**, *123*, 1638–1644. [CrossRef]

37. Bhargava, K.; Ghosh, A.; Mori, Y.; Ramanujam, S. Suggested empirical models for corrosion-induced bond degradation in reinforced concrete. *J. Struct. Eng.* **2008**, *134*, 221–230. [CrossRef]
38. Wang, X.; Liu, X. Bond strength modelling for corroded reinforcements. *Constr. Build. Mater.* **2006**, *20*, 177–186. [CrossRef]
39. Tariq, F.; Bhargava, P. Post corrosion bond-slip models for super ductile steel with concrete. *Constr. Build. Mater.* **2021**, *285*, 122836. [CrossRef]
40. Koulouris, K.; Apostolopoulos, C. Study of the residual bond strength between corroded steel bars and concrete—A comparison with the recommendations of fib model code 2010. *Metals* **2021**, *11*, 757. [CrossRef]
41. Lee, H.S.; Noguchi, T.; Tomosawa, F. Evaluation of the bond properties between concrete and reinforcement as a function of the degree of reinforcement corrosion. *Cem. Concr. Res.* **2002**, *32*, 1313–1318. [CrossRef]
42. Blomfors, M.; Zandi, K.; Lundgren, K.; Coronelli, D. Engineering bond model for corroded reinforcement. *Eng. Struct.* **2018**, *156*, 394–410. [CrossRef]

Article

Flexural Behavior of Reinforced Concrete Beams Retrofitted with Modularized Steel Plates

Min Sook Kim and Young Hak Lee *

Department of Architectural Engineering, Kyung Hee University, Deogyeong-Daero 1732, Yongin 17104, Korea; kimminsook@khu.ac.kr
* Correspondence: leeyh@khu.ac.kr; Tel.: +82-31-201-3815

Abstract: Many structural retrofitting methods tend to only focus on how to improve the strength and ductility of structural members. It is necessary for developing retrofitting strategy to consider not only upgrading the capacity but also achieving rapid and economical construction. In this paper, a new retrofitting details and technique is proposed to improve structural capacity and constructability for retrofitting reinforced concrete beams. The components of retrofitting are prefabricated, and the components are quickly assembled using bolts and chemical anchors on site. The details of modularized steel plates for retrofitting have been chosen based on the finite element analysis. To evaluate the structural performance of concrete beams retrofitted with the proposed details, five concrete beams with and without retrofitting were tested. The proposed retrofitting method significantly increased both the maximum load capacity and ductility of reinforced concrete beams. The test results showed that the flexural performance of the existing reinforced concrete beams increased by 3 times, the ductility by 2.5 times, and the energy dissipation capacity by 7 times.

Keywords: retrofitting; modular; finite element analysis; flexural behavior; ductility

Citation: Kim, M.S.; Lee, Y.H. Flexural Behavior of Reinforced Concrete Beams Retrofitted with Modularized Steel Plates. *Appl. Sci.* **2021**, *11*, 2348. https://doi.org/10.3390/app11052348

Academic Editor: Panagiotis G. Asteris

Received: 28 January 2021
Accepted: 3 March 2021
Published: 6 March 2021

Publisher's Note: MDPI stays neutral with regard to jurisdictional claims in published maps and institutional affiliations.

Copyright: © 2021 by the authors. Licensee MDPI, Basel, Switzerland. This article is an open access article distributed under the terms and conditions of the Creative Commons Attribution (CC BY) license (https://creativecommons.org/licenses/by/4.0/).

1. Introduction

Structural members need to be strengthened through retrofitting due to revision in the current design code or change in load. The retrofitting methods using jacketing have mainly been investigated by many scholars to improve the strength and ductility of existing structural members. Steel jacketing or fiber-reinforced polymer (FRP) jacketing is a commonly employed method used to retrofit concrete beams and columns.

Steel jacketing consists of steel angles or plates with different thicknesses, widths, and spacing and they are installed by welding. Cement or epoxy mortar fills the gap between the jacket and concrete column. Several studies have been conducted evaluating the structural performance of steel jacketing. Garzon-Roca et al. [1] experimentally investigated the behavior of beam-column joint strengthened with steel caging under combined bending and axial load. The test results showed that steel caging increases both the failure load and ductility of the columns. Adam et al. [2] presented parametric study using the finite element model and carried out analyzing the behavior of reinforced columns strengthened by steel caging. Wei and Wu [3] tested concrete columns retrofitted with high strength steel wire and proposed stress-strain relationship. Tarabia and Albakry [4] conducted an axially loaded test on ten column specimens. The size of steel angles, strip spacing, and grout material were considered as parameters. They reported that most of the specimens were failed due to buckling of the steel angle by crushing of the columns. Alvarez et al. [5] presented the nonlinear analysis procedures for concrete columns retrofitted with steel jackets. Backbone curves were constructed of circular and rectangular retrofitted columns to evaluate the response of jacketing columns. Chrysanidis and Tegos [6] proposed new type of hybrid jackets composed of metal grid jacket and high strength mortar. It has been proven that the proposed systems can improve strength and ductility than conventional method.

This technique is widely used in construction field due to cost-effective, developing lateral strength, and axial load carrying capacity. However, it is mainly applied to the column due to weight and welding. Additional lifting equipment is required for installation of tube type steel jacket without welding on site because the steel jacket is prefabricated considering member size.

Composite material was used in the jacketing method to improve movement and installation problems. Commonly carbon or aramid fibers with resin or epoxy resin is used for FRP jacketing. Maaddawy [7] examined the effect of eccentrically load of the concrete column retrofitted with FRP jacketing. It was concluded that the strength of columns decreased as eccentricity ratio increased. A number of studies have been evaluated the efficiency of a FRP jacket joining [8–10]. The behavior of concrete columns retrofitted with FRP jacket subjected to seismic loads was investigated based on experimental and analytical results [11,12]. Experimental investigation was performed to explore the performance of eccentrically loaded rectangular RC columns with different CFRP strengthening schemes and preloading levers by Wang et al. [13]. The results indicated that ultimate capacity can be significantly enhanced by full wrapping of CFRP. They also proposed calculation model to predict the axial load and bending moment capacity of RC columns rapped with FRP jackets.

FRP has low tensile strength and poor adhesion to concrete while FRP sheet have several advantages such as ease of installation and being light weight. Especially, structural retrofitting methods mainly focus on columns or beam-to-column connections, but only retrofitting vertical members may cause stiffness differences between the horizontal members. Few studies have dealt with retrofitting method for concrete beams than columns or beam-column joints.

In this paper, a new modularized retrofitting method for concrete beams is proposed and experimentally examined. The proposed retrofitting method is expected to resolve the problems in terms of the weight and welding of steel jacketing and adhesion and low strength of FRP jacketing. In order to reduce the self-weight of retrofitted concrete beams, modularize steel plate consisting of L- and Z-shaped steel plates are used instead of steel tube or steel jacket. The new modularized retrofitting method can offer rapid construction progress and good quality. The finite element modeling and parametric study were performed to optimize and determine the details of modularized steel plate to reduce stress concentration and exercise its full capacity. A total of four retrofitted concrete beams and one control beam specimen were fabricated and tested to prove that the proposed retrofitting details were capable of improving the load carrying capacity and ductility.

2. The Retrofitting Method for Reinforced Concrete Beams

The proposed retrofitting system aims at improving structural performance, reducing self-weight and rapid assembly based on modularization. This study proposes a new type of retrofitting details using modularized steel plates. The proposed retrofitting details consist of Z-shaped side plates, L-shaped lower plates, and bottom plates with vertical grid as shown in Figure 1a. Two L-shaped lower plates are fixed to the bottom of the concrete beam using chemical anchors. L-shaped lower plates help easy and accurate installation of the Z-shaped side plates. The Z-shaped side plate is connected to the L-shaped lower plate using bolts and then bonded to the side of the concrete beam using the chemical anchors. The vertical grid is inserted into the Z-shaped side plate and combined using bolts. The space formed under the concrete beam is filled with mortar. The details of concrete beam retrofitted with modularized steel plates are shown in Figure 1b. By assembling L-shaped lower and Z-shaped side plates with bolts, the depth of the parts being extended can be adjusted, and it is easy to design and construct regardless of the size of the members. In addition, grid is installed vertically to resist tensile stress, so no subsequent process is required to install additional flexural reinforcement on the site.

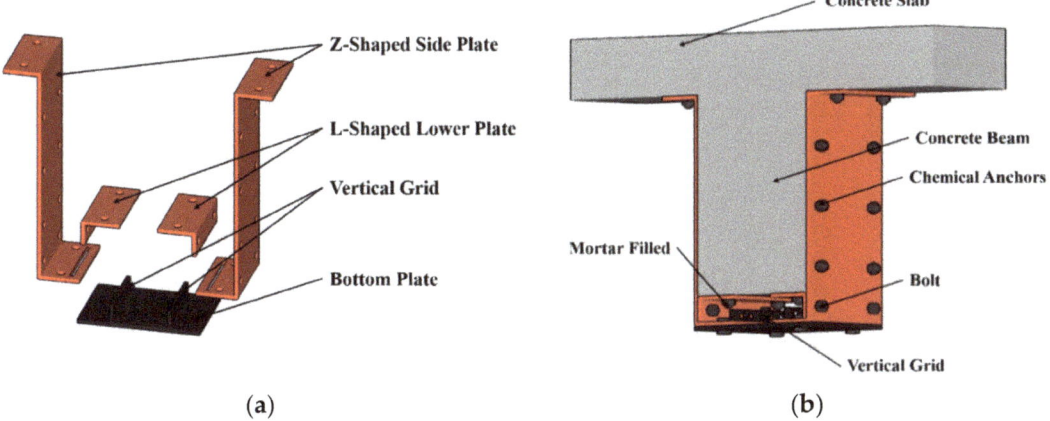

Figure 1. Proposed retrofitting system: (**a**) The components of proposed retrofitting system; (**b**) The retrofitted beam with the proposed system.

3. Parametric Studies

3.1. Analytical Model

Finite element analysis was performed to determine the details of the proposed retrofitting systems. The commercial FE analysis software ANSYS 16.0 was employed to analyze stress for beams. A three-dimensional and eight-node solid element was applied to the model. The steel plates are fixed to the concrete surface using bolts. Automatic surface to surface contact option was used to model the contact interfaces between concrete and steel plates. A surface element was placed and attached to the beam inside the steel plates and around the steel plates. The thickness, stiffness, and strength were not considered for the surface element. The surface of concrete and steel plates are designated as master and slave surface, respectively. Contact interaction was defined between the concrete and steel plates. The friction contact between the parts was defined using Coulomb friction with friction coefficient of 0.3 [14].

The steel plates and concrete were modelled as isotropic materials. Material and geometric nonlinearity are considered. The elastic-perfectly plastic uniaxial material model with bilinear isotropic hardening is used for the concrete model. The yielding stress is equal to the concrete compressive strength and tangent modulus equals zero. The steel plates and bolts are considered to be elastic-perfectly plastic. Steel are assumed to be linear elastic until the yield stress, and after that, the stress remains constant. The yield strength and ultimate strength of the steel plates were 250 MPa, and 460 MPa, respectively. The compressive strength of the concrete was 24 MPa. The material properties are listed in Table 1.

Table 1. Material properties used in the analytical models.

	Compressive Strength (MPa)		Modulus of Elasticity (MPa)	Poisson's Ratio
Concrete	24		24,500	0.18
	Yield strength (MPa)	Tensile strength (MPa)	Modulus of elasticity (MPa)	Poisson's Ratio
Steel plate	250	460	200,000	0.3

The thickness of the bottom plate (A), the thickness of the L-shaped lower plate (B), the thickness of the Z-shaped side plate (C), and the spacing of bolts (D) were considered as variables as shown in Figure 2. T-shaped beam is designed with a 1000 mm × 150 mm slab, 300 mm × 650 mm beam, and 4 m length. The concrete beam model had simply supported condition. The displacement was applied at the node of top surface center of beam in the downward direction. The purpose of the analysis is to compare the stress distribution according to each component details, the displacement of 10 mm was applied for efficiency of analysis. The element mesh size was determined to be 1.5 mm × 1.5 mm considering that the smallest dimension among the components, L-shaped lower plate and Z-shaped side plate, was 2.5 mm. The finite element model is shown in Figure 3.

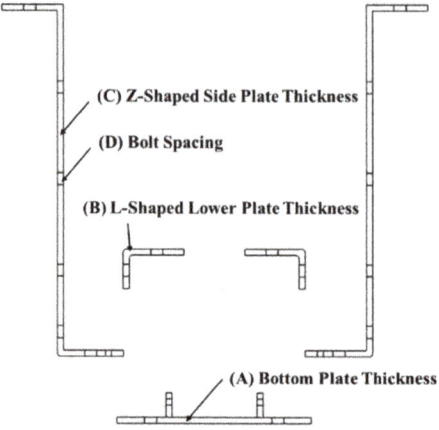

Figure 2. Analytical variables of the proposed retrofitting system.

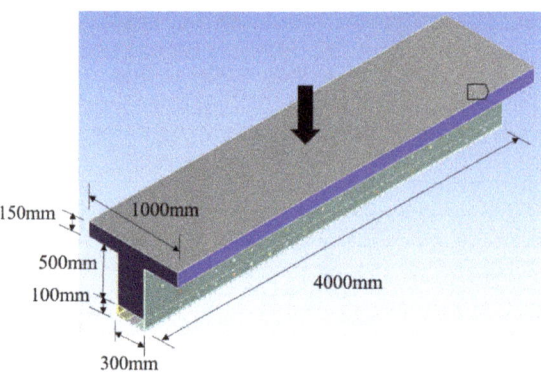

Figure 3. Finite element model.

Four models with different mesh sizes, 20 mm, 10 mm, 5 mm, 1.5 mm, were performed to determine a reasonable mesh size that could be reasonably evaluated for the stress of each component. As a results, the stress converged from the mesh size of 5 mm as shown in Figure 4. A sufficient number of elements are needed to observe the stress distribution of each component. Therefore, the mesh size of 1.5 mm would be applicable to this study.

3.2. Determinations of the Details of the Proposed Retrofitting System

The bottom plate is located at the bottom of the beam and is the component that occurs high tensile stresses. An appropriate thickness of the bottom plate must be determined. As the thickness of the bottom plate increases, the maximum load capacity increases, which may cause compression failure of the concrete beams. Thickness of the bottom plates are 5 mm, 10 mm, 15 mm, 20 mm, and 30 mm to propose a thickness with a stable stress distribution for the analytical model. The results of the analysis are summarized in Table 2.

Table 2. The maximum stress of components.

	(1) Bottom Plate				
Thickness (mm)	5	10	15	20	30
Concrete (MPa)	4.1	4.0	4.0	5.3	6.8
Side plate (MPa)	14.9	15.7	17.3	17.7	18.4
Lower plate (MPa)	1.7	1.8	2.1	2.1	2.8
Bottom plate (MPa)	5.1	2.8	2.9	2.4	2.2
	(2) L-shaped lower plate				
Thickness (mm)	2.5	5	7.5	10	
Concrete (MPa)	4.1	4.1	4.1	4.0	
Side plate (MPa)	15.7	15.7	15.6	15.6	
Lower plate (MPa)	3.46	1.84	4.3	2.0	
Bottom plate (MPa)	2.05	2.03	2.03	2.02	
	(3) Z-shped side plate				
Thickness (mm)	2.5	5	7.5	10	
Concrete (MPa)	3.0	4.1	4.4	5.1	
Side plate (MPa)	11.6	15.7	11.3	9.4	
Lower plate (MPa)	2.9	1.8	2.3	2.2	
Bottom plate (MPa)	1.7	2.0	2.0	2.1	
	(4) Bolt spacing				
Spacing (mm)	150	200	250	300	400
Concrete (MPa)	4.1	4.1	4.0	3.4	2.8
Side plate (MPa)	15.7	15.3	16.7	16.8	20.7
Lower plate (MPa)	2.5	1.8	5.0	6.6	6.0
Bottom plate (MPa)	1.7	2.1	2.5	2.8	2.8

The deflection of the beam decreased as the thickness of the bottom plate increased. As the thickness of the bottom plate increased, the stress of the compressive zone and bolt increased. The thickness of the bottom plate is greater than 10 mm, the stress was concentrated on the bolt hole. This may lead to the crushing of concrete before the flexural failure. The thickness of the L-shaped lower plate was alternatively set to 2.5 mm, 5 mm, 7.5 mm, and 10 mm. As the thickness of the L-shaped lower plate increased, the stress on concrete, Z-shaped side plates, and bottom plates decreased slightly, but did not significantly affect the stress distribution. As Z-shaped side plate thickness increases, the maximum stress is occurred in the compressive zone of the concrete beam. It is concluded that the thickness of the Z-shaped side plate can induce the brittle failure on the compressive zone of the concrete beam. Bolt spacing should be examined in terms of the composite behavior between the concrete and the modular steel plates. As the bolt spacing increased, the stress of the concrete decreased and the stress of the bolt hole increased. The optimal bolt spacing was determined to be 200 mm, which provided a relatively uniform stress distribution.

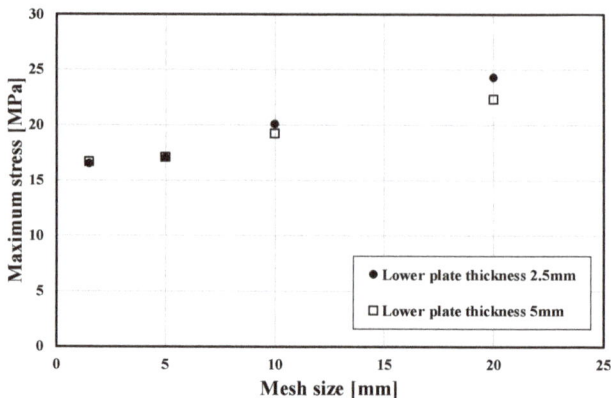

Figure 4. Meshing convergence study.

4. Experimental Program

4.1. Sequence of Construction

The specimens were manufactured in the order of steel rebar placement and formwork installation, concrete placement, combination of modularize steel plate, and mortar injection. As shown in Figure 5, L-shaped lower plates were fixed to the bottom of the concrete specimen, the process is to bolt the lower and side plates and then insert the bottom plates to tighten the bottom plate and the side plate bolts. After integrating the side plates with chemical anchors, the specimen was manufactured by injecting mortar, and the specimen was manufactured upside down to prevent filling defects in mortar.

Figure 5. The process of proposed retrofitting system application: (**a**) The installation of the L-shaped lower plate; (**b**) Assembly of Z-shaped side plate and vertical grid; (**c**) The process of connecting the side plate used in chemical anchors; (**d**) The process of grouting of mortar.

4.2. Specimen Details and Test Parameters

In this paper, the total of five specimens were manufactured, including one reinforced concrete beam and four reinforced concrete beams retrofitted with the modularized steel plates. The finite element analysis to optimize the dimensions of each component was performed assuming the real in-situ situations that beams are supporting the slabs. In this study, the slab (flange) parts in T-shaped section only provide the spaces to fix the steel plates by using chemical bolts. Therefore, flanges were not considered when the specimens were manufactured as shown Figure 6. The concrete beams of all specimens were designed to be 300 mm wide, 350 mm high and 4500 mm long. The thickness of Z-shaped side plate was 2.5 mm, the thickness of the L-shaped lower plates was 5 mm, and the spacing of chemical anchors and bolts was 300 mm, using the same in zigzag format. The thickness of the bottom plate, and the number of vertical grids were considered as experimental variables, which are expected to affect the flexural behavior, and represent the corresponding portion of each variable in Figure 5. The depths of the new beams were 100 and 150 mm, the thicknesses of the bottom plates were 5 and 10 mm, and the number of vertical grid were 0, 2, and 4. For the specimen with two vertical grids, the spacing of each grid was 200 mm, and for the specimen with four vertical grids, each grid was installed at spacing of 65 mm. The specimen illustration is organized in Table 3, and the details of the specimen, according to each variable, are shown in Figure 7. The 28-day compressive strength of the concrete used in this experiment was 24 MPa. For the tensile and compressive reinforcement, deformed rebar was used with a diameter of 19 mm and it's yield strength is 400 MPa. Steel plate was used with a yield strength of 275 MPa. The test results of compressive strength of concrete and steel plate are shown in Figure 8. The high-strength bolts which is F10T M16 bolts with a diameter of 16 mm was used to connect the concrete beams. The material properties are summarized in Table 4.

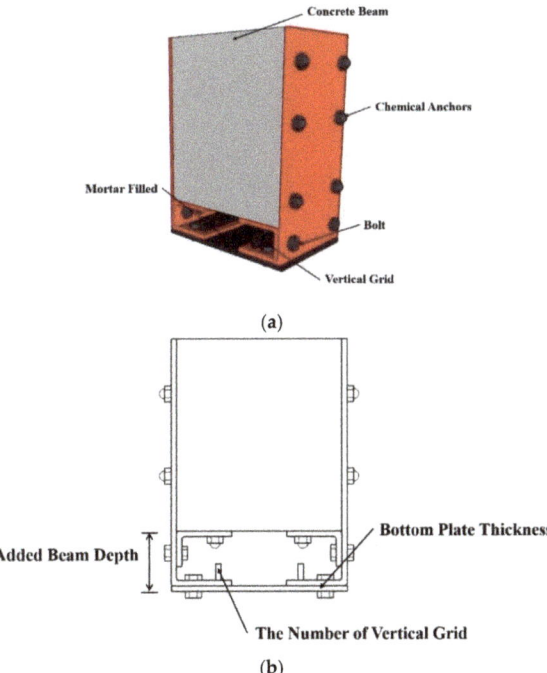

Figure 6. The test parameters: (**a**) The fabrication of proposed retrofitting system in the experiment; (**b**) test parameters.

Table 3. The test specimen parameters.

No.	Name	Added Beam Depth (mm)	Bottom Plate Thickness (mm)	Vertical Grid (EA)
1	RC	-	-	-
2	D100-A [1] 2 [2]	100	5	2
3	D100-B2	100	10	2
4	D100-B4	100	10	4
5	D100-B0	100	10	-

[1] A: Bottom plate thickness (A: 5 mm, B: 10 mm). [2] 2: The number of steel plate openings (2: 2EA, 4: 4EA, 0: 0EA).

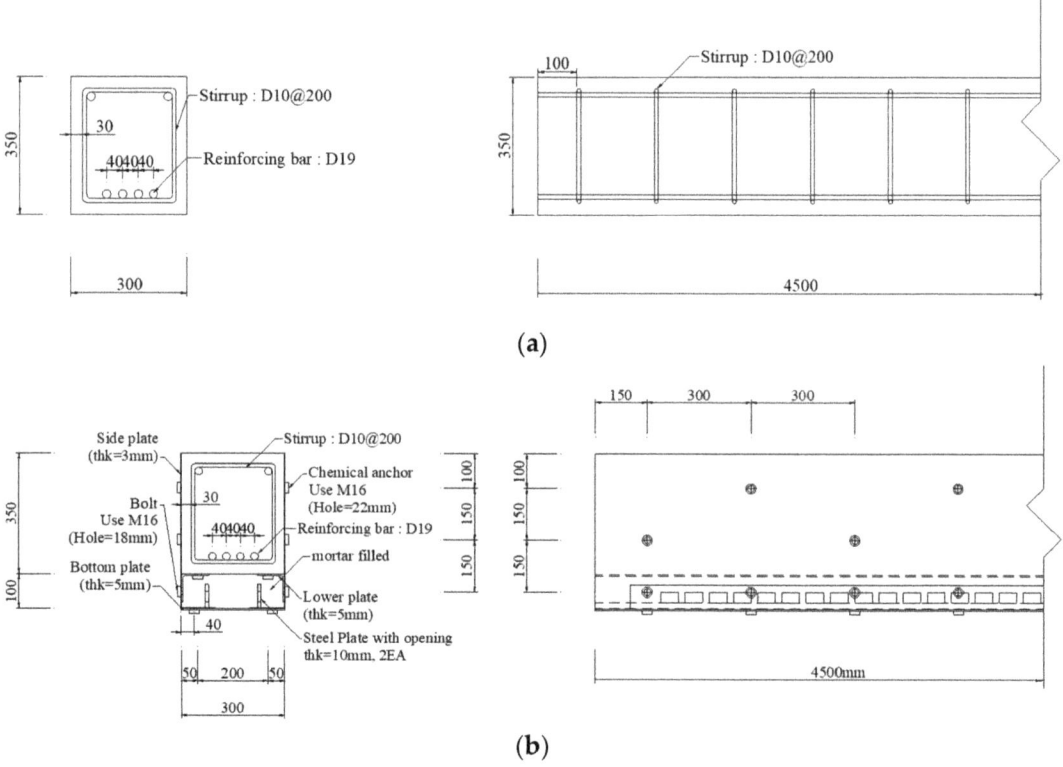

Figure 7. Specimen details: (**a**) RC; (**b**) D100-A2.

4.3. Test Set Up

All specimens were tested using a Universal Testing Machine (UTM) with a 5000 kN load capacity as shown in Figure 9. The specimens were simply supported and subjected to four-point bending loading. The distance from support to loading point was 1750 mm and the distance between the two central point loads was 600 mm. Displacement was applied to each specimen a rate of 2 mm/min. During the flexural test, load and midspan deflection were measured. Deflection was measured with three linear variable displacement transducers (LVDT's) mounted onto the bottom of the specimens. For concrete, the strain gage is attached in the compression zone C1. In order to evaluate the yield by strain of the tensile reinforcing bar, the strain gage was attached in the middle of the tensile reinforcing bar before concrete placement (R1). The strain gage (R2) was attached to the center of the vertical grid's vertical strip to evaluate the flexural behavior as tensile reinforcement. A total of 7 strain gages (S1 ~ S7) were attached to the front of the beam to evaluate the local buckling and deformation of the modularized steel plates. The installation of the LVDTs and strain gages is shown in Figure 10. All specimens were continuously observed to mark the crack pattern and note any signs of failure during the loading process. Loading, displacement and strain were recorded using a data acquisition system. The specimens were tested until the load decreased at the level of approximately 70% to 80% of the maximum load.

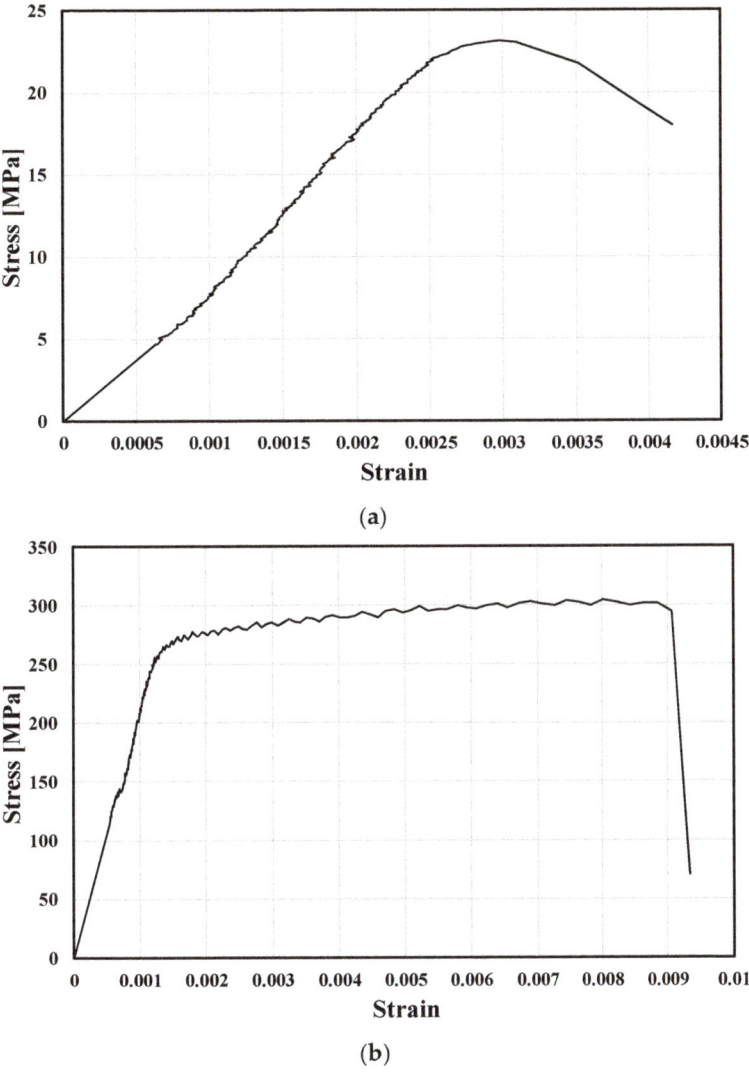

Figure 8. (a) Compressive strength test of concrete; (b) Tensile test of steel plate.

Table 4. Material properties.

Material		Compressive Strength (MPa)	
Concrete		24	
Material	Diameter (mm)	Yield Strength (MPa)	Modulus of Elasticity (MPa)
Rebar	19	400	200,000
Stirrup	10	400	200,000
Material	Yield Strength (MPa)	Tensile Strength (MPa)	Modulus of Elasticity (MPa)
Steel plate	275	410–550	205,000

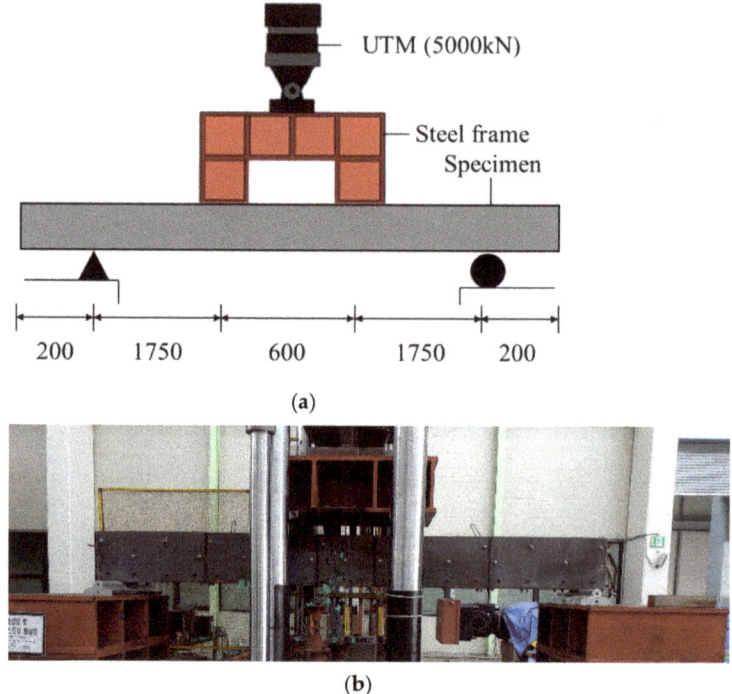

Figure 9. Test setup: (**a**) A schematic of the test setup; (**b**) A photograph of the test setup.

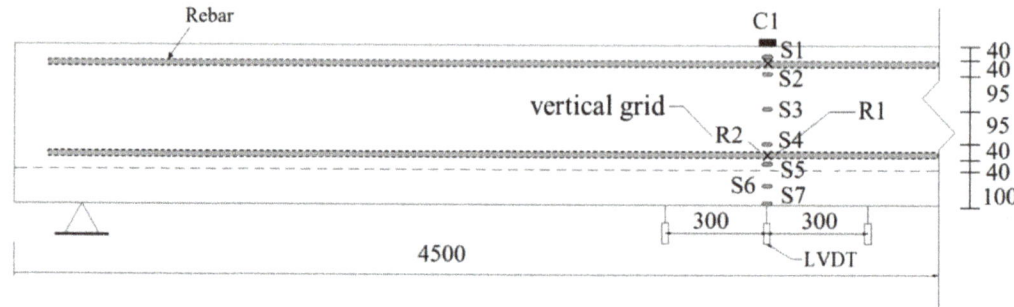

Figure 10. The locations of strain gauges.

5. Test Results and Discussion

5.1. Cracking Behavior and Modes of Failure

Table 5 shows the crack pattern at maximum load for each specimen tested. For the D100-A2 and D100-B2 specimens with variable thicknesses of the bottom plate, it has been confirmed that the crushing of compression zone widens as the thickness of the bottom plate increases. As the number of vertical grid increased, the crushing and spalling of compression zone were more localized.

Table 5. Crack propagation examples.

Name		Crack Propagation	Observed Damage
Bottom plate thickness	D100-A2		-Concrete crushing at the top of the concrete
	D100-B2		-As the bottom plate thickness increases, cracks propagate widely
The number of vertical grid	D100-B0		-Concrete crushing at the top of the concrete
	D100-B2		-As the number of steel plates with openings increases, the concrete crushing occurs more locally.
	D100-B4		

The loading was terminated when the load was reduced 80% after the maximum load was reached. All the specimens showed flexural behavior as shown in Figure 11, cracks started appearing in the compression zone. As the load increased, the cracks of the compression zone became wider, and some spalling was observed on the mortar grouted on the bottom plate. The spalling of the mortar was observed differently according to the vertical grid. More spalling occurred in the D100-B0 specimen without the vertical grid. Finally, as the concrete crushing and spalling increased, the load gradually decreased. Local buckling of the Z-shape side plates and bolt fracture were not observed in all retrofitted specimens as shown in Figures 12 and 13. It is concluded that sliding was not observed at the end of the beam because the steel plates and concrete were perfectly bonded.

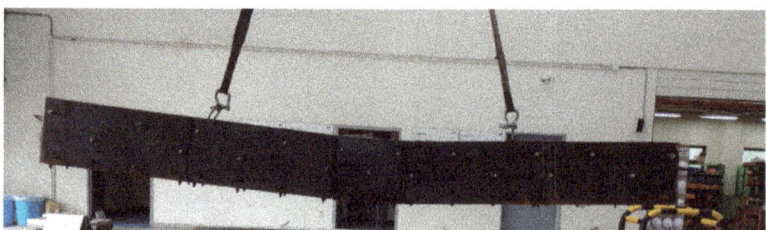

Figure 11. The shape of the test specimen after a flexural test.

Figure 12. The side plate and bottom plate after a flexural test.

Figure 13. Load-Deflection Relationship.

The Load-displacement curve is shown in Figure 14. Specimens retrofitted with modularized steel plates showed greater stiffness and maximum load than RC specimen. Retrofitted specimens showed similar initial stiffness. Remarkable differences were not observed in terms of the number of vertical grid between D100-B2 and D100-B4. The maximum load for D100-B2 was 602 kN and that for D100-B4 was 598 kN, but the maximum load for D100-B0 was 514 kN, which was increased by approximately 1.17 times depending on the presence of vertical grid. This is to achieve the restraining effect of reinforcing parts on the mortar injected into reinforcing areas, and the flexural strength due to reinforcement is improved depending on whether or not it is installed.

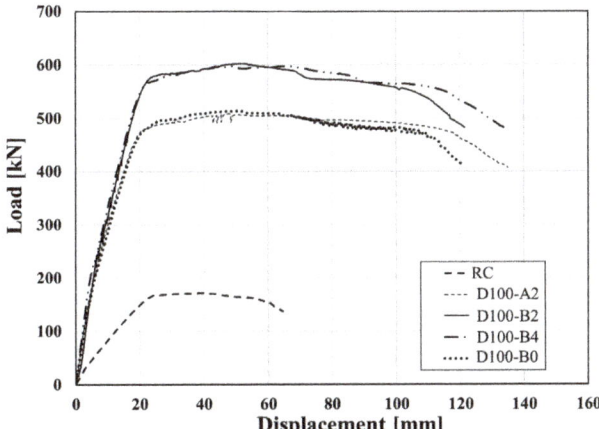

Figure 14. The load-deflection curve test results.

The maximum load increased as the thickness of bottom plate increased. A comparison of the load-displacement curves of the two specimens confirmed that the maximum load for the D100-A2 specimen was 508.48 kN, the maximum load for the D100-B2 specimen was 602.62 kN, and that the flexural performance was approximately 1.19 times higher for the D100-B2 specimen.

The test results are summarized in Table 6. The nominal flexural strength P_n is calculated from plastic stress distribution method suggested by AISC 360-10 [15]. The yield load P_y defines a point as the yield load where a line parallel to the line connecting the origin to the maximum load meets the load-displacement curve as shown in Figure 15. The proposed retrofitting method has shown an increase in the strength, with increasing thicknesses of bottom plates and the number of vertical grids. The experiment resulted in an average increase of 3.2 times the maximum load and an average increase of 1.3 times the maximum displacement.

Table 6. Flexural test results.

No.	Name	P_n [1] (kN)	$P_{y,test}$ [2] (kN)	$P_{u,test}$ [3] (kN)	P_u/P_n	P_u/P_y	δ_y [4] (mm)	δ_u [5] (mm)
1	RC	138.06	137.55	171.65	1.24	1.08	18.09	38.72
2	D100-A2	535.19	370.86	508.48	0.95	1.11	13.48	47.89
3	D100-B2	585.03	466.03	602.62	1.03	1.08	16.27	52.21
4	D100-B4	585.03	423.40	597.55	1.02	1.07	13.90	49.25
5	D100-B0	556.73	382.13	514.00	0.92	1.10	14.61	50.86

[1] P_n: The nominal force calculated by the nominal flexural strength applied to the beam. [2] $P_{y,test}$: The yield load of each specimen. [3] $P_{u,test}$: The maximum load of each specimen. [4] δ_y: The displacement when the yield load applied. [5] δ_u: The displacement when the maximum load applied.

5.2. Load-Strain Relationship

Figure 16a shows the load-strain relationship of the C1 gauge located in the upper of the concrete specimen. The strains of all the specimens were similar and after exceeding the maximum compressive strain of the concrete under the maximum load of 0.003, the gauge was found to have been damaged due to the compressions on the top of the concrete. Figure 16b shows the load-strain relationship for the R2 gauge located at the vertical grid. Although there were differences in the slopes of the initial linear depending on the specimen, the strain of the specimens increased in a similar, and the yield strain reached

0.002 under the maximum load and then yielded. It is judged that the specimens exhibited sufficient flexural behavior because vertical grid as tensile reinforcement yielded.

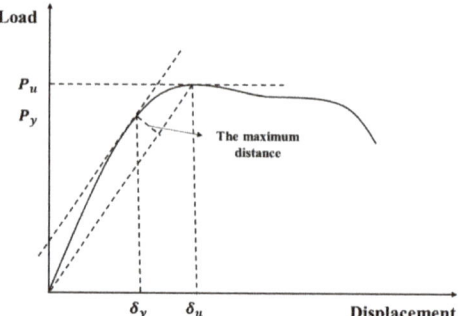

Figure 15. Determination of the yield strength.

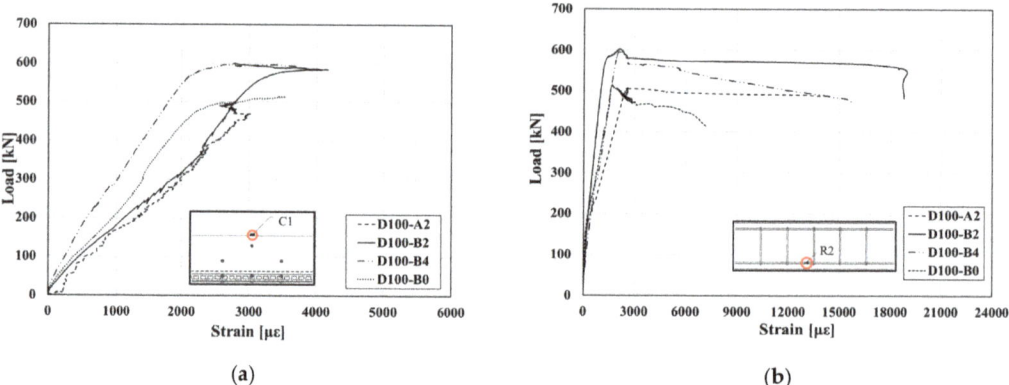

Figure 16. The Load-strain relationships: (a) C1 Guage; (b) R2 Gauge.

5.3. Ductility Capacity

The ductility of flexural members, such as beams, can be calculated by the ratio of the maximum displacement and yield displacement after the maximum load is generated, without a sudden reduction of the load when the load is reduced to 70–80%. In this study, the ductility was calculated by dividing the displacement $\delta_{0.8Pu}$ when the maximum load was reduced to 80% of the maximum load after the maximum load was applied by the displacement δ_y when the yield load was applied. The ductility capacities of each specimen are summarized, as shown in Table 7. The ductility capacity of specimens retrofitted steel plates was shown to be at least 7.47 and up to 10.01, averaging 8.85. Compared to the RC specimen, the ductility capacity is about 2.5 times higher. The ductility of each D100-A2 and D100-B2 specimens, with variable thicknesses of the bottom plate, was 10.01 and 7.47, respectively, indicating that the ductility decreases as the thickness of the bottom plate increases. In addition, the comparison of specimens according to the number of vertical grid resulted in improved ductility as the number of vertical grid increased to 7.47 for D100-B4, 9.67 for D100-B0, and 8.26 for the D100-B0 specimen. However, due to the thickness of the bottom plate and the presence or absence of vertical grid, excessive reinforcement on the tensile zone is considered to be possible due to the crushing of concrete. The ductility capability can be improved by the application of modularized steel plates.

Table 7. Analysis of the ductility capacity for each specimen.

No.	Name	$\delta_{0.8Pu}$ [1]	δ_y [2]	$\delta_{0.8Pu}/\delta_y$ [3]
1	RC	64.78	18.09	3.58
2	D100-A2	134.88	13.48	10.01
3	D100-B2	121.52	16.27	7.47
4	D100-B4	134.43	13.90	9.67
5	D100-B0	120.75	14.61	8.26

[1] $\delta_{0.8Pu}$: The displacement when the maximum load reduced to 80%. [2] δ_y: The displacement when the yield load applied. [3] $\delta_{0.8Pu}/\delta_y$: The ductility capacity of each specimen.

5.4. Energy Dissipation Capacity

Energy dissipation is one of the structural performance evaluation elements of a member that measures the ability of the structure and member to absorb the energy generated when deformation is applied to the structure and components. An analysis of the energy dissipation capacity of specimens is shown in Figure 17.

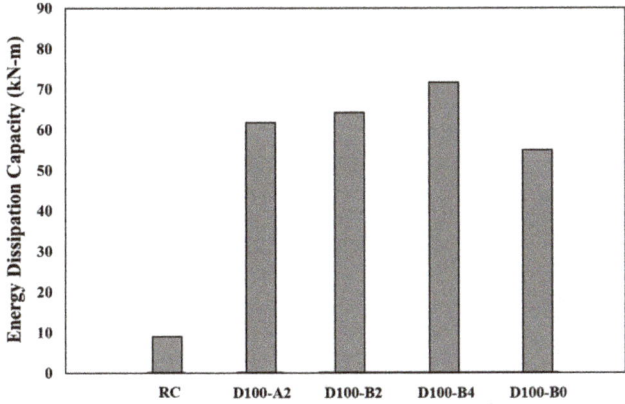

Figure 17. Analysis of energy dissipation capacities.

For the energy dissipation with a thickness increase of the bottom plate, the D100-A2 specimen measured 61.6 kN-m, and the D100-B2 specimen measured 64 kN-m, and the difference with varying thickness of the bottom plate was minimal. The D100-B2 specimen, D100-B4 specimen, and D100-B0 specimen, with variable numbers of vertical grid, are shown to improve their energy dissipation capacity as the number of vertical grid increases to 64 kN-m, 71.6 kN-m, and 55 kN-m, respectively. It has been confirmed that the energy dissipation capability was approximately 7 times higher than that RC specimen.

5.5. Model Validation and Discussion

The results of the numerical model are validated and confirmed by comparing the results of the experiments and the numerical model. Analysis conditions were set the identical as for parametric study. For comparison with the experimental results, a displacement of 120 mm was applied. Figure 18 shows the comparison of the numerical predictions with the experimental results for the D100-B4 specimen. The comparison indicated that numerical results could reasonably predict the overall structural performance in terms of initial stiffness and maximum load capacity. The numerical models slightly became stiffer than the experimental results. This may be due to the numerical model not considering micro-crack or the construction error of the specimen, but it is considered to be acceptable. However, the numerical model shows a different behavior after yielding due to the limitation of the numerical models in implementing the failure mechanism. This is a topic for further investigation by the authors.

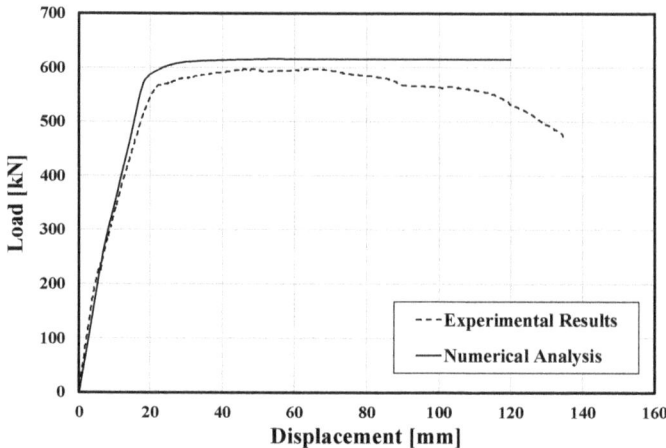

Figure 18. Comparison between the numerical and the experimental results of load versus midspan deflection.

6. Conclusions

The purpose of this study is to propose and examine a modularized steel plate retrofitting method resulting in improved structural performance and constructability. The details of retrofitting method were proposed based on the finite element analysis. The flexural test was conducted to evaluate the structural performance of concrete beams retrofitted with the proposed method. The main conclusions obtained from the study are as follows:

(1) In this study, the details and components of the modularized retrofitting method are proposed, with Z-shaped side plates, L-shaped lower plate, and bottom plate with vertical grid. Finite element analyses were carried out to investigate the effect of the bottom plate thickness, L-shaped lower plate thickness, Z-shaped side plate thickness, and the bolt spacing on the stress distribution. Details of the modularized steel plate retrofitting system were proposed based on the analysis results.

(2) The maximum load increased as the thickness of the bottom plate increased, but it was found to be relatively lacking in ductility. Although the number of vertical grid did not significantly affect stiffness, the increase in that number improved the ductility. In addition, the flexural performance increased by about 1.17 times, depending on the presence of the vertical grid. The vertical grid attached to the bottom plate increased the confinement effect of the mortar and reduced spalling by acting as a flexural reinforcement.

(3) The ductility and energy dissipation capabilities were compared and analyzed. Compared to non-retrofitted reinforced concrete beams, the ductility capacity increased by about 2.5 times, and the energy dissipation capacity increased by about seven times. This confirms that the application of the proposed retrofitting method can improve the structural performance of existing concrete members.

(4) The experiment in this paper focused on the evaluation of the flexural capacity of retrofitted concrete beams. Further experimental research is needed on the shear behavior of concrete beams retrofitted with modularized steel plates. Moreover, a finite element model that can predict structural capacity is required for the design stage.

Author Contributions: Conceptualization, M.S.K. and Y.H.L.; methodology, M.S.K.; validation, Y.H.L.; formal analysis, M.S.K.; investigation, M.S.K. and Y.H.L.; writing—original draft preparation, M.S.K.; writing—review and editing, Y.H.L. All authors have read and agreed to the published version of the manuscript.

Funding: This study was supported by the National Research Foundation of Korea (NRF) grant funded by the Korean government (MSIT) (No. 2019R1A2C1090033).

Institutional Review Board Statement: Not applicable.

Informed Consent Statement: Not applicable.

Data Availability Statement: The results presented in this study are available on request from the corresponding author.

Conflicts of Interest: The authors declare no conflict of interest.

References

1. Garzon-Roca, J.; Ruiz-Pinilla, J.; Adam, J.M. An experimental study on steel caged RC columns subjected to axial force and bending moment. *Eng. Struct.* **2011**, *33*, 580–590. [CrossRef]
2. Adam, J.M.; Ivorra, S.; Pallares, F.J.; Gimenez, E.; Calderon, P.A. Axially loaded RC columns strengthened by steel caging. Finite element modeling. *Constr. Build. Mater.* **2009**, *23*, 2265–2276. [CrossRef]
3. Wei, Y.; Wu, Y.F. Compression behavior of concrete columns confined by high strength steel wire. *Constr. Build. Mater.* **2014**, *54*, 443–453. [CrossRef]
4. Tarabia, A.M.; Albakry, H.F. Strengthening of RC columns by steel angles and strips. *Alex. Eng. J.* **2014**, *53*, 615–626. [CrossRef]
5. Alvarez, J.C.; Brena, S.F.; Arwade, S.R. Nonlinear backbone modeling of concrete columns retrofitted with fiber-reinforced polymer or steel jackets. *Aci Struct. J.* **2018**, *115*, 53–64. [CrossRef]
6. Chrysanidis, T.; Tegos, I. Axial and transverse strengthening of R/C circular columns: Conventional and new type of steel and hybrid jackets using high-strength mortar. *J. Build. Eng.* **2020**, *30*, 101236. [CrossRef]
7. EL Maaddawy, T. Strengthening of eccentrically loaded reinforced concrete columns with fiber-reinforced polymer wrapping system: Experimental investigation and analytical modeling. *J. Compos. Constr.* **2009**, *13*, 13–24. [CrossRef]
8. Fu, B.; Teng, J.G.; Chen, J.F.; Chen, G.M.; Guo, Y.C. Concrete cover separation in FRP plated RC Beams mitigation using FRP U-Jackets. *J. Compos. Constr.* **2017**, *21*, 04016077. [CrossRef]
9. Fu, B.; Chen, G.M.; Teng, J.G. Mitigation of intermediate crack debonding in FRP-plated RC beams using FRP U-jackets. *Compos. Struct.* **2017**, *176*, 883–897. [CrossRef]
10. Mohammed, A.A.; Manalo, A.C.; Maranan, G.B.; Zhuge, Y.; Vijay, P.V.; Pettigrew, J. Behavior of damaged concrete columns repaired with novel FRP jacket. *J. Compos. Constr.* **2019**, *23*, 04019013. [CrossRef]
11. Xu, C.X.; Peng, S.; Deng, J.; Wan, C. Study on seismic behavior of encased steel jacket-strengthened earthquake-damaged composite steel-concrete columns. *J. Build. Eng.* **2018**, *17*, 154–166. [CrossRef]
12. Mohammed, A.A.; Manalo, A.C.; Ferdous, W.; Zhuge, Y.; Vijay, P.V.; Pettigrew, J. Experimental and numerical evaluations on the behaviour of structures repaired using prefabricated FRP composites jacket. *Eng. Struct.* **2020**, *210*, 110358. [CrossRef]
13. Wang, J.; Lu, S.; Yang, J. Behavior of eccentrically loaded rectangular RC columns wrapped with CFRP jackets under different preloading levels. *J. Build. Eng.* **2021**, *34*, 101943. [CrossRef]
14. Lam, D.; Dai, H.T.; Han, L.H.; Ren, Q.X.; Li, W. Behavior of inclined, tapered and STS square CFST stub columns subjected to axial load. *Thin-Walled Struct.* **2012**, *54*, 94–105. [CrossRef]
15. AISC. *Specification for Structural Steel Building*; ANSI/AISC 360-10; AISC: Chicago, IL, USA, 2010.

Article

Strengthening Design of RC Columns with Direct Fastening Steel Jackets

Zhiwei Shan [1], Lijie Chen [2], Kun Liang [2], Ray Kai Leung Su [2,*] and Zhaodong Xu [1]

[1] Key Laboratory of Concrete and Prestressed Concrete Structures of the Ministry of Education, Southeast University, Nanjing 210096, China; shanzw@foxmail.com (Z.S.); zhdxu@163.com (Z.X.)
[2] Department of Civil Engineering, The University of Hong Kong, Pokfulam Road, Hong Kong, China; chenlj@connect.hku.hk (L.C.); kunliang556324@outlook.com (K.L.)
* Correspondence: klsu@hku.hk; Tel.: +852-28592648

Abstract: For non-seismically designed columns with insufficient strength and flexural stiffness, intense inter-story drift can be incurred during a strong earthquake event, potentially leading to the collapse of the entire building. Existing strengthening methods mainly focus on enhancing axial or flexural strength but not the flexural stiffness of columns. In response, a novel direct fastening steel jackets that can increase both flexural strength and stiffness is introduced. This novel strengthening method features straightforward installation and swift strengthening as direct fastening is used to connect steel plates together to form a steel jacketed column. This new connection method can quickly and stably connect two steel components together by driving high strength fasteners into them. In this paper, the design procedure of RC columns strengthened with this novel strengthening method is originally proposed, which includes five steps: (1) estimating lateral load capacity of damaged RC columns; (2) determining connection spacing of steel jacket; (3) estimating the lateral load capacity of strengthened RC column; (4) evaluating the axial load ratio (ALR) of strengthened RC columns; and (5) estimating effective stiffness of strengthened RC columns. Lastly, an example is presented to illustrate the application of the proposed design procedure.

Keywords: RC columns; strengthening; direct fastening; steel jackets; design procedure

Citation: Shan, Z.; Chen, L.; Liang, K.; Su, R.K.L.; Xu, Z. Strengthening Design of RC Columns with Direct Fastening Steel Jackets. *Appl. Sci.* **2021**, *11*, 3649. https://doi.org/10.3390/app11083649

Academic Editor: Pier Paolo Rossi

Received: 1 April 2021
Accepted: 14 April 2021
Published: 18 April 2021

Publisher's Note: MDPI stays neutral with regard to jurisdictional claims in published maps and institutional affiliations.

Copyright: © 2021 by the authors. Licensee MDPI, Basel, Switzerland. This article is an open access article distributed under the terms and conditions of the Creative Commons Attribution (CC BY) license (https://creativecommons.org/licenses/by/4.0/).

1. Introduction

Reinforced concrete (RC) moment-resisting frame buildings are widely used in schools, hospitals, and residential buildings. RC columns are the principal structural component in the resisting of lateral and gravity loads in frame buildings. Based on post-earthquake investigations [1–3], the stability of frame buildings is known to be critically dependent on the seismic performance of RC columns. The structural deficiencies identified in outmoded non-seismically designed RC columns [1–10] are (1) insufficient lap splice length of longitudinal reinforcement at column ends, (2) insufficient transverse reinforcement at the plastic hinge, (3) strong beam-weak column arrangement, (4) insufficient corrosion resistance, (5) insufficient strength due to new functional use of the building, (6) fire-induced damage, and (7) earthquake-induced damage. The first five above mentioned deficiencies may result in insufficient flexural strength, which can be mitigated by various available strengthening techniques such as RC jacketing [11,12], steel jacketing [13,14] and fiber reinforced polymer (FRP) jacketing [15–17]. However, in the case of fire or earthquake damaged columns, both flexural strength and stiffness can be reduced [18–21]. Figure 1 illustrates the seismic displacement demands of structurally damaged and non-damaged buildings within a demand spectrum. Any reduction in lateral strength and stiffness within a structural system can then induce higher seismic displacement demand during subsequent earthquake events. To restore the seismic capacity of columns to their undamaged state, both flexural stiffness and strength must be improved.

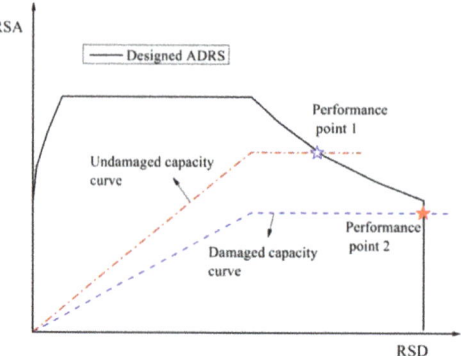

Figure 1. Design spectrum and capacity curves for damaged and undamaged structures.

In response, a novel steel jacket that deviates from traditional steel jacketing (assembled by either welding or bolting) is developed (see Figure 2). This new strengthening method offers straightforward installation and rapid strengthening, as direct fastening is used to connect steel plates together to form a steel jacketed column. The strengthening process is briefly described herein: (1) the prefabrication of steel jackets is achieved by welding the end angles to the end of the steel plates (welding can be conducted in-shop); (2) four steel plates are fastened by anchor bolts to the top and base beams; (3) the steel plates are tightly clamped to minimize gaps between them and the RC columns, and steel angles are temporarily fixed to the adjacent steel plates; and (4) the adjacent steel plates and steel angles are tightly and quickly joined by high strength nail using a powder-actuated gun (see Figure 3).

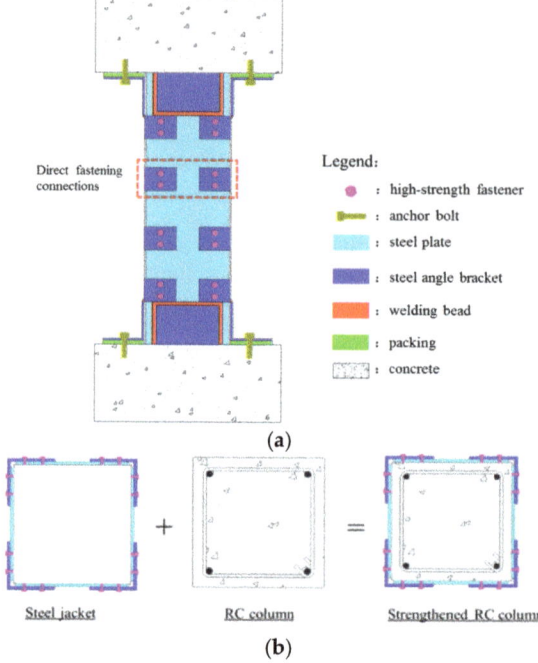

Figure 2. Proposed strengthening scheme: (**a**) front view; (**b**) plane view.

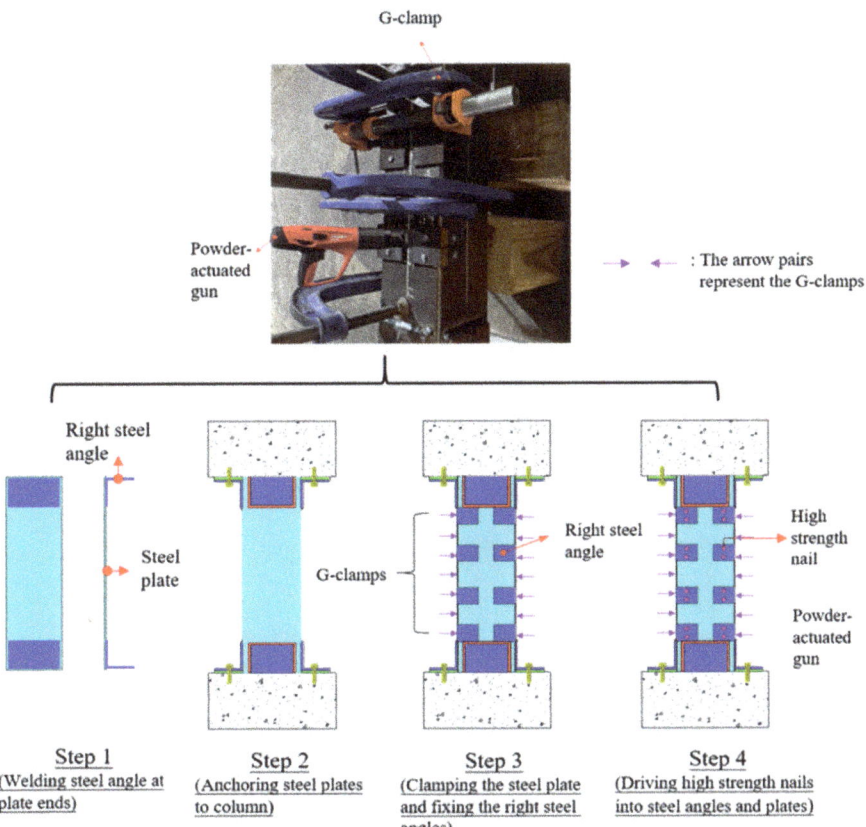

Figure 3. Strengthening procedure.

In this paper, previous experimental and theoretical studies on shear connections joined by direct fastening, axial load strengthening and the seismic strengthening of RC columns strengthened by direct fastening steel jackets are first reviewed. Based on the findings in previous studies [22–25], a design procedure of RC columns strengthened by direct fastening steel jackets is developed. To illustrate the application of the proposed design procedure, a worked example is presented.

2. Review of Previous Work

2.1. Mechanical Behavior of Shear Connectors Joined by Direct Fastening

The connections used in this novel steel jacket resist shear loading incurred as a result of concrete expansion or lateral load. The load transfer process of these connections is similar to that of single lap joints subjected to tensile force at the ends. However, the behaviors of this type of connection have rarely been studied and relevant design specifications are thus not available. Therefore, the experimentation on the single lap joints shown in Figure 4 was carried out to study the mechanical behavior of direct fastening connections [22].

Two kinds of failure modes—bearing and shear fracture—were examined in the tests. Similar to the bearing failure seen in bolted or screwed connections, this failure occurs with enlarged nail holes and bulging of the material around the fastener holes due to large plastic deformation, leading to desirable ductile behavior (see Figure 5).

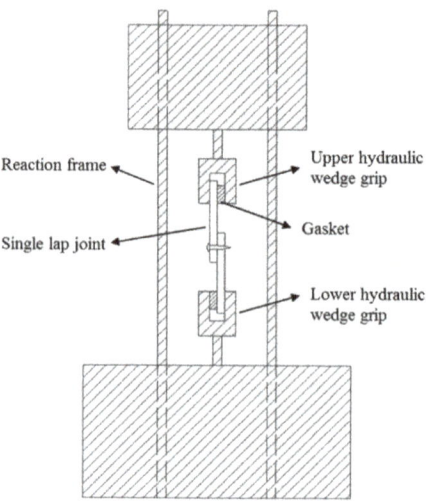

Figure 4. Illustration of connection test joined by direct fastening.

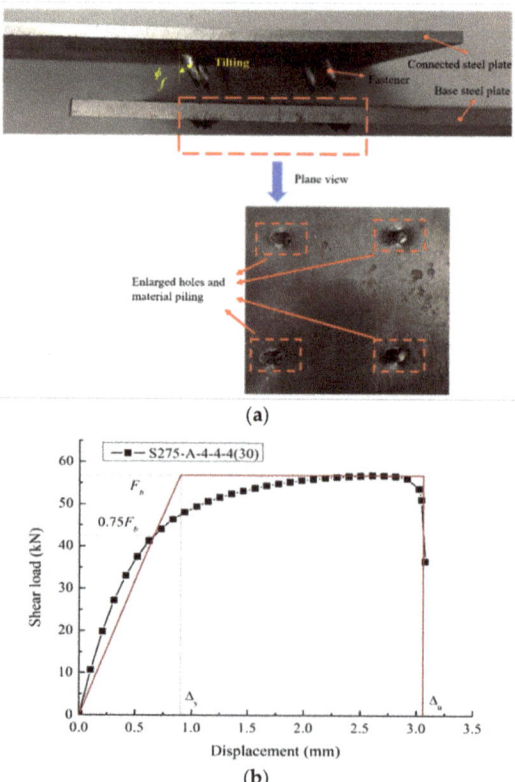

Figure 5. Mechanical behavior of connections: (**a**) bearing failure; (**b**) shear load versus displacement curve.

Although equations for predicting the maximum bearing resistance of bolted and screwed connections are available in current specifications EN-1993-1-8 [26], ANSI/AISC 360-10 [27], and AS 4100 [28], these equations cannot be applied to estimate the maximum bearing resistance (yield strength) of connections joined by direct fastening without modification. Therefore, a unified design equation is developed to evaluate the maximum bearing resistance of the connections:

$$F_b = \psi_{fp}\psi_{fk}\alpha_{br}d_n t_p f_{pu} \tag{1}$$

where α_{br} is the bearing resistance factor, which is taken as 1.6; ψ_{fp} is the factor that considers the effect of protuberance; ψ_{fk} is the factor that takes the effect of knurling into account; d_n is the nominal fastener diameter; t_p is the thickness of the connected steel plate; and f_{up} is the ultimate strength of the connected steel plate.

The two factors caused by the effects of protuberance and knurling are given as:

$$\psi_{fp} = \begin{cases} 1.35 & \text{without pre-drilled holes on connected steel plate} \\ 1.0 & \text{with pre-drilled holes on connected steel plate} \end{cases} \tag{2}$$

$$\psi_{fk} = \begin{cases} 1.17 & \text{fastener with knurling} \\ 1.0 & \text{fastener without knurling} \end{cases} \tag{3}$$

2.2. Axial Strengthening of RC Columns by Direct Fastening Steel Jackets

From the experimental study on the axial strengthening of RC columns by the direct fastening of steel jackets [23], the reliability and effectiveness of the proposed method were observed. Critical parameters (i.e., vertical spacing between adjacent connections, thickness of the steel plate and number of fasteners in each connection) affecting load bearing performance and deformation behavior were identified.

In the proposed strengthening method, steel plates can directly sustain axial load. The axial load contribution of the steel plates is determined by buckling strength as buckling occurs prior to the yield strength of steel plates. The buckling strength is presented in the following section.

Direct fastening connections behave in the manner of transverse reinforcement (i.e., stirrups). Due to the transverse dilation of the column, passive confinement stress is mobilized in stirrups and direct fastening connections, which enhances the strength of the column. To determine confined concrete strength, the equivalent passive confinement stress shown in Equation (3) should be determined.

$$f_{est} = \frac{m\alpha_{st} f_{yst} A_{st}}{s_{st} l_{st}} \tag{4}$$

$$f_{ed} = \frac{2\alpha_d n_f F_b}{(s_d + d_d)d_c} \tag{5}$$

where m is the stirrup legs; l_{st} is the center-to-center distance of the peripheral stirrup; s_{st} is the stirrup spacing; f_{yst} is the yield strength of the stirrup; A_{st} is the cross-sectional area of the stirrup; s_d is the clear vertical spacing of the adjacent connections; d_d is the length of the right steel angle bracket; d_c is the column width; α_{st} is the stress ratio of the stirrup; and α_d is the shear force ratio of the direct fastening connection.

A theoretical study was carried out to examine the stress ratio of the stirrup and the shear force ratio of the direct fastening connection [24]. Based on an extensive parameter study, a lower bound value of 0.34 is used to represent the stress ratio of the stirrups with a yield strength of 400 MPa. To consider the effect of the yield strength, a yield strength factor of γ_{fyst} was proposed. The yield strength factors for yield strengths of 500 MPa and 600 MPa are 0.8 and 0.66, respectively. The shear force ratio of the connections is obtained using:

$$\alpha_d = \gamma_{nf}(0.82 - 0.64\lambda_d) \quad (6)$$

where λ_d is the normalized connection spacing ($\lambda_d = s_d/d_c$); and γ_{nf} represents a factor of the number of fasteners on the shear force ratio of the connections. When the number of fasteners is 2 and 4, the factor is equal to 0.72 and 0.47, respectively.

The failure criterion proposed in [29] is used to derive the confined concrete strength of f_{cc}'. The failure criterion is given by:

$$\tau_{oct}^* = 6.9638 \left(\frac{0.09 - \sigma_{oct}^*}{c - \sigma_{oct}^*} \right)^{0.9297} \quad (7)$$

$$c = 12.2445(\cos 1.5\theta)^{15} + 7.3319(\sin 1.5\theta)^2 \quad (8)$$

$$\cos\theta = \frac{\sqrt{3}}{2} \frac{s_1}{\sqrt{J_2}} \quad (9)$$

$$\tau_{oct}^* = \frac{\tau_{oct}}{f_c'} \quad (10)$$

$$\sigma_{oct}^* = \frac{\sigma_{oct}}{f_c'} \quad (11)$$

where τ_{oct} and σ_{oct} are the octahedral shear and normal stress, respectively; θ is the direction of the deviatoric stress on the deviatoric plane; s_1 is the first deviatoric stress; J_2 represents the second invariant of the deviatoric stress tensor; and f_c' is the compressive strength of the unconfined concrete.

The key parameters in the applied failure criterion are summarized as follows:

$$s_1 = \frac{f_{cc}' - f'}{3} \quad (12)$$

$$J_2 = \frac{1}{3}(f_{cc}' - f')^2 \quad (13)$$

$$\cos\theta = \frac{f_{cc}' - f'}{3\sqrt{2}\tau_{oct}} \quad (14)$$

$$\tau_{oct} = \frac{\sqrt{2(f_{cc}' - f')}}{3} \quad (15)$$

$$\sigma_{oct} = -\frac{2f' + f_{cc}'}{3} \quad (16)$$

where f_{cc}' is the axial compressive strength of the confined concrete and f' is the equivalent passive confinement stress.

2.3. Seismic Strengthening of RC Columns by Direct Fastening Steel Jackets

To investigate the seismic performance of strengthened RC columns using direct fastening steel jackets, an experimental study was conducted [25]. Attention was given to the enhancement of flexural stiffness and strength.

On the basis of comparisons between predicted and measured effective flexural stiffness, the expressions recommended in EN 1994-1-1 [30] are advised to calculate effective flexural stiffness:

$$K_i = (EI)_s + 0.6(EI)_c \quad (17)$$

where $(EI)_s$ is the flexural stiffness provided by the steel jacketing; and $(EI)_c$ is the flexural stiffness provided by the RC column.

Buckling of the steel plate on the compressive side was observed. The corresponding buckling stress of steel plate under a compressive state should thus be embedded within the theoretical model used to predict lateral capacity. Using the Rayleigh–Ritz method and assuming the initial bowing and subsequent deflection as a trigonometric function, the buckling strength of steel plate is given as:

$$\sigma_{p,critical} = \frac{4\pi^2 D d_p}{s_d^2 A_p}(1-\alpha_i) \quad (18)$$

where D is the bending stiffness of the steel plate per unit width; d_p is the plate depth; E_p and μ_p are elastic modulus and Poisson's ratio of steel plate; A_p is the cross-sectional area of steel plate; and α_i is the imperfection factor.

The imperfection factor α_i is introduced to consider the initial bowing effect induced by the welding and is calibrated using the experimental results:

$$\alpha_i = 1.046 - \frac{0.73 \lambda_{sr}}{100} \quad (19)$$

where $\lambda_{sr} = s_d/t_p$ is the slenderness ratio of steel plate. As the empirical imperfection factor was calibrated with limited available data, it is useful to assume a slenderness ratio of the steel plate between 14 and 39.

It is worth noting that the steel plate detached from the RC column at the bottom of the tension fiber. This was largely due to the strain incompatibility between RC column and the steel jacket on the tension side during bending. As a result, the assumption that the plane section would remain plane after deformation throughout the entire length of the column is inaccurate. Hence, a reduction factor of $\eta_i = 0.6$ was introduced for the tensile component of the steel plate when the lateral resistant capacity was estimated.

3. Design Procedure

In this section, the critical parameters (i.e., steel plate thickness, fastener number and connection spacing) are first roughly determined. The feasibility of these parameters are further examined by satisfying four conditions: (1) desirable flexural failure occurs prior to brittle shear failure; (2) lateral load capacity is larger than lateral load demand; (3) ALR does not exceed the limit stipulated in current specification; and (4) effective flexural stiffness should be comparable to that of the undamaged RC column.

3.1. Estimating Lateral Load Capacity of Damaged RC Columns

The deficiency of the RC column may occur for a variety of reasons (e.g., fire). It is necessary to estimate the lateral load capacity of the damaged RC column and compare this with the load demand. According to the assumption that the plane section remains plane after deformation, the strain profile and the stress of the longitudinal rebar can be defined (Figure 6).

$$\begin{cases} -f_{yl} \leq \sigma_{l1} = E_s(d_c - c - \phi_{st} - \frac{\phi_l}{2} - x_c)\frac{\varepsilon_{cu}}{x_c} \leq f_{yl} \\ -f_{yl} \leq \sigma_{lj} = E_s(d_c - c - \phi_{st} - \frac{\phi_l}{2} - (j-1)s_l - x_c)\frac{\varepsilon_{cu}}{x_c} \leq f_{yl} \\ -f_{yl} \leq \sigma_{lm} = E_s(d_c - c - \phi_{st} - \frac{\phi_l}{2} - (m-1)s_l - x_c)\frac{\varepsilon_{cu}}{x_c} \leq f_{yl} \end{cases} \quad (20)$$

where s_l is the longitudinal rebar spacing; subscript lj represents the jth row longitudinal rebar counting from the tensile sides ($1 \leq j \leq m$); m is the total rows of longitudinal rebars which is equal to the stirrup legs; ϕ_l is the diameter of the longitudinal rebar; ϕ_{st} is the diameter of the stirrup; c is the cover thickness of the column; x_c is the compressive depth of the column; ε_{cu} is the ultimate strain of concrete and is taken as 0.003 in accordance with ACI 318 [31]; f_{yl} is the yield strength of the longitudinal reinforcement; and σ_{lj} is the stress of the jth row longitudinal rebar.

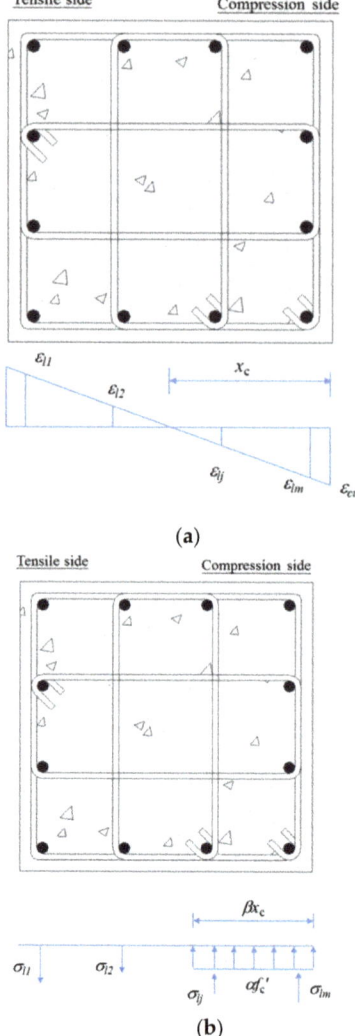

Figure 6. Vertical strain and stress profiles of RC column: (**a**) strain profile; (**b**) stress profile.

The stress profile of the concrete is quantified using the equivalent rectangular stress block. The compressive depth is then determined by the axial equilibrium:

$$N_0 + \frac{\sum_{j=1}^{m} k_j A_l \sigma_{lj}}{\gamma_s} - \frac{\alpha f_c' d_c \beta x_c}{\gamma_c} = 0 \qquad (21)$$

where k_j is the number of the jth row longitudinal rebar; γ_s is the partial safety factor of rebar; γ_c is the partial safety factor of concrete; N_0 is the axial load; A_l is the cross-sectional area of longitudinal rebar; f_c' is the concrete strength; and α and β are the two factors used to determine the equivalent rectangular stress block.

By taking moment to the extreme compressive fiber of the RC column, the moment capacity is given as:

$$M_{cap} = N_0 \frac{d_c}{2} + \frac{\sum_{j=1}^{m} k_j A_l \sigma_{lj}(d_c - c - \phi_{st} - \frac{\phi_l}{2} - (j-1)s_l)}{\gamma_s} - \frac{\alpha f'_c d_c \beta x_c \frac{\beta x_c}{2}}{\gamma_c} \tag{22}$$

Under the assumption that the stiffness of the upper and lower joints is comparable, the lateral load resistance is then given by:

$$V_{cap} = \frac{M_{cap}}{0.5L} \tag{23}$$

where L is the length of the column.

If lateral load capacity cannot satisfy lateral load demand, the RC column can be strengthened using the direct fastening steel jackets.

3.2. Determining Connection Spacing of Steel Jacket

Flexural failure is the desirable failure mode. To ensure that desirable flexural failure occurs, the following relationship should be satisfied in light of ASCE 41-13 [32].

$$V \leq V_{cap,stren} \leq 0.6 V_{stren} \tag{24}$$

where V is the lateral load demand acting on the RC column; $V_{cap,stren}$ is the lateral load corresponding to the moment capacity of the strengthened RC column; and V_{stren} is the shear strength of the strengthened RC column which is determined by the following expression:

$$\begin{aligned} V_{stren} &= V_d + V_c + V_s \\ &= 0.5 \frac{d_c}{s_d + d_d} \frac{2n_f F_b}{\gamma_s} + \frac{0.17(1 + \frac{N_0}{14 A_c})\sqrt{f'_c} d_c d_w}{\gamma_c} + \frac{m A_{st} f_{yst} d_w}{\gamma_s s_{st}} \end{aligned} \tag{25}$$

where n_f is the fastener number in a connection; A_c is the cross-sectional area of the column; and d_w is the depth of the column.

The first term represents the shear resistance produced from the connections. To avoid an overestimation of the shear resistance from the steel jacket, a factor of 0.5 is introduced by following EN 1998-3:2005 [33]. As the partial safety factor for direct fastening connections is absent, the partial safety factor of steel rebar is used.

Combining the above two expressions, the connection spacing of the steel jacket can be determined.

3.3. Estimating Lateral Load Capacity of Strengthened RC Column

According to the assumption that the plane section remains plane after deformation, the strain profile and stress profile of the longitudinal rebar and steel plates of the strengthened RC column are defined, respectively (see Figure 7). It should be noted that the side plate is divided into n equal small parts to accurately define the stress in the side plate. Moreover, the buckling of the steel plate under a compressive state is embedded.

$$\begin{cases} \sigma_{pt} = E_p(d_c + \frac{t_p}{2} - x_c)\frac{\varepsilon_{cu}}{x_c} \leq f_{py} \\ \sigma_{pc} = E_p(\frac{t_p}{2} + x_c)\frac{\varepsilon_{cu}}{x_c} \leq \sigma_{p,critical} \\ -\sigma_{p,critical} \leq \sigma_{pside\_i} = E_p(d_c - \frac{d_c - d_p}{2} - \frac{\Delta_i}{2} - (i-1)\Delta_i - x_c)\frac{\varepsilon_{cu}}{x_c} \leq f_{py} \\ -f_{yl} \leq \sigma_{l1} = E_s(d_c - c - \phi_{st} - \frac{\phi_l}{2} - x_c)\frac{\varepsilon_{cu}}{x_c} \leq f_{yl} \\ -f_{yl} \leq \sigma_{lj} = E_s(d_c - c - \phi_{st} - \frac{\phi_l}{2} - (j-1)s_l - x_c)\frac{\varepsilon_{cu}}{x_c} \leq f_{yl} \\ -f_{yl} \leq \sigma_{lm} = E_s(d_c - c - \phi_{st} - \frac{\phi_l}{2} - (m-1)s_l - x_c)\frac{\varepsilon_{cu}}{x_c} \leq f_{yl} \end{cases} \tag{26}$$

where i represents the ith partitioned part of the steel plate parallel to the lateral load counting from the tensile side; σ_{pside_i} is the stress of the ith partitioned part of the steel plate parallel to the lateral load; Δ_i is the length of the equal small part and is equal to d_p/n; and $\sigma_{p,critical}$ is the ultimate compressive stress of the steel plate, which can be defined by Equation (8).

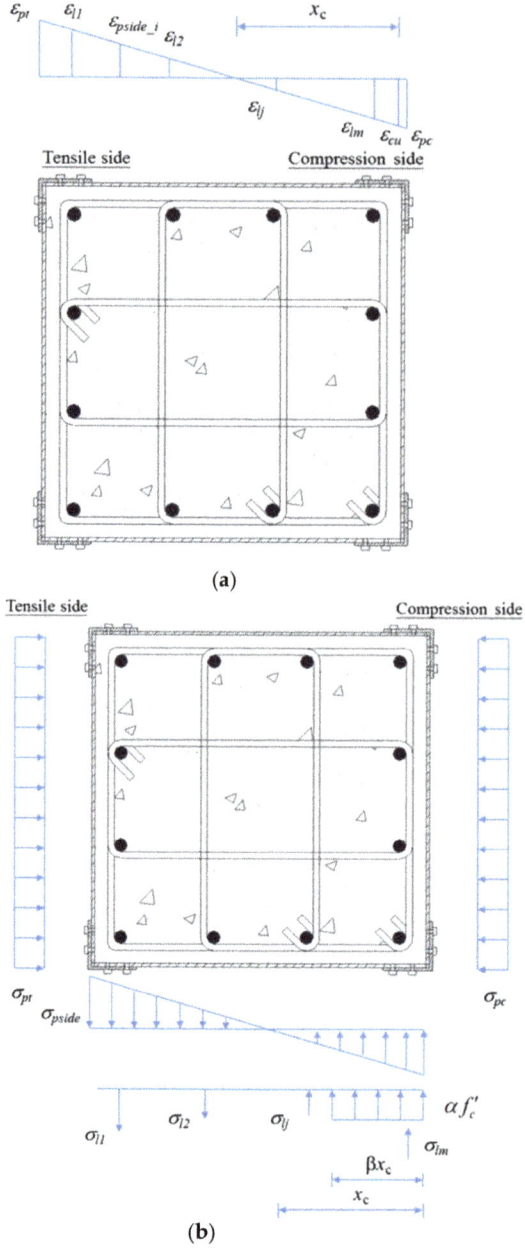

Figure 7. Vertical strain and stress profiles of strengthened RC column: (**a**) strain profile; (**b**) stress profile.

The compressive depth of the strengthened RC column is then determined by the axial equilibrium.

$$0 = N_0 + \frac{\sum_{j=1}^{m} k_j A_l \sigma_{lj}}{\gamma_s} + \frac{\eta_i \sigma_{pt} A_p - A_p \sigma_{pc} + 2\sum_{i=1}^{n} t_p \Delta_i \sigma_{pside\_i}}{\gamma_s} - \frac{\alpha f'_c d_c \beta x_c}{\gamma_c} \quad (27)$$

By taking moment to the extreme compressive fiber of the strengthened RC column, the moment capacity is given as:

$$M_{cap,stren} = N_0 \frac{d_c}{2} - \frac{\alpha f'_c d_c \beta x_c \frac{\beta x_c}{2}}{\gamma_c} + \frac{\sum_{j=1}^{m} k_j A_l \sigma_{lj} (d_c - c - \phi_{st} - \frac{\phi_l}{2} - (j-1)s_l)}{\gamma_s} + \frac{\eta_i \sigma_{pt} A_p (d_c + \frac{t_p}{2}) + A_p \sigma_{pc} \frac{t_p}{2} + 2\sum_{i=1}^{n} t_p \Delta_i \sigma_{pside\_i} (d_c - \frac{d_c - d_p}{2} - \frac{\Delta_i}{2} - (i-1)\Delta_i)}{\gamma_s} \quad (28)$$

Under the assumption the stiffness of the upper and lower joints is comparable, the lateral load resistance is then given by:

$$V_{cap,stren} = \frac{M_{cap,stren}}{0.5L} \quad (29)$$

Then, the relationship between the ultimate lateral load and the lateral load resistance of the strengthened RC column can be checked.

3.4. Estimating ALR of Strengthened RC Columns

The confined concrete divisions are shown in Figure 8a. Confined concrete 1 is confined by the erected steel jacket. Confined concrete 2 is confined by the stirrup. Confined concrete 3 is confined by the stirrup and erected steel jacket. Because the area of confined concrete 1 and confined concrete 2 is relatively smaller than that of confined concrete 3, and these two areas are close, the confined concrete divisions can be simplified in the design calculation. The simplified concrete divisions are shown in Figure 8b, in which the confined effect of confined concrete 2 is imposed on confined concrete 1. Thus, the initial three varieties of confined concrete can be equivalent to one confined concrete that is confined by a stirrup and steel jacket. For the confined concrete in Figure 8b, the imposed equivalent passive confinement stress is $f_l = f_{est} + f_{ed}$, which can be determined according to (3). Using the confined concrete strength model in (5), the confined concrete strength (f_{cc}') can be determined. Thus, the axial load capacity is given as:

$$N_{c,stren} = \frac{A_{cc} f'_{cc} + A_{c0} f'_c}{\gamma_c} + \frac{(4m-4) A_l f_{yl} + 4P_{p,critical}}{\gamma_s} \quad (30)$$

where A_{cc} is the area of confined concrete in Figure 8b; and A_{c0} is the area of the unconfined concrete in Figure 8b.

The confined concrete area A_{cc} and unconfined area A_{c0} are respectively given as:

$$A_{cc} = d_c^2 - 4\frac{d_l^2}{6} - (4m-4)A_l \quad (31)$$

$$A_{c0} = 4\frac{d_l^2}{6} \quad (32)$$

where d_l is the edge-to-edge horizontal distance of the connections, as shown in Figure 8b.

The ALR is given as:

$$\text{ALR} = \frac{N_0}{N_{c,stren}} \quad (33)$$

According to the recommendation in EN 1998-1:2004 [34], the ALR must necessarily be less than 0.65.

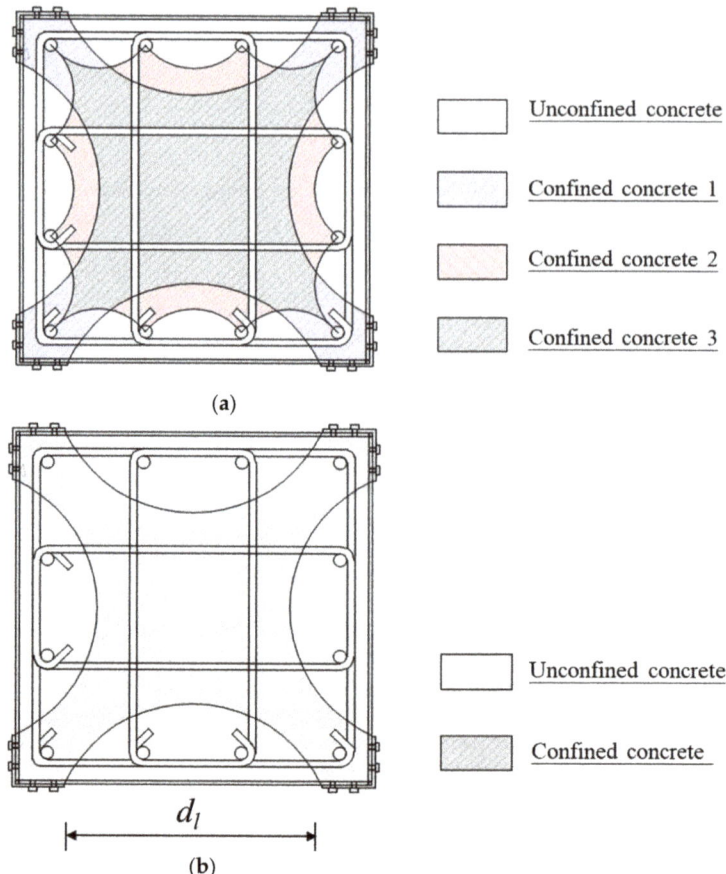

Figure 8. Divisions of the confined concrete: (**a**) exact divisions; (**b**) simplified divisions.

3.5. Estimating Effective Stiffness of Strengthened RC Columns

In light of the findings relating to the seismic strengthening of RC columns by direct fastening steel jackets, the effective flexural stiffness of the strengthened RC column can be estimated by (7). The effective flexural stiffness of the strengthened RC column should be comparable to that of the undamaged RC column, with the aim of the deformation capacity and seismic load carrying capacity remaining unchanged.

4. Parameter Study

A parameter study is conducted in this section to investigate the enhancement ratio ($\eta_M = V_{cap,stren}/V_{cap}$) of the lateral load capacity of strengthened RC columns and roughly estimate the required thickness of steel plate. Four critical variables (i.e., thickness of steel plate, cross-sectional dimension of column, longitudinal rebar ratio ρ_l and ALR) are investigated to plot the design curves. The range of these four variables is set out in Table 1. The yield strength and elastic modulus of the longitudinal rebar are 500 MPa and 200 GPa, respectively. The yield strength, ultimate strength and elastic modulus of the steel plate are 300 MPa, 400 MPa and 200 GPa, respectively. The cylinder strength of the concrete is 30 MPa. The compressive strength of the compressive steel plate is greatly affected by its slenderness ratio (λ_{sr}). For conservative consideration, λ_{sr} is taken as 38, which is the upper bound value of the slenderness ratio of steel plate determined from the experimental study [23,25].

Table 1. Range of four variables in parametric study.

t_p (mm)	ALR	d_c (mm)	ρ_l
3, 4, 5	0.15, 0.3	400, 500	0.015, 0.03

It can be determined from Figure 9 that the enhancement ratio is significantly affected by the thickness of the steel plate and longitudinal rebar ratio. The enhancement ratio increases with the thickness of the steel plate, which can be seen from Equation (17). The enhancement ratio declines by 20% when the longitudinal rebar ratio increases from 0.015 to 0.03, as RC columns comprised of more longitudinal rebar possess a higher lateral load capacity, causing the decreased enhancement ratio. Compared with the effect of the longitudinal rebar ratio, the effect of the ALR on the enhancement ratio is marginal, especially for RC columns possessing higher longitudinal rebar ratios.

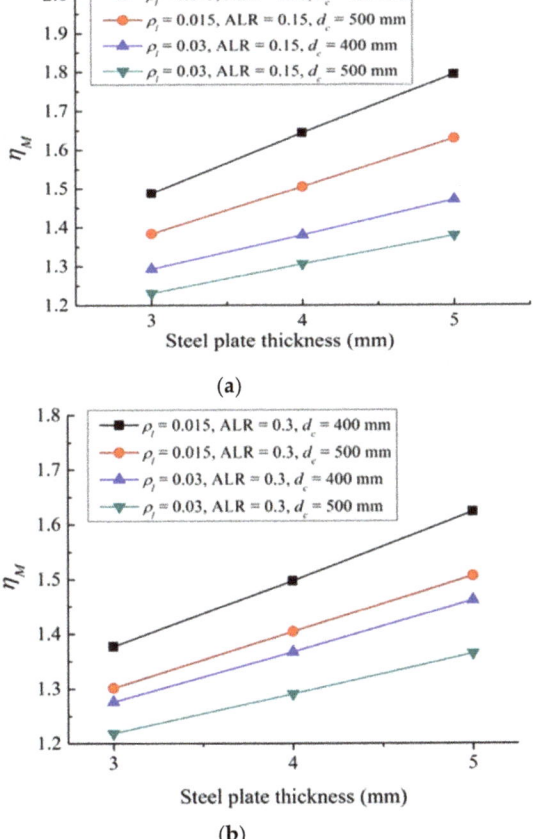

Figure 9. Enhancement of lateral load capacity: (**a**) ALR = 0.15; (**b**) ALR = 0.3.

5. Worked Example

The prototype of the RC column is shown in Figure 10. The height of the column is 3000 mm. The cross-sectional dimension is 500 mm × 500 mm. The reinforcement cage is formed by 12T20 and R10@150. At the ultimate limit state, the factored axial load (N_0) and factored lateral load (V) are 1400 kN and 300 kN, respectively. The elastic modulus of the rebar is 200 GPa. The yield strength of the stirrup and the longitudinal rebar is 400 MPa and 500 MPa, respectively. The cylinder strength (f_c'), cube strength (f_{cu}), and elastic modulus (E_c) of the concrete are 30 MPa, 40 MPa, and 25 GPa, respectively. The cover thickness (c) is 20 mm. It is postulated that the column experienced a fire event up to 570 °C. The strength and stiffness of the RC column can be greatly impaired after exposure to high temperature. The proposed strengthening method of directly fastening steel jackets can be used to restore fire-damaged RC columns. The elastic modulus, yield strength and ultimate strength of the steel plate are 200 GPa, 300 MPa and 400 MPa, respectively. The steel angle bracket used to connect the two adjacent steel plates measures 75 mm × 75 mm × 5 mm with a length of 50 mm. The 4 mm diameter high strength knurled fastener is used in this strengthening system. The partial safety factor of steel and concrete is taken as 1.2 and 1.5.

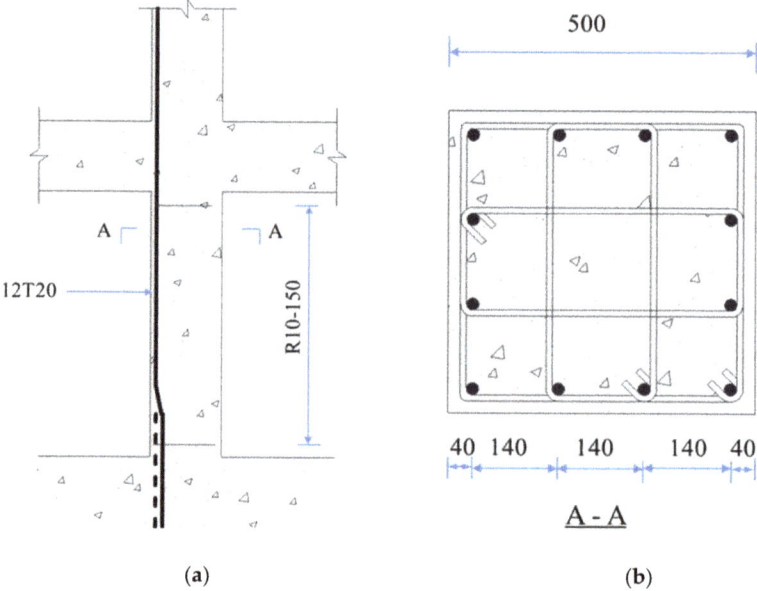

Figure 10. Reinforcement details of column: (**a**) front view; (**b**) plane view.

5.1. Evaluating Lateral Load Capacity of Fire-Damaged Column

After the concrete is exposed to fire, the concrete strength is impaired. The concrete strength post fire can be estimated according to the recommendation in EN-1992-1-2 [35]. The residual cylinder strength ($f_{c',pf}'$) of the RC column following exposure to 570 °C fire is $0.5 f_c'$.

According to the study on the mild steel post fire in [22], the residual strength and elastic modulus of rebar can be readily determined. The residual strength of longitudinal rebar is given as:

$$\begin{aligned} f_{yl,pf} &= f_{yl} R s_y \\ &= 500 \times (1.0 - 0.2 \times \tfrac{570-400}{600}) \\ &= 470 \text{ MPa} \end{aligned}$$

The residual strength of the stirrup is given as:

$$\begin{aligned} f_{yst,pf} &= f_{yst} Rs_y \\ &= 400 \times (1.0 - 0.2 \times \tfrac{570-400}{600}) \\ &= 376 \text{ MPa} \end{aligned}$$

The residual elastic modulus of rebar is given as:

$$\begin{aligned} E_{s,pf} &= E_s Rs_E \\ &= 200 \times (1.0 - 0.07 \times \tfrac{570-350}{500}) \\ &= 194 \text{ GPa} \end{aligned}$$

According to (20) and (21), the stresses of the four rows of longitudinal rebar and thus the compressive height of the plane section can be determined.

$$N_0 + \frac{4A_l\sigma_{l1} + 2A_l\sigma_{l2} + 2A_l\sigma_{l3} + 4A_l\sigma_{l4}}{1.2} - \frac{\alpha f'_{c,pf} d_c \beta x_c}{1.5} = 0$$

The compressive height of the plane section is solved as:

$$x_c = 304 \text{ mm}$$

The moment capacity is derived by (22).

$$\begin{aligned} M_{cap} &= N_0 \tfrac{d_c}{2} - \tfrac{\alpha f'_{c,pf} d_c \beta x_c \tfrac{\beta x_c}{2}}{1.5} \\ &+ \tfrac{4A_l\sigma_{l1}(d_c - c - \phi_{st} - \tfrac{\phi_l}{2}) + 2A_l\sigma_{l2}(d_c - c - \phi_{st} - \tfrac{\phi_l}{2} - s_l)}{1.2} \\ &+ \tfrac{2A_l\sigma_{l3}(d_c - c - \phi_{st} - \tfrac{\phi_l}{2} - 2s_l) + 4A_l\sigma_{l4}(d_c - c - \phi_{st} - \tfrac{\phi_l}{2} - 3s_l)}{1.2} \\ &= 311 \text{ kN} \cdot \text{m} \end{aligned}$$

The lateral load resistance of the fire-damaged RC column is given by (23).

$$V_{cap} = \frac{M_{cap}}{0.5L} = \frac{311 \times 10^3}{0.5 \times 3000} = 207 \text{ kN} < V = 300 \text{ kN}$$

Thus, the fire-damaged RC column is not safe, and the required enhancement ratio of the RC column is around 1.4.

5.2. Evaluating Lateral Load Capacity of Strengthened RC Column

Since the required enhancement ratio is around 1.4 and the longitudinal rebar ratio is 0.015, 4 mm steel plate is utilized to strengthen the fire-damaged RC column based on the findings in Figure 8. Additionally, four fasteners on the connection are used.

(a) Determining connection spacing

Flexural failure is the desirable failure mode. To realize this, Equation (24) should be satisfied.

$$300 \text{ kN} = V \leq V_{cap,stren} \leq 0.6 V_{stren}$$

According to Equation (25), the shear capacity of the strengthened fire-damaged RC column is given as:

$$\begin{aligned} V_{stren} &= V_d + V_{c,pf} + V_{s,pf} \\ &= 0.5 \tfrac{d_c}{s_d + d_d} \tfrac{2n_f F_d}{\gamma_s} + \tfrac{0.17(1 + \tfrac{N_0}{14A_c})\sqrt{f'_{c,pf}} d_c d_w}{\gamma_c} + \tfrac{mA_{st} f_{yst,pf} d_w}{\gamma_s s_{st}} \\ &= 0.5 \times \tfrac{500}{s'_d} \times 2 \times \tfrac{4 \times 1.6 \times 1.35 \times 1.17 \times 4 \times 4 \times 400}{1.2} + \tfrac{0.17(1 + \tfrac{1400 \times 1000}{14 \times 500 \times 500})}{1.5} \\ &\times \sqrt{15} \times 500 \times (500 - c - \phi_{st} - \tfrac{\phi_l}{2}) + \tfrac{4 \times \tfrac{3.14 \times 10^2}{4} \times 376 \times (500 - c - \phi_{st} - \tfrac{\phi_l}{2})}{150 \times 1.2} \end{aligned}$$

Solving the above two equations, the connection spacing is given as:

$$s_d \leq 160 \text{ mm}$$

Here, the connection spacing is taken as 100 mm. The corresponding shear capacity of the strengthened column is 623 kN.

(b) Determining ultimate compressive stress in compressive steel plate

The slenderness ratio of the steel plate is given as:

$$\lambda_{sr} = \frac{s_d}{t_p} = \frac{100}{4} = 25$$

The initial imperfection factor is given by (19).

$$\alpha_i = 1.046 - 0.0073\lambda_{sr} = 1.046 - 0.0073 \times 25 = 0.86$$

The ultimate compressive stress of the steel plate is determined by (18).

$$\begin{aligned} \sigma_{p,critical} &= \frac{4\pi^2 D d_p}{s_d^2 A_p}(1-\alpha_i) = \frac{4\pi^2 E_p t_p^2}{12 s_d^2 (1-\mu_p^2)}(1-\alpha_i) \\ &= \frac{4 \times 3.14^2 \times 200 \times 10^3 \times 4^2}{12 \times 100^2 \times (1-0.3^2)} \times (1-0.86) \\ &= 162 \text{ MPa} < f_{yp} = 300 \text{ MPa} \end{aligned}$$

(c) Determining lateral load capacity of strengthened RC column

The stresses in steel plate and longitudinal rebar are given by Equation (26). The compressive depth of the strengthened fire-damaged RC column is determined by Equation (27).

$$\begin{aligned} 0 = & N_0 + \frac{4A_l\sigma_{l1} + 2A_l\sigma_{l2} + 2A_l\sigma_{l3} + 4A_l\sigma_{l4}}{1.2} \\ & + \frac{0.6\sigma_{pt1}A_p - A_p\sigma_{pc1} + 2\sum_{i=1}^{n} t_p\Delta_i\sigma_{pside\_i}}{1.2} - \frac{\alpha f'_{c,pf} d_c \beta x_c}{1.5} \end{aligned}$$

The compressive height of the plane section of the strengthened fire-damaged RC column is solved as:

$$x_c = 302 \text{ mm}$$

The moment capacity of the strengthened fire-damaged RC column is derived by Equation (28).

$$\begin{aligned} M_{cap,stren} = & N_0 \frac{d_c}{2} - \frac{\alpha f'_{c,pf} d_c \beta x_c \frac{\beta x_c}{2}}{1.5} \\ & + \frac{4A_l\sigma_{l1}(d_c-c-\phi_{st}-\frac{\phi_l}{2}) + 2A_l\sigma_{l2}(d_c-c-\phi_{st}-\frac{\phi_l}{2}-s_l)}{1.2} \\ & + \frac{2A_l\sigma_{l3}(d_c-c-\phi_{st}-\frac{\phi_l}{2}-2s_l) + 4A_l\sigma_{l4}(d_c-c-\phi_{st}-\frac{\phi_l}{2}-3s_l)}{1.2} \\ & + \frac{0.6\sigma_{pt1}A_p(d_c+\frac{t_p}{2}) + A_p\sigma_{pc1}\frac{t_p}{2} + 2\sum_{i=1}^{n} t_p\Delta_i\sigma_{pside\_i}(d_c-\frac{d_c-d_p}{2}-\frac{\Delta_i}{2}-(i-1)\Delta_i)}{1.2} \\ = & 530 \text{ kN} \cdot \text{m} \end{aligned}$$

The lateral load resistance of the strengthened fire-damaged RC column is given by Equation (29).

$$V_{cap,stren} = \frac{M_{cap,stren}}{0.5L} = \frac{530 \times 10^3}{0.5 \times 3000} = 353 \text{ kN} > V = 300 \text{ kN}$$

The condition in (24) is rechecked herein.

$$V_{cap,stren} = 353 \text{ kN} < 0.6 V_{stren} = 374 \text{ kN}$$

5.3. Checking ALR

(a) Estimating equivalent passive confinement stress

The equivalent passive confinement stress originated from the stirrup and the connections are respectively given by (4) and (5).

$$f_{est} = \frac{4\alpha_{st} f_{yst,pf} A_{st}}{s_{st} l_{st}} \qquad f_{ed} = \frac{2\alpha_d n_f F_d}{(s_d + d_d) d_c}$$
$$= \frac{4 \times 0.34 \times 376 \times 3.14 \times \frac{10^2}{4}}{150 \times (500 - 20 \times 2 - 10)} \qquad = \frac{2 \times 0.47 \times (0.82 - 0.64 \times \frac{100}{500}) \times 4 \times 1.6 \times 1.35 \times 1.17 \times 4 \times 400}{150 \times 500}$$
$$= 0.59 \text{ MPa} \qquad = 0.56 \text{ MPa}$$

(b) Estimating axial load capacity

For the confined concrete in Figure 7b, the imposed equivalent passive confinement stress is $f_l = f_{est} + f_{ed}$ = 0.59 + 0.56 = 1.15 MPa. Using the confined concrete strength model in (7)–(16), the confined concrete strength is 22.5 MPa. Thus, the axial load capacity is given by Equations (30)–(32).

$$N_{c,stren} = \frac{A_{cc} f'_{cc,pf} + A_{c0} f'_{c,pf}}{\gamma_c} + \frac{12 A_l f_{yl,pf} + 4 A_p \sigma_{p,critical}}{\gamma_s}$$
$$= \frac{(500 \times 500 - 12 \times \frac{3.14 \times 20^2}{4} - \frac{2}{3}(500 - 70 \times 2)^2) \times 22.5 + \frac{2}{3}(500 - 70 \times 2)^2 \times 15}{1.5}$$
$$+ \frac{12 \times \frac{3.14 \times 20^2}{4} \times 470 + 4 \times 4 \times 495 \times 162}{1.2}$$
$$= 5800 \text{ kN}$$

The ALR is given by (33).

$$\text{ALR} = \frac{N_0}{N_{c,stren}} = \frac{1400}{5800} = 0.24 < 0.65$$

5.4. Checking Effective Flexural Stiffness

Based on the study conducted by Hwang et al. [6], the residual elastic modulus of concrete retains around $0.2E_c$ after exposure to a 570 °C fire event. According to (17), the effective flexural stiffness of the strengthened RC column is given as:

$$K_i = (EI)_s + 0.6(EI)_{c,pf}$$
$$= 2 \frac{E_p t_p d_p^3}{12} + E_p t_p d_p \frac{d_c^2}{2} + 0.6 \frac{0.2 E_c d_c^4}{12}$$
$$= 2 \times \frac{200000 \times 4 \times 495^3}{12} + 200000 \times 4 \times 495 \times \frac{500^2}{2} + 0.6 \times \frac{0.2 \times 25000 \times 500^4}{12}$$
$$= 8.1 \times 10^{13} \text{N} \cdot \text{mm}^2 \approx 7.8 \times 10^{13} \text{N} \cdot \text{mm}^2 = 0.6 \times \frac{25000 \times 500^4}{12} = 0.6(EI)_c$$

Until then, the parameters of the steel jacket (i.e., steel plate thickness, fastener number and connection spacing) are determined. The strengthening scheme is shown in Figure 11.

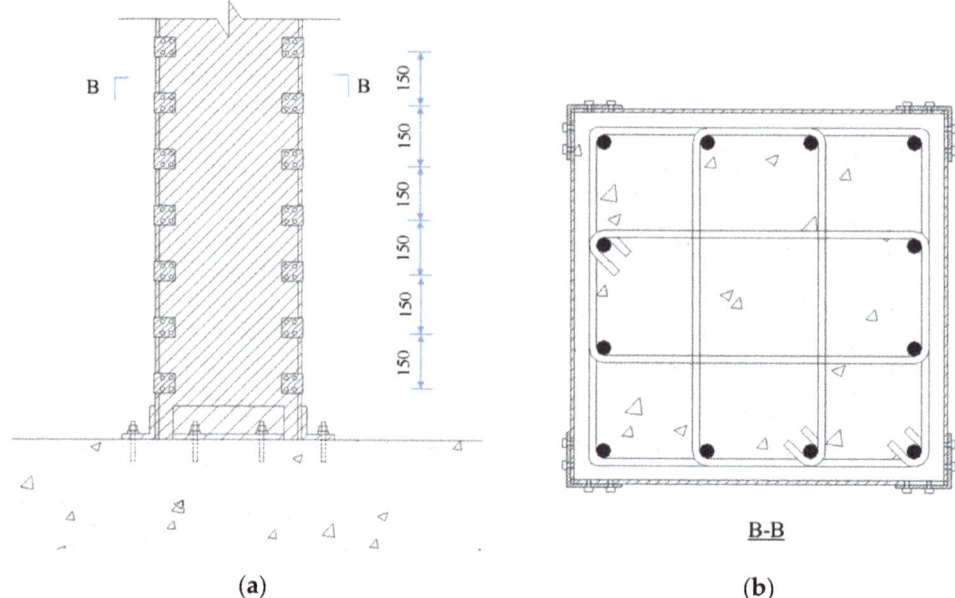

Figure 11. Strengthened RC column: (**a**) front view; (**b**) plane view.

6. Conclusions

In this paper, a novel direct fastening steel jacket is introduced. The main findings of the available experimental and theoretical studies on the proposed strengthening method are reviewed, based on which the design procedure for the proposed strengthening method is presented. The design procedure can be processed by the application of four critical steps: (1) sizing the steel plate thickness based on the required enhancement ratio and Figure 9; (2) sizing the connection spacing and fastener number by avoiding flexural failure; (3) checking lateral load demand and capacity; and (4) checking the ALR limit. If either of the final two checking conditions is not satisfied, the process should be repeated from step (1) until these two final conditions are satisfied. Subsequently, the effective flexural stiffness is examined to ensure it is similar to that of the undamaged RC column.

Author Contributions: Conceptualization, Z.S. and R.K.L.S.; methodology, Z.S. and R.K.L.S.; investigation, L.C. and K.L.; writing—original draft preparation, Z.S.; writing—review and editing, Z.S., R.K.L.S. and Z.X. All authors have read and agreed to the published version of the manuscript.

Funding: This research was supported by the Fundamental Research Funds for the Central Universities.

Institutional Review Board Statement: Not applicable.

Informed Consent Statement: Not applicable.

Data Availability Statement: The results presented in this study are available on request from the corresponding author.

Acknowledgments: This research was supported by the Fundamental Research Funds for the Central Universities. The authors would like to thank for their financial support.

Conflicts of Interest: The authors declare no conflict of interest.

References

1. Free, M.; Rossetto, T.; Peiris, N.; Taucer, F.; Zhao, B.; Koo, R.; Wang, J.; Ma, X.; Verrucci, E. The Wenchuan earthquake of May 12, 2008: Field observations and recommendations for reconstruction. In Proceedings of the 14th World Conference on Earthquake Engineering, Beijing, China, 12–17 October 2008.
2. Li, B.; Wang, Z.; Mosalam, K.M.; Xie, H.P. Wenchuan earthquake field reconnaissance on reinforced concrete framed buildings with and without masonry infill walls. In Proceedings of the 14th World Conference on Earthquake Engineering, Beijing, China, 12–17 October 2008.
3. Sezen, H.; Whittaker, A.S.; Elwood, K.J.; Mosalam, K.M. Performance of reinforced concrete buildings during the August 17, 1999 Kocaeli, Turkey earthquake, and seismic design and construction practise in Turkey. *Eng. Struct.* **2003**, *25*, 103–114. [CrossRef]
4. Bousias, S.; Spathis, A.L.; Fardis, M.N. Seismic retrofitting of columns with lap splices through CFRP jackets. In Proceedings of the 13th World Conference on Earthquake Engineering, Vancouver, BC, Canada, 1–6 August 2004.
5. Su, R.K.L.; Lam, N.T.K.; Tsang, H.H. Seismic drift demand and capacity of nonseismically designed buildings in Hong Kong. *Electron. J. Struct. Eng.* **2008**, *8*, 110–120.
6. Hwang, E.; Kim, G.; Choe, G.; Yoon, M.; Gucunski, N.; Nam, J. Evaluation of concrete degradation depending on heating conditions by ultrasonic pulse velocity. *Constr. Build. Mater.* **2018**, *171*, 511–520. [CrossRef]
7. Anagnostou, E.; Rousakis, T.C.; Karabinis, A.I. Seismic retrofitting of damaged rc columns with lap-spliced bars using frp sheets. *Compos. Part B Eng.* **2019**, *166*, 598–612. [CrossRef]
8. Al-Harthy, A.S.; Stewart, M.G.; Mullard, J. Concrete cover cracking caused by steel reinforcement corrosion. *Mag. Concrete Res.* **2011**, *63*, 655–667. [CrossRef]
9. Shan, Z.W.; Su, R.K.L. Improved uncoupled closed-form solution for adhesive stresses in plated beams based on Timoshenko beam theory. *Int. J. Adhes. Adhes.* **2019**, *96*, 102472. [CrossRef]
10. Shan, Z.W.; Su, R.K.L. Flexural capacity model for RC beams strengthened with bolted side-plates incorporating both partial longitudinal and transverse interactions. *Eng. Struct.* **2018**, *168*, 44–57. [CrossRef]
11. Chang, S.Y.; Chen, T.W.; Tran, N.C.; Liao, W.I. Seismic retrofitting of RC columns with RC jackets and wing walls with different structural details. *Earthq. Eng. Eng. Vib.* **2014**, *13*, 279–292. [CrossRef]
12. Vandoros, K.G.; Dritsos, S.E. Concrete jacket construction detail effectiveness when strengthening RC columns. *Constr. Build. Mater.* **2008**, *22*, 264–276. [CrossRef]
13. Garzon-Roca, J.; Ruiz-Pinilla, J.; Adam, J.M.; Calderón, P.A. An experimental study on steel-caged RC columns subjected to axial force and bending moment. *Eng. Struct.* **2011**, *33*, 580–590. [CrossRef]
14. Wang, L.; Su, R.K.L.; Cheng, B.; Li, L.Z.; Shan, Z.W. Seismic behavior of preloaded rectangular RC columns strengthened with precambered steel plates under high axial load ratios. *Eng. Struct.* **2017**, *152*, 683–697. [CrossRef]
15. Lam, L.; Teng, J.G. Strength models for fiber-reinforced plastic-confined concrete. *J. Struct. Eng.* **2002**, *128*, 612–623. [CrossRef]
16. Demers, M.; Neale, K.W. Confinement of reinforced concrete columns with fibre-reinforced composite sheets—An experimental study. *Can. J. Civil. Eng.* **1999**, *26*, 226–241. [CrossRef]
17. Yaqub, M.; Bailey, C.G.; Nedwell, P. Axial capacity of post-heated square columns wrapped with FRP composites. *Cem. Concr. Comp.* **2011**, *33*, 694–701. [CrossRef]
18. Wang, L.; Su, R.K.L. Repair of fire-exposed preloaded rectangular concrete columns by postcompressed steel plates. *J. Struct. Eng.* **2014**, *140*, 04013083. [CrossRef]
19. Lin, C.H.; Tsay, C.S. Deterioration of strength and stiffness of reinforced concrete columns after fire. *J. Chin. Inst. Eng.* **1990**, *13*, 273–283. [CrossRef]
20. Vieira, J.P.B.; Correia, J.R.; De Brito, J. Post-fire residual mechanical properties of concrete made with recycled concrete coarse aggregates. *Cem. Concr. Res.* **2011**, *41*, 533–541. [CrossRef]
21. Hyland, C.; Smith, A. *CTV Building Collapse Investigation*; Department of Building and Housing: Christchurch Central City, New Zealand, 2012.
22. Shan, Z.W.; Su, R.K.L. Behavior of shear connectors joined by direct fastening. *Eng. Struct.* **2019**, *196*, 109321. [CrossRef]
23. Shan, Z.W.; Su, R.K.L. Axial strengthening of RC columns by direct fastening of steel plates. *Struct. Eng. Mech.* **2021**, *77*, 705–720.
24. Shan, Z.W.; Looi, D.T.W.; Su, R.K.L. Confinement model of RC columns strengthened by direct fastening of steel plate. *Steel Compos. Struct.* **2021**. accepted.
25. Shan, Z.W.; Looi, D.T.W.; Su, R.K.L. A novel seismic strengthening method of RC columns confined by direct fastening steel plates. *Eng. Struct.* **2020**, *218*, 110838. [CrossRef]
26. EN-1993-1-8. *Design of Steel Structure. Part 1.8: Design of Joints*; European Committee for Standardization: Brussels, Belgium, 2005.
27. ANSI/AISC 360-16. *Specification for Structural Steel Buildings*; American Institute of Steel Construction: Chicago, IL, USA, 2016.
28. AS 4100. *Steel Structures*; Standards: Sydney, Australia, 1998.
29. Guo, Z.H.; Wang, C.Z. Investigation of strength and failure criterion of concrete under multi-axial stresses. *Chin. Civil. Eng. J.* **1991**, *24*, 1–14.
30. EN 1994-1-1:2004. *Design of Composite Steel and Concrete Structures. Part 1 1: General Rules and Rules for Buildings*; European Committee for Standardization: Brussels, Belgium, 2004.
31. ACI Committee 318. *Building Code Requirements for Structural Concrete (ACI 318M-14) and Commentary (ACI 318RM-14)*; American Concrete Institute: Farmington Hills, MI, USA, 2015.

32. ASCE/SEI 41-13. *Seismic Evaluation and Retrofit of Existing Buildings*; ASCE Standard: Reston, VI, USA, 2014.
33. EN 1998-3:2005. *Design of Structures for Earthquake Resistance—Part 3: Assessment and Retrofitting of Buildings*; European Committee for Standardization: Brussels, Belgium, 2005.
34. EN 1998-1:2004. *Design of Structures for Earthquake Resistance—Part 1: General Rules, Seismic Actions and Rules for Buildings*; European Committee for Standardization: Brussels, Belgium, 2004.
35. EN 1992-1-2. *Eurocode 2: Design of Concrete Structures—Part 1–2: General Rules—Structural Fire Design*; European Committee for Standardization: Brussels, Belgium, 2004.

Article
Cyclic Tests and Numerical Analyses on Bolt-Connected Precast Reinforced Concrete Deep Beams

Jing Li [1], Lizhong Jiang [1], Hong Zheng [2], Liqiang Jiang [1,*] and Lingyu Zhou [1,*]

[1] School of Civil Engineering, Central South University, Changsha 410075, China; 144806014@csu.edu.cn (J.L.); lzhjiang@csu.edu.cn (L.J.)
[2] School of Civil Engineering, Chang'an University, Xi'an 710061, China; cehzheng@chd.edu.cn
* Correspondence: jianglq2019@csu.edu.cn (L.J.); zhoulingyu@csu.edu.cn (L.Z.)

Abstract: A bolt-connected precast reinforced concrete deep beam (RDB) is proposed as a lateral resisting component that can be used in frame structures to resist seismic loads. RDB can be installed in the steel frame by connecting to the frame beam with only high-strength bolts, which is different from the commonly used cast-in-place RC walls. Two 1/3 scaled specimens with different height-to-length ratios were tested to obtain their seismic performance. The finite element method is used to model the seismic behavior of the test specimens, and parametric analyses are conducted to study the effect on the height-to-length ratio, the strength of the concrete and the height-to-thickness ratio of RDBs. The experimental and numerical results show that the RDB with a low height-to-length ratio exhibited a shear–bending failure mode, while the RDB with a high height-to-length ratio failed with a shear-dominated failure mode. By comparing the RDB with a height-to-length ratio of 2.0, the ultimate capacity, initial stiffness and ductility of the RDB with a height-to-length ratio of 0.75 increased by 277%, 429% and 141%, respectively. It was found that the seismic performance of frame structures could be effectively adjusted by changing the height-to-length ratio and length-to-thickness of the RDB. The RDB is a desirable lateral-resisting component for existing and new frame buildings.

Keywords: precast reinforced concrete deep beam; experimental study; seismic performance; finite element method (FEM)

Citation: Li, J.; Jiang, L.; Zheng, H.; Jiang, L.; Zhou, L. Cyclic Tests and Numerical Analyses on Bolt-Connected Precast Reinforced Concrete Deep Beams. *Appl. Sci.* **2021**, *11*, 5356. https://doi.org/10.3390/app11125356

Academic Editors: Pier Paolo Rossi and Melina Bosco

Received: 10 April 2021
Accepted: 3 June 2021
Published: 9 June 2021

Publisher's Note: MDPI stays neutral with regard to jurisdictional claims in published maps and institutional affiliations.

Copyright: © 2021 by the authors. Licensee MDPI, Basel, Switzerland. This article is an open access article distributed under the terms and conditions of the Creative Commons Attribution (CC BY) license (https://creativecommons.org/licenses/by/4.0/).

1. Introduction

A reinforced concrete (RC) shear wall is commonly used as an infilled lateral load-resisting component in RC frames for high-rise buildings. The RC frames with infill RC shear walls exhibited better performance compared with the traditional bare RC frames because of the insufficient lateral stiffness and resistance of RC frames. The frame columns and beams are used to resist the vertical loads and moments due to earthquakes, while the RC shear walls (RCSW) resisted the majority of shear forces [1]. The initial researches focused on different kinds of RCSW systems and conducted experimental and theoretical investigations on their seismic performance [2–4]. Furthermore, experimental and numerical studies were performed to analyze the mechanical behavior of different types of connections between RC walls and building frames [5–7]. However, the RC shear walls may exhibit oversized lateral stiffness resulting in an insufficient lateral deformation capacity of the structure with an insufficient load-carrying capacity. Besides, the shear force carried by the RC shear walls may be transmitted as tension or compressive force to the frame columns [8]. To prevent severe damage from occurring at the end of frame columns, strong columns must be designed to comply with the design principle of "strong frame, weak wall". Furthermore, to meet the architectural requirements, the effects of different kinds of openings on the seismic performance of RCSWs also need to be investigated, which may make the seismic design of RCSWs more difficult.

Three major issues should be considered for the frames with RCSW systems. Firstly, a few RC shear walls might be placed in the first story for the stores and open lobbies due to the architectural demands. These buildings may have a soft first story if insufficient shear walls are placed in the first story compared to the upper stories [9]. Secondly, rapid repair and retrofit methods of a severely damaged component are desired for structures after an earthquake occurs [10,11]. The RC shear walls are commonly connected with frame columns and beams cast in place; thus, the frame columns and beams may be destroyed if the retrofit work is performed on the damaged walls. Lastly, the dimensions of the infill walls are determined by the dimensions of the surrounding frame. The lateral stiffness and resistance of the structure are mainly adjusted through the thickness of the walls. However, the thickness is a strict rule for the RC walls due to stability requirements [2]. Therefore, the lateral stiffness and resistance of the RC frames with RCSWs are difficult to change on a large scale to meet the optimization design of RC structures.

In the past decades, the RC wall panel (RCWP) was introduced by Kahn and Hanson [12] as a strengthening member to enhance the performance of RC structures. It was also proposed as a retrofit/repair strategy for the RC frames [13] because the member could be connected to structures with precast construction. Engineered cementitious composite materials (ECC) were proposed for RCWP due to their good performance to dissipate energies generated from earthquakes [14], and a series of experimental and theoretical investigations were performed. In recent years, many researchers have suggested that the RC walls should be separated with frame columns, including separating the RC infill wall from the RC frame by slits [15], and slitting the separated RC infill wall [9]. Other types of steel wall panels, such as double-skin wall panels [16] and panel wall strengthened systems [17–20], were proposed and analyzed. Different from the steel wall panels, the RC wall panels were commonly connected to the boundary frames with cast-in-place construction. Thus the RC wall panels are difficult to remove or repair. The precast concrete components or structures should be developed for the rapid construction of new structures and repairing existing damaged structures under seismic hazard, and it is important to guarantee the desired structural performance for the precast structures [21,22]. The connections are the key aspects of obtaining precast structures' desired performance [23–26]. Therefore, a new type of connection could realize the rapid construction and replacement of RC wall panels and could also ensure the connecting performance between RC wall panels and boundary frames.

Based on the researches presented here, the bolt-connected precast reinforced concrete deep beam (RDB) is proposed in this paper. It connects to the frame beam with high-strength bolts. It exhibits some advantages to handle the above issues compared to the other RC walls for the following reasons. Firstly, the RDB can be flexibly placed, and thus it provides architectural space for the installation of doors and windows. Secondly, it can be easily assembled or removed from the bolted connection; thus, it can be rapidly and cost-efficiently fabricated in building repair or retrofit. Lastly, the length of the RDB could be selected in a wide range; thus, it can achieve a wide range of initial stiffness and lateral resistance. In addition, the main deformation of the deep beam is combining bending and shearing deformation; thus, RDB may exhibit better ductility by comparing it with the RC infill walls for relatively sufficient plastic development of materials. It is also a part of a project that uses a deep beam as a lateral resisting system.

In this study, experimental investigations of two precast reinforced concrete deep beam specimens that were 1/3 scaled were performed, and numerical studies of RDBs were conducted and verified by test results. The initial stiffness, as well as the mechanical behavior of the bolt connections, were also discussed and calculated. Such results provide information for the seismic design of bolt-connected precast RDB components.

2. Materials and Methods

2.1. Material Properties

In Table 1, the average cube compressive strength of concrete material, as well as the material performance of reinforcing bars, is presented. The concrete and reinforcement steel bars were designed as C30 and HPB235, respectively. Based on the Chinese standard

GB50081 [27], concrete cube specimens taken from the batches of concrete were tested. Three repeatable specimens were fabricated with 28 days of standard maintenance, and the average values were used. For the high-strength bolts used in this paper, the yield strength was 800 N/mm^2, and the yield-failure ratio was 0.8. The pretension force applied to each bolt was 125 kN.

Table 1. Mechanical properties of concrete and reinforcing bars.

Material	Item	Result
Concrete	Average cube compressive strength, f_{cu}	45.3 MPa
Reinforcing bars	Diameter, D	7.8 mm
	Yield strength, f_y	237.0 MPa
	Tensile strength, f_u	536.7 MPa
	Elastic modulus, E_s	207.2×10^3 MPa
	Elongation ratio, Δ	36.6%

2.2. Test Specimens

In the test program, two bolt-connected precast reinforced concrete deep beams (RDB) were designed on a 1/3 scale and different height-to-length ratios (RDB-A and RDB-B were 2.0 and 0.75, respectively). The dimensions of these specimens and the test setup are shown in Figure 1. The height of RDBs was 900 mm, and the length was 450 mm for RDB-A and 1200 mm for RDB-B, respectively. The clear height of the specimens was 810 mm. The thickness of the specimens was 60 mm. The reinforcing bars were arranged at a distance of 75 mm, and the diameter was 8 mm.

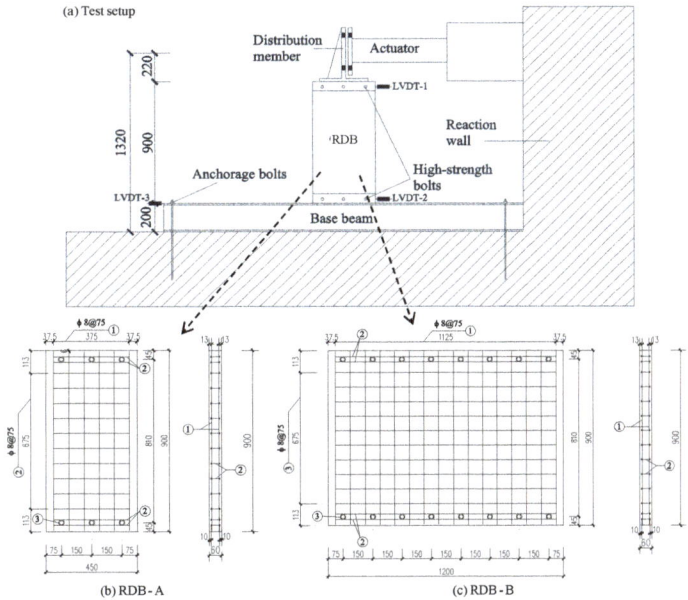

Figure 1. Test setup and test specimens.

2.3. Setup and Measurements of Test Specimens

The construction details of the test device and the specimens are presented in Figure 2. To distribute the load in the RDBs more evenly, a distribution member was designed to connect to the actuator through M45 bolts. Two angles were placed at the two sides of RDBs, and these angles were equipped back-to-back to support the RDBs. M20 high-strength bolts were used to connect the angles and the RDBs. The angles were connected to the base beam and distribution member by M20 bolts, and they were connected to the strong floor by anchorage bolts. To ensure that no out-of-plane deformation developed in the specimens during the tests, two rigid beams were constructed to provide lateral supports. Two pairs of sliding constraints were installed between the rigid beams and the distribution member.

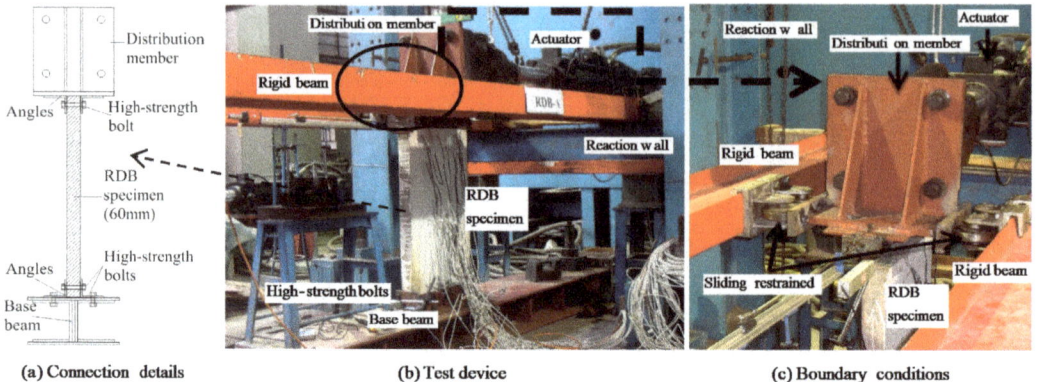

Figure 2. Details of the test device, specimens, connection details and boundary conditions.

Three displacement-based linear transducers were used for measuring the lateral displacement of the specimen. They were placed at the two sides of the RDBs and the base beam, and they were used to measure the relative lateral deformation of the test specimens. The stress distribution of the RDBs was traced by uniaxial strain gauges, and they were placed along the reinforced bars to analyze the mechanical behavior of the RDBs, as presented in Figure 3. The loading pattern of the test specimens was designed based on the Chinese standard JGJ 101 [28], and it was controlled by displacement. The diagrams of the loading pattern are shown in Figure 4. An increment of 1 mm on displacement was used at the elastic mechanical stage, and the displacement did not repeat at the same loading level. The increment was increased at the elastic–plastic stage, and the displacement was repeated three times at the same loading level. Two methods were used to determine the yield displacement of the test specimens: some of the strain gauges exceeded the yield value, and the slope of the load–displacement curves decreased sharply.

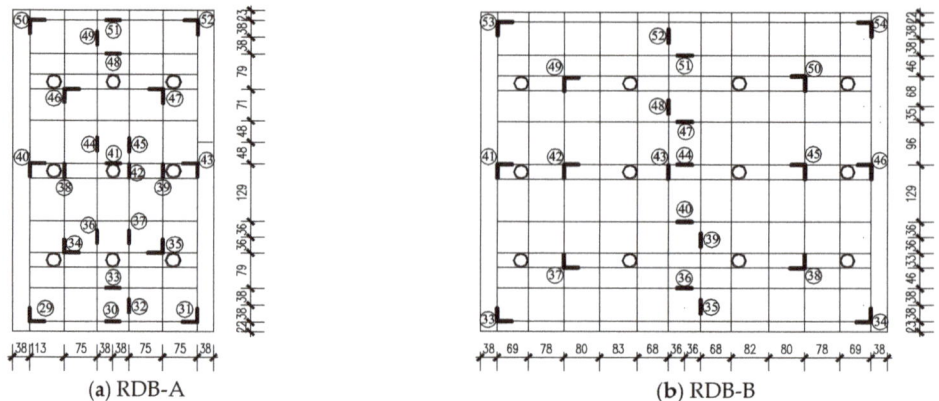

Figure 3. Arrangement of uniaxial strain gauges on the reinforcing bars.

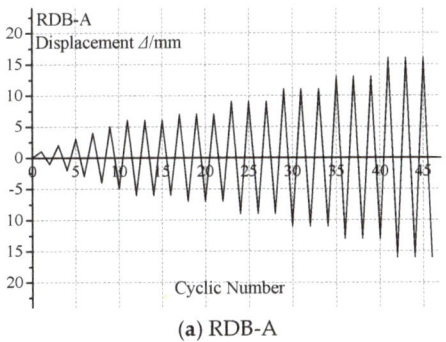 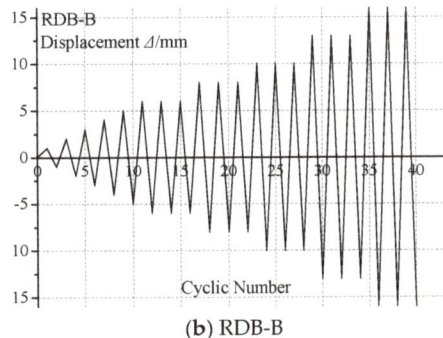

(a) RDB-A   (b) RDB-B

Figure 4. Loading patterns for the test specimens.

3. Test Results

3.1. Failure Mode and Hysteretic Behavior

The RDB-A failed to combine with the bend and shear destroy. An inclined crack was found at the bottom of the deep beam at the 2 mm loading step. At this loading step, concrete spalling was observed at the top side corner of the deep beam, but the reinforcing bars were not exposed, as shown in Figure 5a. New cracks were observed at the corner and bottom of the specimen at the story displacement of 3 mm, and a major crack had obviously developed along the direction of 45° in a horizontal direction. When the displacement reached 5 mm, the concrete at the bottom corners was apparently crushed, as presented in Figure 5b, which resulted in exposing the stirrups. A crisp noise was produced from the high-strength bolts at this step. After that, a relative slip deformation was found, and the cracks developed quickly. The final failure pattern of the RDB-A specimen is depicted in Figure 5c.

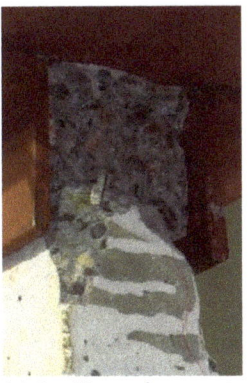 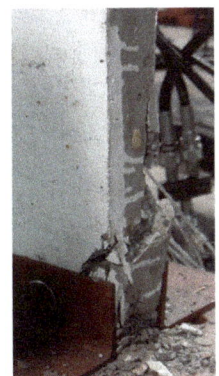 

(a) Spalling of concrete   (b) Crack at the corner   (c) Failure of the RDB-A

Figure 5. Failure modes of the RDB-A.

The RDB-B failed, dominated by the shearing destroy. The first crack was found at the bottom corner nearby, with the first high-strength bolt at the story displacement of 2 mm. When the displacement had reached 3 mm, such cracks could not be closed, as shown in Figure 6a, and a crisp noise was heard from the high-strength bolts. The first diagonal crack was detected; it was formed along the direction of 45° in a horizontal direction when the lateral displacement was 5 mm, as shown in Figure 6b. After that, more other inclined cracks were formed at the center region of the deep beam. The final failure mode is presented in Figure 6c.

(a) Cracks at the corner

(b) Diagonal cracks at the center region

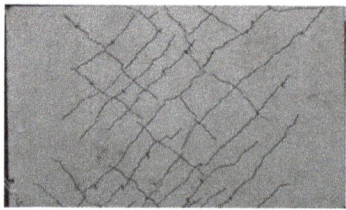
(c) Failure of the RDB-B

Figure 6. Failure modes of the RDB-B.

The relation curves of lateral load and displacement curves of the specimens are depicted in Figure 7. The force and displacement results of some key points are presented in Table 2. When the height-to-length ratio of the RDB specimen decreased from 2.0 to 0.75, the ultimate capacity, initial lateral stiffness and ductility increased about 277%, 429% and 141%, respectively. The results indicated that the height-to-length ratio is an important factor in adjusting the strength and stiffness of the RDBs on a large scale.

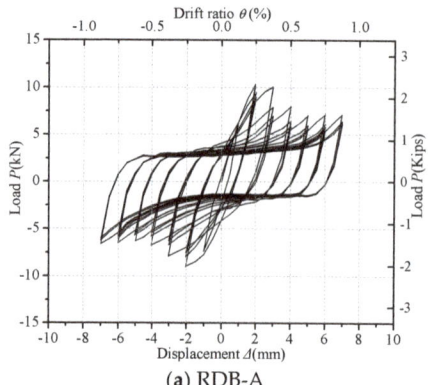
(a) RDB-A

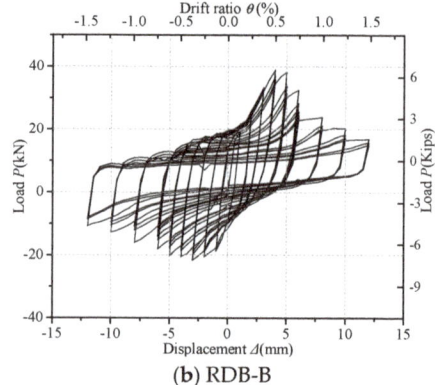
(b) RDB-B

Figure 7. Loading-displacement curves of the specimens.

Table 2. Force, displacement and ductility results.

Test Specimen	Yield Value			Maximum Value			Ductility Value
	P_y (kN)	Δ_y (mm)	φ_y (%)	P_{max} (kN)	Δ_{max} (mm)	φ_{max} (%)	$\mu_\Delta = \Delta_{max}/\Delta_y$
RDB-A	9.23	3.07	0.34	10.35	3.63	0.40	1.18
RDB-B	25.69	2.08	0.23	39.06	5.90	0.65	2.84

3.2. Ductility and Energy Dissipation

In this paper, the displacement-based ductility coefficient μ_Δ is utilized to represent the ductility performance of RDBs. The μ_Δ is determined by the displacements when the specimens yield and fail, and the calculation is $\mu_\Delta = \Delta_{max}/\Delta_y$. The equal energy method is used to obtain the value of Δ_y. The failure displacement is defined as 85% times the maximum value [8]. The ductility results are listed in Table 2. The RDB-B exhibited better ductility than the RDB-A, indicating that the ductility of RDBs increases if the height-to-length ratio reduces.

To evaluate the energy dissipation performance of RDB specimens, the energy dissipation coefficient E is used. The detailed calculation and the diagram are shown in Figure 8a [8]. The energy dissipation coefficients of the test specimens in this paper are presented in Figure 8b. The results show that the energy dissipation coefficient decreases with the increase of the lateral displacement for the most part. The energy dissipation capacity of RDB-A is close to the RDB-B, indicating that the energy dissipation capacity of RDBs changes little by decreasing the height-to-length ratio. Due to the limited number of specimens, it should be highlighted that the effects of the height-to-length ratio on ductility and energy dissipation coefficient are not regular.

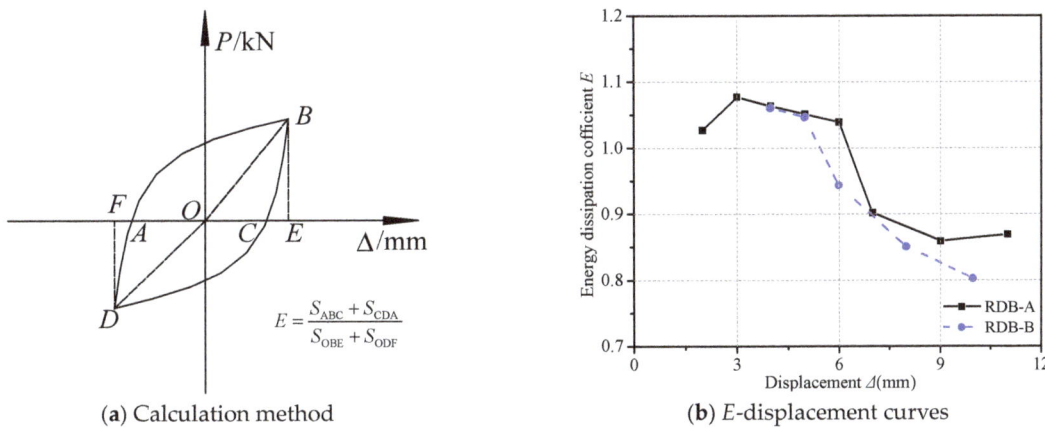

(a) Calculation method (b) E-displacement curves

Figure 8. Energy dissipation results of test specimens.

3.3. Elastic Stiffness

The boundary condition is an important factor to determine the stiffness as well as the ultimate strength of RC deep beams [29]. Thus, the boundary conditions of the RDBs can be simplified as a deep beam with fixed supports [8], and the idealized diagram is presented in Figure 9a. The initial stiffness of the RDB could be obtained by considering the bending deformation and shearing deformation of the RDB. The deformation diagram and force distribution diagram of the RDBs associated with lateral force F are shown in Figure 9b.

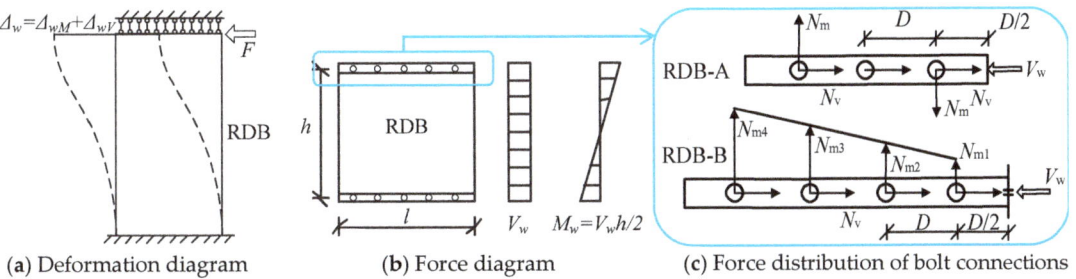

(a) Deformation diagram (b) Force diagram (c) Force distribution of bolt connections

Figure 9. Deformation and force diagrams for RDBs and bolt connections.

The lateral displacement of RDBs Δ_w is a combination of bending deformation Δ_{wM} and shearing deformation Δ_{wV}. The displacement of RDBs can be calculated based on the Graph Multiplication Method in basic theory in structural mechanics:

$$\Delta_w = \Delta_{wM} + \Delta_{wV} = \frac{V_w h^3}{12 E_c I_w} + \frac{1.2 V_w h}{GA} \quad (1)$$

where I_w, V_w and A are the moments of inertia, shear force and section area of RDBs, respectively; E_c and G are the elastic modulus and shearing modulus of concrete. Thus, Equation (1) can be expressed as:

$$\Delta_w = \frac{V_w}{E_c t}\left[\frac{1}{(l/h)^3} + \frac{2.88}{(l/h)}\right] \quad (2)$$

where h is the effective height and l is the effective length. The elastic lateral stiffness K_e is calculated as:

$$K_e = \frac{E_c t}{\left[(h/l)^3 + 2.88(h/l)\right]} \quad (3)$$

Table 3 presents the comparison of the elastic lateral stiffness of the test specimens obtained from Equation (3) and the tests. It was found that the error is no more than 20%, and the values derived from the equation slightly overestimate the lateral stiffness of the test specimens. It can be concluded that the imperfections of the specimens and uncertainty in test procedures were not considered in the equation. However, the overestimation can also be treated as safety reserves in the structural design of RDBs.

Table 3. Comparison of elastic stiffness of the theoretical and experimental test results.

Specimen	Elastic Lateral Stiffness (kN/mm)		Error
	$K_{e\text{-theoretical}}$	$K_{e\text{-experimental}}$	
RDB-A	13.08	11.06	15.4%
RDB-B	69.71	58.48	16.1%

3.4. Mechanical Behavior of the Bolt Connection

The frictional types of high-strength bolts were used in the tests to transfer the shear forces. Thus, it is important to ensure the effectiveness and reliability of the bolt connections. In this paper, the mechanical behavior of the bolt connections is divided into two main phases; they are: (1) phase I, the carried shear force of a bolt is less than its friction force; (2) phase II, the carried shear force of a bolt is larger than its friction force.

A mechanical diagram of RDBs at phase I is presented in Figure 9c; the shear force of bolts generated from the lateral load (N_v) and bending moment (N_m) are included in the diagram. The lateral force is considered as an even distribution force on the bolts; the shear force of each bolt is determined as $N_v = V_w/n$, where n is the number of the bolts in a support of RDBs. According to Figure 9c, $M_w = V_w h/2 = 0.405 V_w$ for the RDB-A, thus the $N_m = M_w/4D = 0.675 V_w$, and $N_v = V_w/3$. The largest shear force is $0.75 V_w$. For the mechanical diagram of RDB-B, $M_w = V_w h/2 = 0.405\ V_w$ and $N_{m4} = 0.225 V_w$, and $N_v = V_w/8$. The largest shear force is $0.26\ V_w$.

At phase II, due to the carried shear force of the bolts that exceeded the friction force, the bending moment is carried by the RDB, and thus the bolts carry the shear force solely. The largest shear force of the bolts is $0.33\ V_w$ and $0.125\ V_w$ for the RDB-A and RDB-B, respectively.

The shear force carried by the bolts in specimens is 10.35 kN and 39.06 kN for the RDB-A and RDB-B (Table 2), respectively. Thus the largest shear force is 7.76 kN and 10.16 kN in phase I for bolt connections in RDB-A and RDB-B, respectively, and the largest shear force is 3.42 kN and 4.88 kN in phase II for the bolt connections in RDB-A and RDB-B, respectively.

The friction force N_{vuI} in phase I and shear resistance N_{vuII} in phase II of the bolts used in this paper are:

$$N_{vuI} = 0.9 n_f \mu P \tag{4}$$

$$N_{vuII} = \frac{\pi d_0^2}{4} f_v \tag{5}$$

where n_f is the number of frictional faces of bolts; μ is the anti-sliding coefficient between bolts and angles, and it is determined as 0.3 according to suggestions [8]; P is the pre-tensile force, it is 125 kN in this study; d_0 is the clear diameter of bolts, and it is 20 mm; f_v is the yield shear strength of bolts, and it is 320 MPa.

Thus N_{vuI} = 67.5 kN and N_{vuII} =100.5 kN can be calculated according to Equations (4) and (5), and the calculated values represent the loading resistance of a high-strength bolt at the two phases. The calculated values are larger than the maximum shear force of the bolts during tests; thus, it can be concluded that high-strength bolts are safe enough in the test specimens. According to test results recorded by the LVDTs, there was also no slipping behavior or bolt-damage failure observed in the test specimens.

4. Numerical Study

4.1. Validation of Finite Element Model

The numerical study was performed by ANSYS, as presented in Figure 10, and the "Solid 65" and "Pipe20" elements were selected out to model the mechanical behavior of reinforced concrete. A bilinear model was used for the reinforcements, and the material test results such as the elastic modulus, yield stress and ultimate stress were derived from material test results. The concrete damage plasticity model in ANSYS was used for the concrete material in RDBs, because it considers the nonlinear behavior of concrete after it cracks or crushes. The compressive strength curve was taken from the compressive test results of the concrete, and the tensile cracking strength was estimated as 5.6% of the compressive strength. The 3D 8-node solid element captured the cracking and crippling behavior of the concrete. The same mesh sizes (50 mm) were used to mesh the model. The intersectant nodes in concrete elements and reinforcing bar elements were coupled together at the same positions so that they could cause consistent deformations. The boundary conditions of the bottom and top side of the concrete panel are considered as a fixed connection (at the bottom side, fixed the X, Y, Z, Rot X, Rot Y and Rot Z; at the top side, fixed the Y, Z, Rot X, Rot Y and Rot Z; the X-direction represents the direction that applied lateral loads). The lateral load is assumed as a uniform load to apply to the top side of the concrete panel in the finite element model, and this mechanical assumption was valid by Jiang et al. [8].

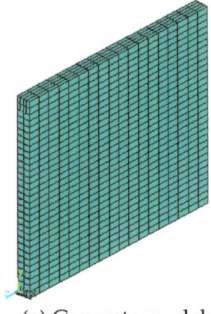
(a) Concrete model

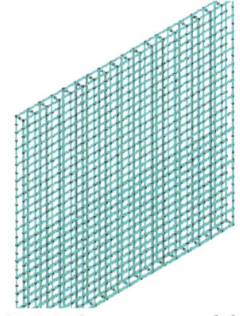
(b) Reinforcement model

Figure 10. Deformation and force diagrams for RDB.

The limitations of the modeling method are: (1) the effects of bolt connections, including the influence on stress concentration and the weakness on RC walls, are not considered in this model; (2) the bond-slipping deformation between reinforcements and concrete is not considered in the model.

The comparison between test and FEM results on lateral–loading curves and failure modes are presented in Figures 11 and 12, respectively. There is a relatively close agreement between the test results and FEM results in the elastic and elastic–plastic stage. For the RDB-A, the peak shear force of RDBs during the tests and FEM analysis is 10.35 kN and 12.22 kN, respectively, and it is 39.06 kN and 45.24 kN for RDB-B specimen, respectively. The difference in maximum shear force value between tests and FE models is no more than 18%. Besides, the failure modes of FEM (Figure 12) can predict the position where concrete cracking is in the test specimens at the estimated boundary conditions. Thus, it can be concluded that the numerical method could effectively simulate the nonlinear behavior of the RDBs, especially for the ultimate lateral resistance of RDBs, even though the largest error is about 18%. However, the loading, deformation and degradation changing processes are basically predicted by the numerical method, and it is assumed that the method can be used for the following parametric study to analyze the effects of different parameters on the mechanical performance of RDBs.

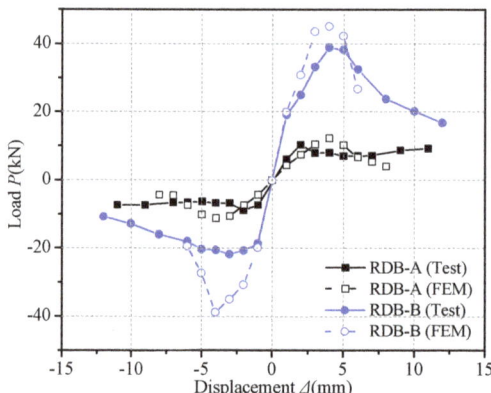

Figure 11. Comparison of the load-displacement relationship between the test and FEM results.

4.2. Parametric Analysis Study

A parametric analysis was performed to summarize the effects on geometrical and material parameters based on the verified numerical modeling method in Section 4.1. The following parameters were considered: height-to-length ratio (α), height-to-thickness (β) and material type of concrete. The height of RDBs is determined as 900 mm, which is the same as the test specimens. The thickness of RDB includes 60 mm, 80 mm and 100 mm. The height-to-length ratio of RDB is ranged from 0.75 to 2.0. The compressive strength of concrete includes three types: 25 N/mm^2 (C25), 30 N/mm^2 (C30) and 35 N/mm^2 (C35). Table 4 presents the detailed information of these RDB models. The effects on lateral loading–displacement curves with parameters of height-to-length ratio, height-to-thickness and material type of concrete are shown in Figure 13.

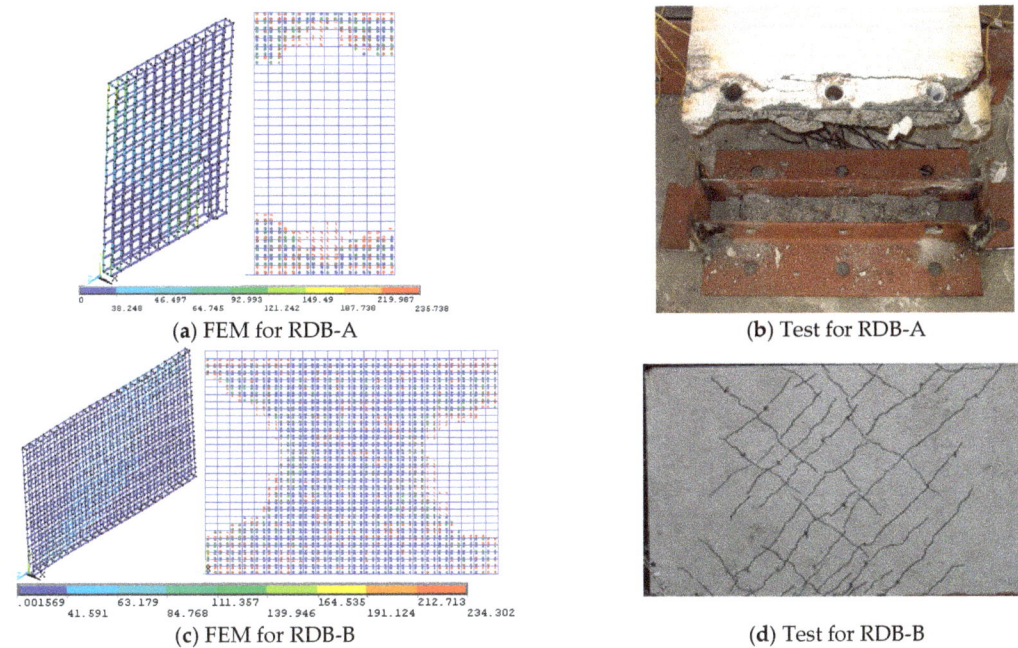

Figure 12. Failure pattern comparison of the tests and FEM predictions.

Table 4. Parameters analyzed for RDBs.

Specimens	Effects	h/mm	t/mm	l/mm	Concrete	Parameters α	β
RDB-1		900	60	450	C30	2.0	15
RDB-2		900	60	600	C30	1.5	15
RDB-3	Height-to-length ratio of RDB ($\alpha = h/l$)	900	60	750	C30	1.2	15
RDB-4		900	60	900	C30	1.0	15
RDB-5		900	60	1050	C30	0.75	15
RDB-6	Strength of concrete material	900	60	600	C25	1.2	15
RDB-7		900	60	600	C35	1.2	15
RDB-8	Height-to-thickness ratio ($\beta = h/t$)	900	80	600	C30	1.5	11.25
RDB-9		900	100	600	C30	1.5	9

Note: h, l and t are the height, length and thickness of the RDB, respectively.

As the results presented in Figure 13a show, it can be concluded that the height-to-length ratio of the RDB is showing a high contribution to the shear resistance of the RDBs. A higher height-to-length ratio of the RDB results in a higher shear resistance throughout all of the loading levels, as demonstrated by the test results. According to Figure 13b, it can be seen that the shear resistance of RDB is increasing with the increase in compressive strength of the concrete material. However, the increment is decreasing with the increase in the compressive strength. The shear resistance of RDB is obviously increasing with the decrease in the height-to-thickness ratio of RDB (see in Figure 13c), and such increment is increasing with the decrease in the height-to-thickness ratio of RDB.

Therefore, it can be concluded that the shear resistance of RDB can be efficiently strengthened by a large margin by decreasing the height-to-length ratio and increasing the height-to-thickness ratio of RDBs. However, the material type of concrete shows little effect on the seismic performance of RDBs.

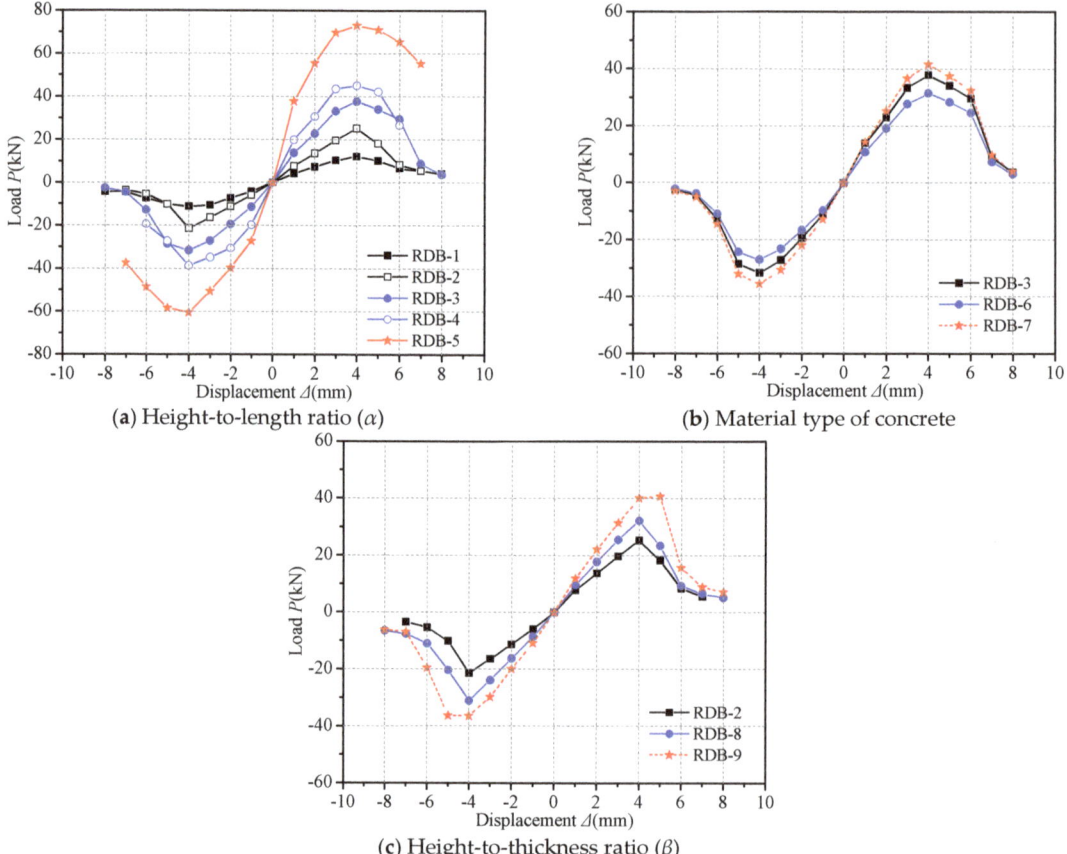

Figure 13. Effects of parameters on seismic performance of RDBs.

5. Conclusions and Discussions

In order to understand the mechanical behavior and seismic performance of RDBs, two scaled specimens were designed and tested under cyclic loads, and the corresponding nonlinear FE models were established and verified. The conclusions are listed as follows:

(1) The first crack formed at the corner of RDB-A due to large normal stress, but the first crack of RDB-B formed at the center region due to large shear stress. Besides, the RDB-A failed with a bending–shear failure mode for both typical horizontal and diagonal cracks that developed in the RDB-A; while the RDB-B failed with a shear dominating failure mode, as only diagonal cracks were observed in RDB-B. This indicates the bending–shear dominated mechanical behavior of RDB-A and the shear dominated mechanical behavior of RDB-B;

(2) By comparing the RDB with the height-to-length ratio of 2.0, the ultimate capacity, initial lateral stiffness and displacement-based ductility of the RDB with a height-to-length ratio of 0.75 increased 277%, 429% and 141%, respectively. This indicates that the height-to-length ratio is a good choice to adjust the capacity and stiffness of RDB to cater to the performance requirement of building structures;

(3) According to the parametric analysis, the shear resistance, elastic lateral stiffness and displacement-based ductility are significantly enhanced when decreasing the height-to-length ratio of RDB, which is similar to the test results. The height-to-thickness

ratio shows a general influence on the seismic performance of RDBs, but the effect of concrete strength on the seismic performance of RDBs is relatively small.

The bolt-connected RC deep beam (RDB) proposed in this paper is quite different from other types of RC shear walls. The bending–shear mechanical characteristic of RDBs is a benefit for the plastic development of RC materials. The length of RDBs could be an effective parameter to adjust the lateral stiffness and resistance of frame structures, and the RDBs could be flexibly arranged to achieve the architecture requirement. According to the failure patterns of RDB specimens, the bolt connection was effective to ensure the performance utilization of RDBs. Thus, the bolt connection could be treated as a reliable connection for RC deep beam and boundary frames. The authors have conducted experimental and numerical analyses on steel frames with steel panel walls, RC panel walls and composite steel panel walls [17–20]. The aim of this paper is to investigate the seismic performance of RDBs solely and to understand the collaborative working relationship between the steel frames and RDBs. We found that the seismic performance of the steel frame with RDB specimens is better than the sum of the steel frame and the RDB based on the test results. Though the numerical model developed in this paper is simplified compared to the macroscopic model [20], the simple model proposed in this paper has relatively good accuracy in predicting the lateral resistance of RDB.

Author Contributions: Conceptualization, H.Z. and L.J. (Liqiang Jiang); methodology, J.L. and H.Z.; software, J.L. and H.Z.; validation, J.L., H.Z. and L.Z.; investigation, J.L., H.Z. and L.Z.; writing—original draft preparation, J.L. and L.J. (Liqiang Jiang); writing—review and editing, L.J. (Lizhong Jiang) and L.Z. All authors have read and agreed to the published version of the manuscript.

Funding: This research was funded by the National Natural Science Foundation of China, grant number 50678025.

Institutional Review Board Statement: Not applicable.

Informed Consent Statement: Not applicable.

Data Availability Statement: The results presented in this study are available on request from the corresponding author. The data are not publicly available due to the reason that the authors are conducting further analysis on the same structural model.

Acknowledgments: We would like to thank the editors and reviewers for the comments on the manuscript.

Conflicts of Interest: The authors declare no conflict of interest.

Nomenclature

P_y	yield strength;
Δ_y	yield displacement;
φ_y	yield drift ratio;
P_{max}	maximum strength;
Δ_{max}	maximum displacement;
φ_{max}	maximum drift ratio;
μ_Δ	displacement ductility coefficient;
E	energy dissipation coefficient;
F	lateral force of the RDBs;
h	effective height of the RDBs;
l	effective length of the RDBs;
t	effective thickness of the RDBs;
Δ_w	lateral displacement of the RDBs;
Δ_{wM}	bending deformation of the RDBs;
Δ_{wV}	shearing deformation of the RDBs;
I_w	moment of inertia of the RDBs;
E_c	elastic modulus of the concrete;

G	elastic shear modulus of the concrete;
A	cross-section area of the RDBs;
K_e	elastic stiffness of the RDBs;
$K_{e\text{-theoretical}}$	theoretical elastic stiffness of the RDBs;
$K_{e\text{-experimental}}$	experimental elastic stiffness of the RDBs;
V_w	shear force of the RDBs;
N_v	shear force of a bolt due to lateral load;
n	the number of bolts;
N_m	shear force of a bolt due to bending moment;
D	distance between bolt connections;
N_{vuI}	friction force of the high strength bolts at phase I;
N_{vuII}	ultimate shear resistance of the high strength bolts at phase II;
n_f	the number of frictional faces;
μ	anti-sliding coefficient;
P	pre-tensile force of the high strength bolts;
d_0	clear diameter of the bolts;
f_v	yield shear strength of the bolts;
α	height-to-length ratio;
β	height-to-thickness ratio.

References

1. Tong, X.; Hajjar, J.F.; Schultz, A.E.; Shield, C.K. Cyclic behavior of steel frame structures with composite reinforced concrete infill walls and partially-restrained connections. *J. Constr. Steel Res.* **2005**, *61*, 531–552. [CrossRef]
2. Wallace, J.W. Behavior, design, and modeling of structural walls and coupling beams—Lessons from recent laboratory tests and earthquakes. *Int. J. Concr. Struct. Mater.* **2012**, *6*, 3–18. [CrossRef]
3. Xu, W.; Yang, X.; Wang, F. Experimental investigation on the seismic behavior of newly-developed precast reinforced concrete block masonry shear walls. *Appl. Sci.* **2018**, *8*, 1071. [CrossRef]
4. Sun, L.; Guo, H.; Liu, Y. Experimental Study on Seismic Behavior of Steel Frames with Infilled Recycled Aggregate Concrete Shear Walls. *Appl. Sci.* **2019**, *9*, 4723. [CrossRef]
5. Magliulo, G.; Ercolino, M.; Cimmino, M.; Capozzi, V.; Manfredi, G. Cyclic shear test on a dowel beam-to-column connection of precast buildings. *Earthq. Struct.* **2015**, *9*, 541–562. [CrossRef]
6. Fischinger, M.; Kante, P.; Isakovic, T. Shake-Table Response of a Coupled RC Wall with Thin T-Shaped Piers. *J. Struct. Eng.* **2017**, *143*, 04017004. [CrossRef]
7. Brunesi, E.; Peloso, S.; Pinho, R.; Nascimbene, R. Cyclic tensile testing of a three-way panel connection for precast wall-slab-wall structures. *Struct. Concr.* **2019**, *20*, 1307–1315. [CrossRef]
8. Jiang, L.; Zheng, H.; Hu, Y. Experimental seismic performance of steel- and composite steel-panel wall strengthened steel frames. *Arch. Civ. Mech. Eng.* **2017**, *17*, 520–534. [CrossRef]
9. Ju, R.-S.; Lee, H.-J.; Chen, C.-C.; Tao, C.-C. Experimental study on separating reinforced concrete infill walls from steel moment frames. *J. Constr. Steel Res.* **2012**, *71*, 119–128. [CrossRef]
10. Gribniak, V.; Misiūnaitė, I.; Rimkus, A.; Sokolov, A.; Šapalas, A. Deformations of FRP–Concrete Composite Beam: Experiment and Numerical Analysis. *Appl. Sci.* **2019**, *9*, 5164. [CrossRef]
11. Salih, R.; Zhou, F.; Abbas, N.; Mastoi, A.K. Experimental Investigation of Reinforced Concrete Beam with Openings Strengthened Using FRP Sheets under Cyclic Load. *Materials* **2020**, *13*, 3127. [CrossRef]
12. Kahn, L.F.; Hanson, R.D. Infilled Walls for Earthquake Strengthening. *J. Struct. Div.* **1979**, *105*, 283–296. [CrossRef]
13. Horii, H.; Kabele, P.; Takeuchi, S. *On the Prediction Method for the Structural Performance of Repaired/Retrofitted Structures, Fracture Mechanics in Concrete Structures: Proceedings of FRAMCOS-3*; AEDIFICATIO Publishers: Freiburg, Germany, 1998.
14. Kesner, K.E.; Bilington, S.L. Investigation of infill panels made from engineered cementitious composites for seismic strengthening and retrofit. *J. Struct. Eng.* **2015**, *131*, 1712–1720. [CrossRef]
15. Yanagisawa, M. An experimental study on structural performance of RC frame with non-structural slit wall-part 1 result of test. *Proc. Architect. Inst. Jpn.* **2008**, *48*, 261–264. (In Japanese)
16. Hong, S.-G.; Cho, B.-H.; Chung, K.-S.; Moon, J.-H. Behavior of framed modular building system with double skin steel panels. *J. Constr. Steel Res.* **2011**, *67*, 936–946. [CrossRef]
17. Jiang, L.; Zheng, H.; Liu, Y.; Yuan, X. Experimental investigation of composite steel plate deep beam infill steel frame. *Int. J. Steel Struct.* **2014**, *14*, 479–488. [CrossRef]
18. Jiang, L.; Zheng, H.; Hu, Y. Effects of configuration parameters on seismic performance of steel frames equipped with composite steel panel wall. *Struct. Des. Tall Spéc. Build.* **2018**, *27*, e1542. [CrossRef]
19. Jiang, L.; Jiang, L.; Hu, Y.; Ye, J.; Zheng, H. Seismic life-cycle cost assessment of steel frames equipped with steel panel walls. *Eng. Struct.* **2020**, *211*, 110399. [CrossRef]

20. Jiang, L.; Jiang, L.; Ye, J.; Zheng, H. Macroscopic modelling of steel frames equipped with bolt-connected reinforced concrete panel wall. *Eng. Struct.* **2020**, *213*, 110549. [CrossRef]
21. Lim, W.-Y.; Hong, S.-G. Cyclic loading tests for precast concrete cantilever walls with C-type connections. *Earthq. Struct.* **2014**, *7*, 753–777. [CrossRef]
22. Singhal, S.; Chourasia, A.; Chellappa, S.; Parashar, J. Precast reinforced concrete shear walls: State of the art review. *Struct. Concr.* **2019**, *20*, 886–898. [CrossRef]
23. Zoubek, B.; Fischinger, M.; Isaković, T. Cyclic response of hammer-head strap cladding-to-structure connections used in RC precast building. *Eng. Struct.* **2016**, *119*, 135–148. [CrossRef]
24. Jiang, H.; Qiu, H.; Sun, J.; Yang, Y. Behavior of steel–concrete composite bolted connector in precast reinforced concrete shear wall. *Adv. Struct. Eng.* **2019**, *22*, 2572–2582. [CrossRef]
25. Seifi, P.; Henry, R.S.; Ingham, J.M. In-plane cyclic testing of precast concrete wall panels with grouted metal duct base connections. *Eng. Struct.* **2019**, *184*, 85–98. [CrossRef]
26. Kinnane, O.; West, R.; Hegarty, R.O. Structural shear performance of insulated precast concrete sandwich panels with steel plate connectors. *Eng. Struct.* **2020**, *215*, 110691. [CrossRef]
27. *GB/T 50081—Standard for Test Method of Mechanical Properties on Ordinary Concrete*; Chinese Architecture & Building Press: Beijing, China, 2002.
28. *JGJ 101—Specification of Testing Methods for Earthquake Resistant Building*; Chinese Architecture & Building Press: Beijing, China, 1997.
29. Mansour, M.; El-Ariss, B.; El-Maaddawy, T. Effect of Support Conditions on Performance of Continuous Reinforced Concrete Deep Beams. *Buildings* **2020**, *10*, 212. [CrossRef]

Article

3D FEA of Infilled RC Framed Structures Protected by Seismic Joints and FRP Jackets

Theodoros Rousakis *, Vachan Vanian, Theodora Fanaradelli and Evgenia Anagnostou

Department of Civil Engineering, Democritus University of Thrace, 67100 Xanthi, Greece; vvanian@civil.duth.gr (V.V.); tfanarad@civil.duth.gr (T.F.); eanagno@civil.duth.gr (E.A.)
* Correspondence: trousak@civil.duth.gr

Abstract: This study focused on characteristic cases of recently tested real-scale RC framed wall infilled structures with innovative seismic protection through polyurethane joints (PUFJ) or polyurethane-impregnated fiber grids (FRPU). The frames revealed a highly ductile response while preventing infill collapse. Herein, suitable 3D pseudo-dynamic FE models were developed in order to reproduce the experimental results. The advanced Explicit Dynamics framework may help reveal the unique features of the considered interventions. Externally applied double-sided FRPU jackets on OrthoBlock infills may maintain an adequate bond with the surrounding RC frame as well as with the brick infill substrate at up to a 3.6% drift. In a weak four-column RC structure, the OrthoBlock infills with PUFJ seismic joints may increase the initial stiffness remarkably, increase the base shear by three times (compared with the bare structure) and maintain a high horizontal drift of 3.7%. After this phase, the structure may receive FRPU retrofitting, reveal the redistribution of stress over broad infill regions, including predamaged parts, and still develop a higher initial stiffness and base shear (compared with the bare RC). The realization of a desirable ductile behavior of infilled frames through PUFJ of only 20 mm thickness, as well as through FRPU jacketing, may remarkably broaden the alternatives in seismic protection against the collapse of structures.

Keywords: RC frames; brick infills; prior damage; finite element analyses; seismic joint; FRP

Citation: Rousakis, T.; Vanian, V.; Fanaradelli, T.; Anagnostou, E. 3D FEA of Infilled RC Framed Structures Protected by Seismic Joints and FRP Jackets. *Appl. Sci.* **2021**, *11*, 6403. https://doi.org/10.3390/app11146403

Academic Editor: Pier Paolo Rossi

Received: 20 June 2021
Accepted: 8 July 2021
Published: 11 July 2021

Publisher's Note: MDPI stays neutral with regard to jurisdictional claims in published maps and institutional affiliations.

Copyright: © 2021 by the authors. Licensee MDPI, Basel, Switzerland. This article is an open access article distributed under the terms and conditions of the Creative Commons Attribution (CC BY) license (https://creativecommons.org/licenses/by/4.0/).

1. Introduction

The behavior of infilled reinforced concrete (RC) frames under in-plane and out-of-plane loading is an open issue as there is a variability of existing RC frames designed and constructed according to different design codes over the years, with or without modern seismic-resistant provisions or suitable reinforcement detailing. Further, the infills may include solid or hollow clay bricks (usually with a horizontal direction of the holes of the bricks), concrete solid or hollow blocks and several other types of infill wall units. Therefore, there is a high variation in the modulus of elasticity, strength and modes of failure of the infills when interacting with the RC frames. In some cases, the effects of the infills on the RC frames may be detrimental and lead to building collapses during severe earthquakes. On the other hand, collapsed infills alone may lead to human injuries or loss. Numerous experimental campaigns have addressed the variable behavior of infilled RC frames. The high scatter in the prediction of several critical parameters—especially the ones related to displacement ductility—reveals the complexity and uncertainty involved (see the studies by [1–7], among others). Furthermore, the out-of-plane response of the infills is equally crucial to predict (see, e.g., [4,8–10], among others). The abovementioned studies suggest that, in most cases, common clay brick infills reveal an accumulation of severe damage for the lateral drift of the infilled RC frame that ranges between 0.5 and 1.5%. Hence, they lead to reduced displacement ductility levels with respect to the case of a bare RC frame, despite the fact they increase the elastic stiffness and maximum base shear capacity of the infilled frame.

Several investigations at the building level, following rigorous assessment approaches, suggest that, in most cases, brick wall infills may enhance the performance of RC frames at low-intensity earthquake excitations (increased base shear capacity). Further, at higher intensities, shear-related failures of columns or out-of-plane collapse of damaged infills (especially if they have openings) may lead to premature exhaustion of buildings' structural capacity (i.e., at lower drifts) with respect to the bare structure at different serviceability or ultimate limit states (see, e.g., dynamic experiments in [11] or analytical investigations in [12–14], among others). Wall infills are very crucial to take into account, especially in the damage assessment at serviceability limit states [14]. In any case, an increased uncertainty in building behavior should be considered as the dynamic excitation intensity is increased and/or multiple structural deficiencies are identified.

Innovative repair and strengthening solutions for RC frames with infill walls involved externally applied sprayable ductile fiber-reinforced cementitious composites [15], or fiber-reinforced polymer sheets [3,16,17], among others. Koutas et al. [18] tested RC frames infilled with brick walls, strengthened with textile-reinforced mortar. The RC frames were of old-type detailing (with inherent deficiencies). In most of the cases of such RC frames subjected to a constant vertical load and cyclic horizontal loading, the infill walls revealed limited ductility before their retrofitting. Even after their strengthening, the structural behavior and failure mechanism were similar to the as-built infilled frames. However, the retrofitted infilled RC frames exhibited a significant increase in the shear resistance and dissipated energy [19,20]. Again, usually, the accumulation of severe damage occurs for a lateral drift range between 0.5 and 1.5%. Such values for the infill performance limit the potential of the retrofitted RC frames [18]. This early damage accumulation may pose a threat, in some cases, to the targeted ductile behavior of strengthened RC frames with infill walls. For example, detrimental effects may be introduced by the highly uncertain and undesirable infill wall–RC frame interaction. Such effects may include premature shear-related failure of the columns in contact with the infills, or partial collapse of the infills, among others, or trigger an undesirable soft-story mechanism [21] and even whole-building collapse (if seismically induced overloads or out-of-plane infill collapses occur).

In general, the deterioration of the bond between commonly used cement-based materials at the boundary between the RC frame and the infill, as well as the severe damage accumulation within the fragile brick infills (subjected, mainly in plane, to variably imposed diagonal compressive deformation and suffering diagonal cracks or corner crushes), led to the development of different seismic protection solutions. Marinkovic and Butenweg [22,23] proposed a system to protect the infilled frame against in-plane and out-of-plane loads. Specially designed, three-sided elastomer joints at the infill–RC boundary interface were placed with a total thickness of 62.5 mm (25 mm plus 37.5 mm). A bottom elastomer was used as well. The composite frame sustained in-plane and out-of-plane loading, reaching a 3% ultimate drift for in-plane loading. Morandi et al. [6] proposed sliding joints inserted in the masonry at different levels with 30 mm thickness ($3 \times 30 = 90$ mm) and deformable joints at the wall–frame interface (25 mm thickness). Similarly, the seismic protection was remarkable up to a 3% frame drift.

Real-scale shake table experiments are of great significance for the validation of novel seismic protection systems. INMASPOL ("INfills and MASonry structures protected by deformable POLyurethanes in seismic areas"—INMASPOL SERA Horizon 2020 project) involved a real-scale RC frame building subjected to multiple ground excitations on a shake table. The excitation was suitable for the validation of the out-of-plane performance of the OrthoBlock brick wall infills of the structure, protected by innovative seismic joints, as well as for their in-plane response. The joints were made of highly deformable polyurethane (polyurethane flexible joints, PUFJ) of 20 mm thickness. The structure was driven to a high lateral drift of up to 3.7% without base shear degradation and no brick infill collapse. Then, the brick infills were retrofitted with an innovative glass grid, impregnated with high-deformability polyurethane (fiber-reinforced polyurethane, FRPU) and retested (for the experimental validation, see [20,24]). The techniques of seismic joints and externally bonded fiber grid jackets have already been validated separately in real-scale RC frames subjected to pseudo-static cycles of horizontal top displacements [25]. Different frames with boundary polymer seismic joints on three or four sides were tested with a 20 mm thickness. Further, an infilled frame without seismic joints was retrofitted with a fiber grid jacket impregnated with highly deformable polyurethane. In all cases of the tested frames, the ductility of the composite frames was remarkable, and the corresponding drift ranged from 3.5% to 4.4%. However, during experimental campaigns involving large-scale frames, only a few parameters can be investigated due to the high demand of resources and time. Three-dimensional (3D) finite element (FE) modeling and analyses may help address the effects of critical design parameters (see [26–34], among others).

This paper developed suitable 3D pseudo-dynamic FE models for a real-scale brick infilled frame receiving external fiber grid retrofitting. Further, this study developed suitable 3D FE models of the real-scale RC INMASPOL frame building with innovative polymer seismic joints at the boundary of the infills. The analyses cover the first phase of successive shake table tests as well as the second phase of shake table tests, after retrofitting of the damaged structure with externally bonded fiber grids. The structure sustained a series of scaled modified real earthquake recordings of increasing maximum ground acceleration during tests, in accordance with the dynamic pushover approach. The analyses presented herein concern the inelastic pseudo-dynamic pushover response of the structure through imposed top displacements. Bare structure analysis is also included for comparative investigations. The 3D FE analyses' results compare well with the experimental ones and may provide a unique insight on the variable interaction of the brick infill–RC frame with seismic polymer joints and retrofitting through FRP jacketing. The results help to elucidate the unique features of the interventions that enable the investigated RC framed structures to reveal high top displacement ductility. Suitable combined retrofits are also examined.

2. Brief Presentation of Existing Experimental Results
2.1. Seismic Protection of Infilled RC Frames with FRP Jackets

Herein, the experimental results of the real-scale RC frame A2R from the study by [25,35] were considered for the analytical elaborations. The RC frame had plane dimensions of a 270 cm-long beam with extensions of 30 cm on both sides and a height of 245 cm (245 cm to the level of the diaphragm). The columns had cross-section dimensions of 25 × 25 cm, 8Φ16 longitudinal continuous rebars (without lap splices) and Φ10/100 mm closed stirrups (one peripheral and one rhombic per 100 mm). The columns had two top extensions of 30 cm, as well as a common foundation beam with dimensions of 30 × 40 × 355 cm. The top beam had cross-section dimensions of 25 × 25 cm and was reinforced with 8Φ14 longitudinal rebars and Φ10/100 mm closed stirrups (peripheral). The dimensions of the RC frame and the detailing of the internal steel reinforcement are presented in Figure 1a and in Table 1. Specimen A2R included an infill wall with special thermal insulating hollow clay units, KEBE OrthoBlocks K100 (KEBE, Greece), with dimensions of 100/240/250 mm and a weight of about 100 kg/m^2 (infill wall with vertical holes in the bricks), built with mortar. The mortar for the building of the walls (with nominal strength class M10) formed 3 mm-thick layers at the head or bed joints as well as at the infill–RC frame interface (no seismic joint). The infill was retrofitted with an external glass grid (of type SikaWrap 350 G Grid, with a real weight of 360 g/m^2) impregnated with highly deformable PS (Sika) polymer (fiber-reinforced polyurethane jacket, FRPU, Figure 1b). The corresponding properties are cited in Table 2.

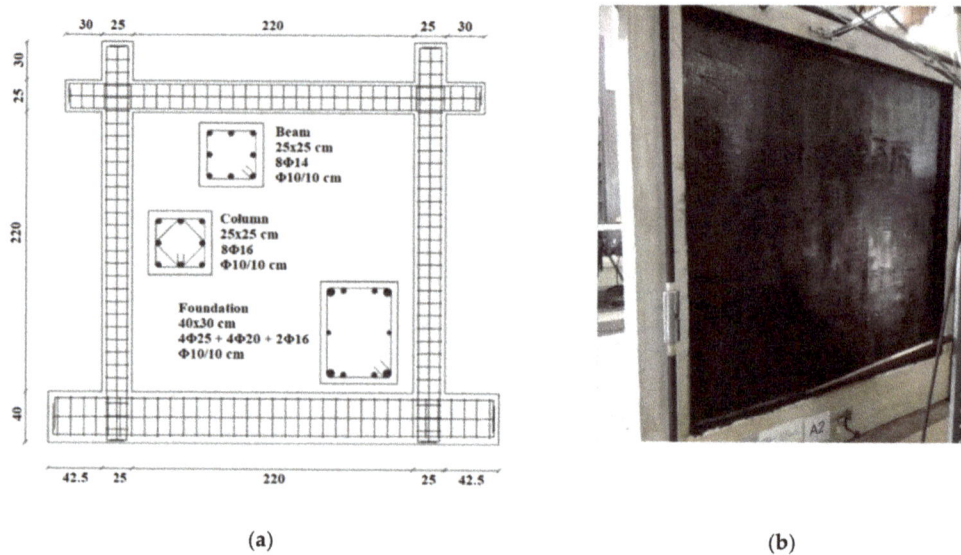

Figure 1. (a) Dimensions of the RC frame and detailing of the internal steel reinforcement and (b) specimen A2R with FRPU jacket on the infill surfaces.

Table 1. The dimensions of the RC frame and the detailing of the internal steel reinforcement.

	b (mm)	h (mm)	H (mm)	Longitudinal Steel Bars	Stirrups
Columns	250	250	2450 [1]	8Φ16	Φ10/100
Foundation beam	300	400	3550	4Φ25 + 4Φ20 + 2Φ16	Φ10/100
Top beam	250	250	3550	8Φ14	Φ10/100

[1] To the level of the diaphragm.

Table 2. Material properties of the infilled RC structure.

Material	Density (kg/m^3)	Structural Element	Property	Value
Concrete	2380	Foundation	Compressive strength [1]	34.1 MPa
		Columns		27.1 MPa
		Slab		34.2 MPa
Steel	7850	Reinforcement	Characteristic yield strength	500 MPa
Clay	1800	OrthoBlock	Weight	100 kg/m^2
Glass Fiber	2600	In FRPU jacket	Elastic modulus	80 GPa
			Strength	2600 MPa
			Ultimate elongation	4%
Polyurethane	910	PUFJ	Elastic modulus	4 MPa
			Strength	1.4 MPa
			Ultimate elongation	110%
		In FRPU jacket	Elastic modulus	16 MPa
			Strength	2.5 MPa
			Ultimate elongation	40%

[1] Cubic strength at 28 days.

The two columns of the frame received a constant concentric axial load of 375 kN each. The horizontal actuator imposed successive pseudo-static cycles of gradually increased displacements at the top beam. More details on the materials used, the experimental setup, the instrumentation and test results of the A2R can be found in [25,35].

2.2. Seismic Protection of Infilled RC Structure with Highly Deformable Joints

2.2.1. Material Properties

A real-scale fully symmetrical one-story RC structure, consisting of masonry wall infilled RC frames (4 columns and 4 beams with a top slab), was subjected to shake table tests (see also Figure 2). The structure was designed according to current Eurocodes 2 and 8 [36,37]. The plane dimension of the structure was 3.8 × 3.8 m, and it was 3.3 m high. The material properties for the concrete, the reinforcing steel bars, the masonry (made of OrthoBlocks identical to the ones in A2R), the glass fiber grid and the PS polymer in the FRPU composite (FRPU identical to the one in A2R) are cited in Table 2. The structure was protected with 20 mm-thick polyurethane flexible joints (PUFJ) at the infill–RC frame interfaces made of PM polymer (mechanical properties are cited in Table 2). More details on the material properties can be found in [24].

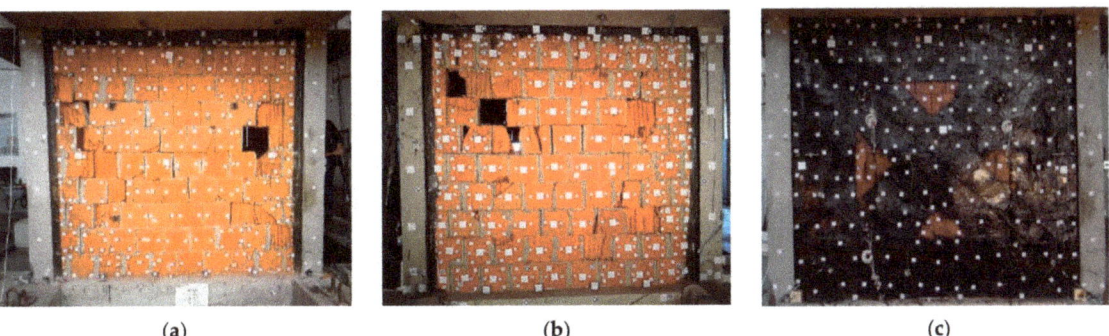

Figure 2. Structure significantly damaged after PHASE 1: (**a**) north side—type B infill; (**b**) south side—type B infill; and (**c**) FRPU-strengthened infill of type B after PHASE 1.

2.2.2. Structure Detailing

The frames had floor plan dimensions of 2.7 × 2.7 m—based on the columns' position—and a height of 2.5 m (to the top of the slab). The dimensions and longitudinal and transverse reinforcement detailing are cited in Tables 3 and 4, respectively. The beam was hidden inside the slab. The 4 columns supported, on their top, an RC slab that was 200 mm thick. The slab carried carefully anchored steel ingots as additional masses (total 7200 kg) for the tested structure. The slab had a central square opening for access to the inside of the structure. The welded mesh top and bottom reinforcements of the slab are cited in Table 4. An additional slab reinforcement was placed at the perimeter edges and at the edges of the opening. The structure had a special foundation, with holes and hooks for attaching it to the shake table and for lifting and manipulating the structure (Figure 2a).

The infills in the building had a height of 2.3 m. Therefore, the ratio of the vertical (or horizontal) interface joint thickness to the infill wall height was 20/2300 = 0.0087. The infills were 100 mm thick, with holes in the vertical direction (identical to the ones in A2R). Out of the total of four infills, the two pairs of parallel walls had different seismic joint configurations to experimentally investigate the implementation of PUFJ in existing and new buildings. The two parallel infills of type B (Figure 2b) were constructed directly on the RC foundation with a 3 mm-thick mortar bed joint, while there was a PUFJ of 20 mm thickness, produced in situ through injection of polyurethane Sika PM. The polymer was injected between the infill and the RC columns, and between the infill and the RC slab (three-sided joint). The adhesiveness of Sika PM to brick substrates was determined equal to 0.63 MPa by a pull-off test in [38], followed by an adhesive brick/polymer failure mode. The two parallel infills of type C were constructed on a prefabricated 20 mm-thick PUFJ bottom joint (bonded to the RC foundation beam), while identical prefabricated joints were bonded at the sides and top of the infill (four-sided joint). The adhesiveness of Sika PS to brick substrates (used in FRPU) was determined by a pull-off test in [38] and was equal to 1.48 MPa, followed by an adhesive brick/polymer failure mode. More details on the detailing of the structure can be found in [24].

Table 3. Dimensions of structure and of different parts.

	Height (mm)	Length (mm)	Width (mm)
Infill	2300	2300	100
Brick Unit	240	250	100
Bed Joint	-	-	3 (thickness)
Head Joint	-	-	3 (thickness)
Wall B FRPU	-	-	20 (thickness) 3 sides
Wall C FRPU	-	-	20 (thickness) 4 sides
Beam	200	2300	200
Columns	2500	200	200
Beam–Column Joint	200	200	200

Table 4. Reinforcement detailing.

	Columns	Beam
Longitudinal	8Φ10	8Φ10
Transverse peripheral	Φ8/50	Φ8/50
Transverse rhombic	Φ8/50	-
	Additional Information	
Clear concrete cover	42 mm	
Slab thickness/Reinforcement	200 mm/Q503	
	Total 7200 kg mass anchored on the top of the slab	

2.2.3. Loading Phases

Only PHASE 1 and PHASE 2 were considered for the numerical analysis. The original position of the model was such that both walls of type B were loaded in plane, and both walls of type C were loaded out of plane. In this position (PHASE 1), the structure suffered at least 10 dynamic earthquake runs of increasing maximum acceleration. The structure sustained a top slab relative displacement of 88.9 mm, corresponding to a 3.7% drift during the last run of PHASE 1. The damage to the infills of type B after the last run is depicted in Figure 2a,b. However, it should be noted that during that run, both brick infills of type B (tested in plane), protected by the PUFJ seismic joint, did not exhibit any significant damage up to a 2.5% drift of the structure. Both brick infills of type C (tested out of plane) had no significant damage after the last run of PHASE 1. After the first PHASE, the damaged brick infills of type B received FRPU external strengthening bonded on both sides of the damaged wall using a flexible adhesive of type Sika PS without any special treatment of the infill face and no crack repair (see Figure 2c). Then, three additional runs were performed during PHASE 2 up to a 1.62% drift of the structure. Further excitation during PHASE 2 was avoided in order to secure the integrity of the structure for PHASE 3 of the experiments (not presented herein) as the RC columns had already suffered significant damage. No additional significant damage to the infill walls of type B or to the infill walls of type C or any damage to the FRPU retrofit was observed after PHASE 2.

3. 3D FE Models

The numerical models were developed, and the material parameters were suitably calibrated based on the experimental parameters, using the Explicit Dynamics framework of the ANSYS Workbench [39] (see also [33,34]).

3.1. Geometry and Elements

RC members were modeled with solid elements for the concrete and linear elements for the reinforcement. Longitudinal steel bars and steel stirrups for columns, beams and slabs were modeled using the concept of "reinforcement" body interaction, not permitting separate reinforcement behavior. The parts for PUFJs were solid bodies with a 20 mm thickness, surrounding the 3 sides of the brick infill–RC frame interface for the type B infill and the 4 sides for the type C infill. Due to the small thickness of FRPUs, they were modeled as surface bodies to eliminate numerical instabilities and enhance the mesh quality. The body interaction between different components was considered as "bonded". The model used the 8-node reduced integration hexahedral elements. These elements are suitable for dynamic applications including large deformations, large strains, large rotations and complex contact conditions in Explicit Dynamics. Linear 4-node tetrahedron elements are available for use in Explicit Dynamics analysis whenever hexahedral element meshing fails. Figure 3 shows the geometry of the FE models and mesh density.

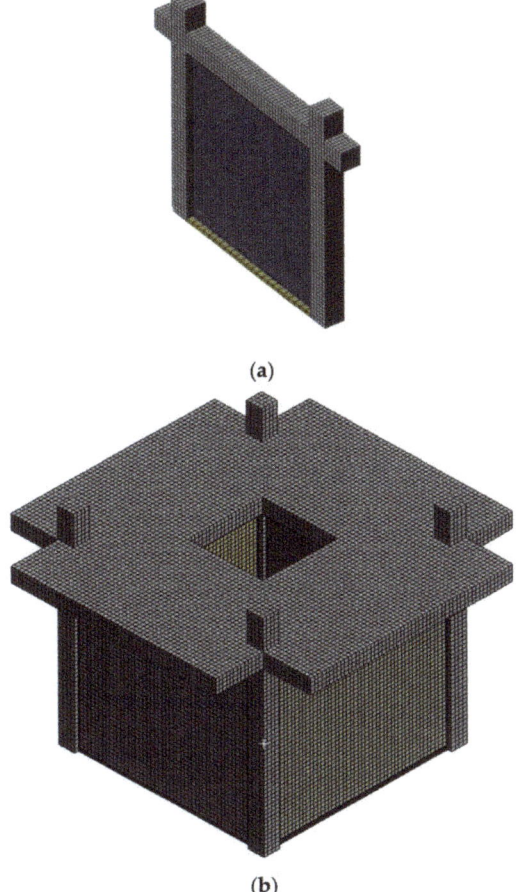

Figure 3. Geometry and mesh of the (**a**) A2R frame and of (**b**) 4-column model of INMASPOL structure.

3.2. Material Models

The properties of the materials satisfied the experimental properties. PUFJs and glass FRPU were modeled as elastic materials. Concrete was of type RHT concrete, an advanced plasticity model for brittle materials developed by Riedel et al. [40–42]. The failure function for the RHT constitutive model, its fracture surface and the strain softening and damage relations of the model are gathered in Table 5.

Table 5. RHT concrete model.

RHT Concrete Model Brief Presentation		
Equation	No	Definition
$f(P, \sigma_{eq}, \theta, \dot{\varepsilon}) = \sigma_{eq} - Y_{TXC(P)} \times F_{CAP(P)} \times R_{3(\theta)} \times (F)_{RATE(i)}$	(1)	Generalized failure surface
$Y_{TXC} = f'_c \left[A_{Fail} \left(P^* - P^*_{spall} F_{RATE} \right)^{N_{Fail}} \right]$	(2)	Fracture surface
$F_{RATE} = \begin{cases} 1 + \left(\frac{\dot{\varepsilon}}{\dot{\varepsilon}_0}\right)^{\alpha} & \text{for } P > \frac{1}{3} f_c \text{ (compression)} \\ 1 + \left(\frac{\dot{\varepsilon}}{\dot{\varepsilon}_0}\right)^{\delta} & \text{for } P < \frac{1}{3} f_t \text{ (tension)} \end{cases}$ where $\dot{\varepsilon}_0 = 3e - 6$ in tension and $30e - 6$ in compression.	(3)	Rate-dependent enhancement factor
$R_3 = \frac{2(1-Q_2^2)\cos\theta + (2Q_2-1)\sqrt{4(1-Q_2^2)\cos^2\theta - 4Q_2}}{4(1-Q_2^2)\cos^2\theta + (1-2Q_2)^2}$ $\cos(3\theta) = \frac{3\sqrt{3} J_3}{2 \sqrt[3]{J_2}}$ and $Q_2 = Q_{2.0} + BQ.P^*$ with $0.5 < Q_2 < 1$, $BQ = 0.0105$	(4)	Third invariant dependence term
$Y^* = Y_{elastic} + \frac{\varepsilon_{pl}}{\varepsilon_{pl(pre-softening)}} (Y_{fail} - Y_{elastic})$ where $\varepsilon_{pl(pre-softening)} = \frac{Y_{fail} - Y_{el}}{3G} \left(\frac{G_{elastic}}{G_{elastic} - G_{plastic}} \right)$	(5)	Bilinear strain hardening function for the case of uniaxial compression
$D = \sum \frac{\Delta \varepsilon_{pl}}{\varepsilon_p^{failure}}$ where $\varepsilon_p^{failure} = D_1 \left(P^* - P^*_{spall} \right)^{D_2}$	(6)	Damage is assumed to accumulate due to inelastic deviatoric straining (shear-induced cracking)
$Y^*_{fractured} = (1-D) Y^*_{failure} + D Y^*_{residual}$ where $Y^*_{residual} = \text{Min} \left[B(P^*)^M, Y_{TXC} \times SFMAX \right]$	(7)	Strain softening
$G_{fractured} = (1-D) G_{elastic} + D G_{residual}$	(8)	The shear modulus reduction
$P = \max[D \times P_{min}, P(\rho, \theta)]$	(9)	Maximum tensile pressure in the material is limited

σ_{eq} is the uniaxial compressive strength; $Y_{TXC(p)}$ is the fracture surface; $F_{CAP(P)}$ is a dimensionless cap function which activates the elastic strength surface within the RHT material model at high pressures; $R_{3(\theta)}$ is the third invariant dependence term; $(F)_{RATE(i)}$ is the strain rate effect represented through fracture strength with plastic strain rate; f'_c is the cylindrical compressive strength; A_{Fail}, N_{Fail} are user-defined parameters; P^* is pressure normalized to f'_c; P^*_{spall} is the normalized hydrodynamic tensile limit; F_{RATE} is the rate-dependent enhancement factor in Equation (3); f_c and f_t are static uniaxial compressive and tensile strengths; $\dot{\varepsilon}$, $\dot{\varepsilon}_0$ are the static strain rates; α is the compressive strain rate factor; δ is the tensile strain rate factor; $Y_{elastic}$ is the initial elastic surface; Y_{fail} is the failure surface; G is the shear modulus; η_{pl} is the plastic strain before the failure strength; $\eta_{pl(pre-softening)}$ is the total plastic strain, and it can be determined by the secant modulus between the elastic limit surface and the failure surface; D_1 and D_2 are material constants used to describe the effective strain to fracture as a function of pressure. Damage accumulation can have two effects in the model.

3.3. Boundary Conditions and Loads

The supports of the columns and infill had a fixed boundary condition. Pseudo-dynamic analyses of controlled monotonic displacement, imposed at the top of the structure (beam in A2R or at the top slab), were conducted to obtain the inelastic behavior of the A2R frame or of the structure for PHASE 1 and PHASE 2. The pseudo-dynamic pushover approach followed herein utilizes the advanced Explicit Dynamics features and minimizes the analysis time for demanding cyclic or dynamic tests (see also [33,34]).

4. 3D FE Analyses Results

4.1. A2R Frame

Figure 4 shows, in black and black-gray curves, the experimental envelope base shear–top displacement curves of A2R for (+) and (−) push, respectively, based on the results reported in [25,35]. The external application of the FRPU jackets on the OrthoBlock infill resulted in a remarkable top displacement ductility of the infilled frame A2R, with a negligible base shear reduction up to a 3.6% drift for push (+) and (−). Table 1 cites the experimental and numerical results of the A2R frame. The parameter used for the comparisons in this study is the absolute divergence (AD), also known as the absolute error (AE), and it is defined in Equation (10).

$$AD = \frac{|a_{anal.} - a_{exp.}|}{a_{exp.}} \qquad (10)$$

where a is the compared value (analytical or experimental).

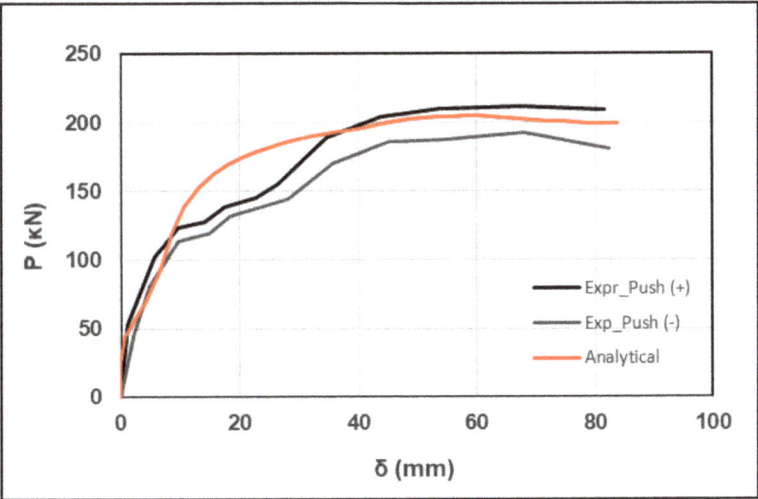

Figure 4. Analytical base shear force–displacement curve for A2R, compared against the experimental ones.

The pseudo-dynamic analytical FE red curve coincides with the experimental ones up to 10 mm of top displacement, and then it falls in between the experimental curves for top displacement higher than 35 mm. The analytical maximum base shear is 204.7 kN (between the experimental push (−) and push (+)) at a top displacement of 59.1 mm, which is 9.1% lower than the experimental one. Satisfactory is the prediction of the ultimate base shear as well (Table 6). The divergence at the range of 10–35 mm top displacement may be attributed to the relatively rough modeling of the infill wall as a 3D panel.

Table 6. Comparative results for A2R frame.

	A2R		
	Exp.	Anal.	AD (%)
Pmax (+) (kN)	214.1	204.7	4.4
δPmax (+) (mm)	65.0	59.1	9.1
Pmax (−) (kN)	190.6	204.7	7.4
δPmax (−) (mm)	65.0	59.1	9.1
Pu (+) (kN)	210.4	199.1	5.4
δPu (+) (mm)	81.4	81.4	-
Pu (−) (kN)	178.7	199.1	11.4
δPu (−) (mm)	81.4	81.4	-

4.2. INMASPOL Structure

Based on the FE developments in RC frames and on the characteristics of the INMASPOL structure, three numerical models were developed and analyzed. The bare RC structure with no infills (model 1) was analyzed for comparison reasons. The response of analytical model 1 was compared against the response of the brick wall infilled structure with the innovative PUFJs (model 2, INMASPOL structure), as tested in PHASE 1. Then, the damaged brick wall infilled structure received external double-sided glass FRPU jacketing (model 3, corresponding to PHASE 2 shake table tests). The damage in model 3 was implemented by removing the continuous material from the infill walls, based on the observed infill type B wall damage after the end of the experimental PHASE 1 (see Figure 2a). That is, in model 3 the infill had selected 3D finite elements removed to simulate the locally damaged parts of the infill. Such approaches can also be found in the analytical investigation of the residual capacity of buildings with damaged infills or RC members in [43].

Figure 5a presents the diagrams of the experimental top slab accelerations (equivalent to the base shear divided by the excited mass of the structure) versus the relative drift of the top slab (relative top slab–column bottom displacement divided by the respective vertical distance) for each dynamic excitation of the structure on the shake table for PHASE 1, while Figure 5b shows these for PHASE 2. Herein, it should be noted that the results are reported in acceleration–drift terms, as in [20], and can be easily transformed to the equivalent base shear–top displacement terms.

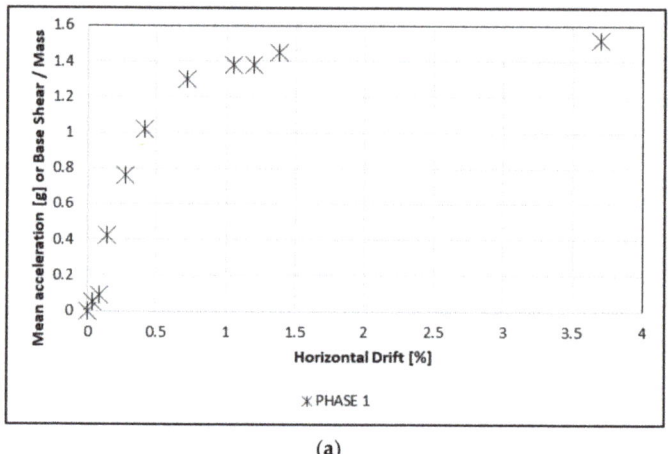

(a)

Figure 5. Cont.

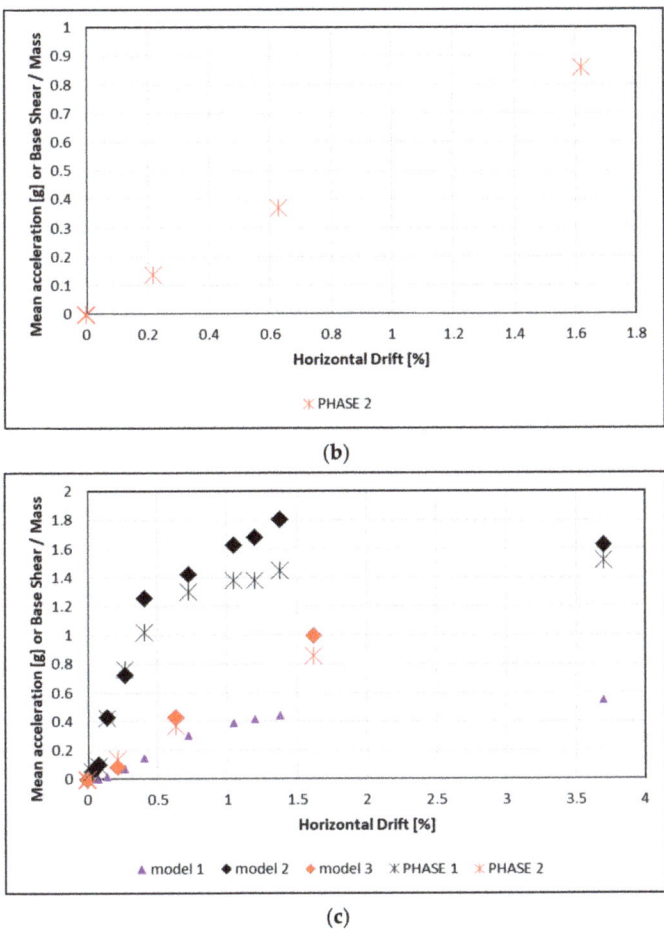

Figure 5. Mean slab acceleration (or base shear/mass) versus horizontal drift diagram (**a**) for experimental PHASE 1; (**b**) for experimental PHASE 3; and (**c**) for numerical model 1, model 2 and model 3, against the experimental results of PHASE 1 and PHASE 2.

Figure 5c shows the results obtained from the pushover analyses in terms of top slab acceleration (base shear divided by the excited mass of the structure) versus the drift of the structure (relative top slab–column bottom displacement divided by the respective vertical distance).

The black star marks in Figure 5a,c show the experimental results from the shake table tests for the INMASPOL real-scale RC structure with OrthoBlock infills and PUFJ highly deformable seismic joints. The structure exhibits a ductile top acceleration–drift behavior which is equivalent to a ductile base shear–drift behavior. It reaches a 3.7% drift without degradation of the top acceleration (base shear). Figure 2a suggests there is no collapse of the infills tested in plane or out of plane after a 3.7% drift. Further, Figure 2c shows that the damage in infills of type B could be repaired with emergency FRPU external jackets without any special treatment of the infill (more details can be found in [20,24]).

The red star marks in Figure 5b,c show the experimental results from the shake table tests for the INMASPOL real-scale RC structure with OrthoBlock infills and PUFJ highly deformable seismic joints after the repair with FRPU (PHASE 2). The structure exhibits a rather linear elastic-like top acceleration–drift behavior which is equivalent to a linear elastic base shear–drift behavior. During retesting of the retrofitted structure, the achieved drift was 1.62% without degradation of the top acceleration (base shear). The FRPU maintained the integrity of the damaged infills, and collapse was also avoided in PHASE 2 (more details can be found in [20]). Interestingly, the stiffness of the repaired structure seems relatively increased, owing to the effects of the externally bonded FRPU jackets (more details can be found in [20]).

The abovementioned dynamic experimental results validate the advanced seismic protection provided by the 20 mm-thick PUFJ to the infill and the RC frame. An undesirable interaction is avoided, and the base shear drift behavior is remarkably ductile, while the potential of the structure is higher (tests at higher accelerations were avoided to prevent the structure from collapse and continue with PHASES 2 and 3). Similarly, the results in PHASE 2 validate the advanced retrofit of the infills by the FRPU jackets, and the potential of the structure is again higher.

Figure 5c also includes the FE analysis results for model 1 (triangles), for model 2 (black rhomb) and for model 3 (red rhomb). The developed models 2 and 3 compare well with the experimental results of the PHASE 1 and PHASE 2 tests. For model 2, the INMASPOL structure, in PHASE 1, at the initial stage of the acceleration–drift diagram, the behavior of the structure is elastic, and the experimental (black star marks) and analytical curves (black rhomb marks) coincide. After the stage of yielding of the steel bars of the concrete columns, the analytical curves start to overestimate the experimental points. The analytical acceleration value at a 3.7% drift slightly overestimates the corresponding experimental one. Overestimation of the experimental values by the analyses may be attributed to the higher damage accumulation in the RC columns and in the brick infills of type B during the experiments (the structure suffered a total of 10 earthquake excitations of increasing severity).

At PHASE 2, the analytical model was updated with the observed damage of the brick infill and with the retrofit with the FRPU (see the mesh in Figure 6a). The analytical values of model 3 (red rhomb) slightly overestimate the experimental ones (red star). The model reproduces the FRPU repair effects despite the local brick damage. Figure 6c reveals the engagement of broad infill regions (predamaged regions included) through the interaction with the highly deformable FRPU jacket and the RC frame.

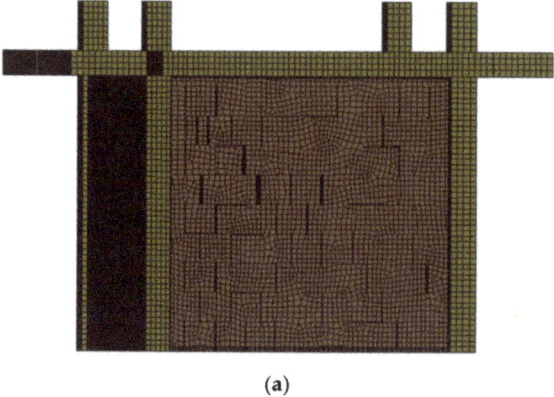

(a)

Figure 6. *Cont.*

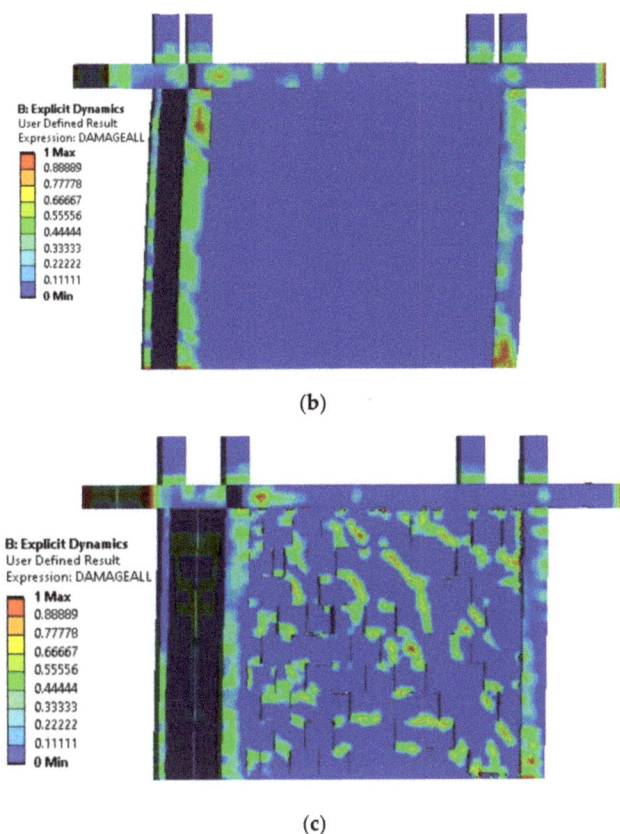

Figure 6. (**a**) Mesh of the RC structure, of the damaged brick infill and of the external FRPU retrofit; (**b**) model 2 at the end of PHASE 1; and (**c**) model 3 at the end of PHASE 2.

Furthermore, the advanced FE analyses allow for a comparative investigation of the analytical response of the bare RC structure (model 1) against analytical models 2 and 3. The analyses in Figure 6 show that the bare structure sustains a maximum acceleration of around 0.55 g, while model 2's and model 3's maximum accelerations occurred at the end of PHASE 1 at around 1.63 g and 0.99 g, respectively. Therefore, the infilled structure with seismic protection developed a three times higher maximum acceleration (or base shear) than the bare structure at PHASE 1. The corresponding increase in the initial stiffness was similarly remarkable while maintaining the high displacement ductility (3.7% horizontal drift). At PHASE 2, the retrofitted structure still revealed a higher initial stiffness when compared with the bare structure, while it could develop, at least, around double the maximum acceleration (base shear). The area under the acceleration–drift curve (denoting a measure of the energy dissipation potential) is higher for model 2 (RC structure with infills protected with PUFJ joints) than for model 1 (bare RC structure). Therefore, in the case that the OrthoBlock infill with PUFJ protection is considered in the intervention for a weak bare RC frame, it can ensure a remarkable increase in the elastic stiffness and base shear while providing a desirable ductile P–δ behavior, with a drift as high as 3.7%.

4.3. FRPU Retrofitted INMASPOL Structure with the Addition of Steel Dowels

The developed 3D FE models allow for the analytical investigation of the FRPU retrofitted INMASPOL structure with additional mild intervention with steel dowels. Only five steel dowels were used at the top infill–RC beam boundary interface in order to increase the initial stiffness of the damaged structure, based on the approach proposed by Facconi and Minelli [44]. The steel connectors had a diameter of 16 mm and a length of 90 mm and consisted of a smooth-surface round bar made of structural steel having a nominal yield strength of 235 MPa. The distance between the five dowels was 410 mm (horizontal direction), while the middle dowel was placed at the middle of the beam span (no dowels in the vertical interfaces). This detailing avoids the position of the dowels within the corner beam–column regions that will increase their efficiency but will also increase the damage within the brick infill. As it is depicted in model 3, FRPU is efficient enough to maintain the integrity of the corner regions. The anchorage of the steel dowels in the reinforced concrete beam is 40 mm, and in the masonry wall, it is 50 mm. To increase the cohesiveness of the connection and to prevent early-stage damage, the area around dowels in the masonry was modeled with thixotropic mortar, as in [44].

Indeed, Figure 7 shows that at a 0.22% drift, the developed acceleration is 0.135 g for the revised model 3DOW with the additional steel dowels. Model 3 without dowels developed only 0.08 g acceleration (Figure 5c). The increase in the developed acceleration is 68% and reveals the earlier engagement of the infill. After a 0.6% drift, the accelerations for the two models almost coincide. Figure 8 reveals that the proposed dowel detailing leads to damage accumulation around the region within the brick infill in which the dowels are anchored. A vertical mode in controlled damage accumulation within the infill is denoted that coincides with the position of the five dowels. However, the interaction with the FRPU maintains the integrity of the infill. Based on the developed 3D FE models, the structure may receive a suitable combination of retrofits that could manipulate (in this case increase) the engagement of the infill and, correspondingly, the initial stiffness, if required. Further, the damage accumulation may be controlled in order not to lead to degradation in the performance of the structure.

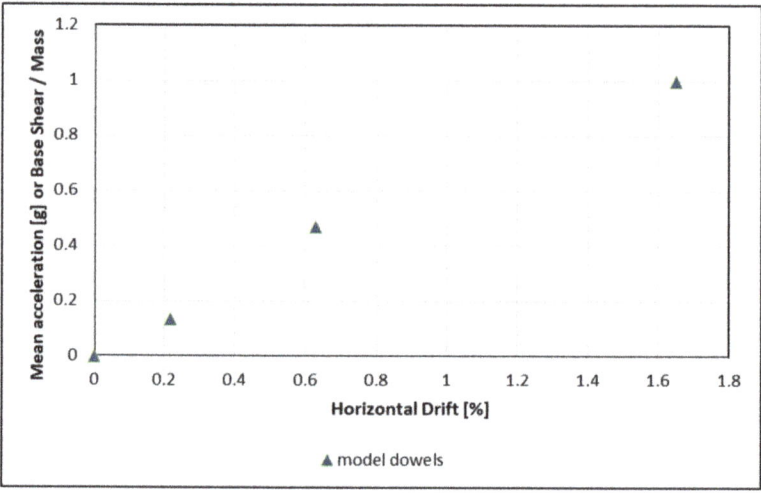

Figure 7. Mean slab acceleration (or base shear/mass) versus horizontal drift diagram for the model 3DOW with additional steel dowels.

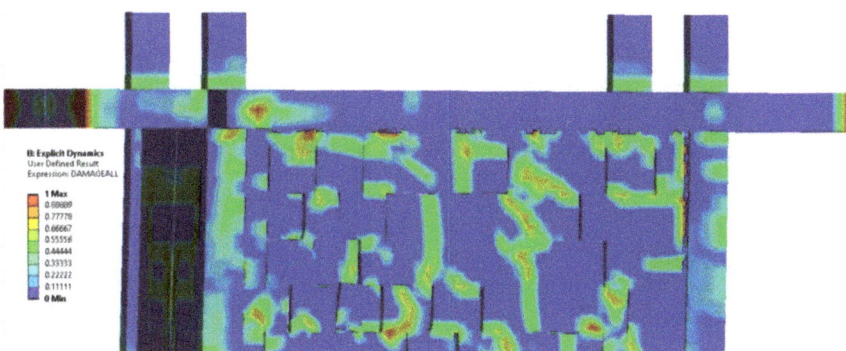

Figure 8. Damage accumulation in model 3DOW at the end of PHASE 2 (upper part).

5. Conclusions

The developed 3D FE analytical models provide satisfactory predictions of the experimental P–δ response of the RC frame A2R with infills and FRPU jacketing. Similarly, satisfactory are the predictions for the experimental top acceleration–drift response of the INMASPOL structure having seismic protection through PUFJs in PHASE 1 and then retrofitting with FRPU jackets in PHASE 2.

The external application of double-sided FRPU jackets on the OrthoBlock infill resulted in a remarkable top displacement ductility of the infilled RC frame A2R, with a negligible base shear reduction up to a 3.6% drift for push (+) and (−). FRPU jackets with highly deformable polyurethane impregnation resin may maintain an adequate bond with the surrounding RC frame. Further, they maintain an adequate bond with the brick infill substrate. In all cases, FRPU jacketing ensured the integrity of the infill and of the jacket, excluding total or partial collapse of the infill or of smaller parts.

The INMASPOL real-scale RC frame possesses a low base shear capacity and a high ductility to satisfy the aims of the shake table tests: (a) carry enough mass to lead the real-scale infilled structure to in-plane infill failure and allow for the investigation of the infill contribution at high ductility levels and up to failure; (b) suffer specially designed severe excitation for the out-of-plane performance of the infills. The real-scale RC structure with OrthoBlock infills and PUFJ highly deformable seismic joints exhibited a ductile top acceleration–drift behavior (which is equivalent to a ductile base shear–drift behavior) up to a 3.7% drift, without degradation of the top acceleration (base shear). No collapse of the infills tested in plane or out of plane after a 3.7% drift was evidenced. The damage to infills of type B could be repaired with emergency FRPU external jackets without any special treatment of the infill. After its emergency repair with FRPU (PHASE 2), the structure exhibited a rather linear elastic-like top acceleration–drift behavior (equivalent to a linear elastic base shear–drift behavior) up to 1.62%, without degradation of the top acceleration (base shear). The FRPU maintained the integrity of the damaged infills, and collapse was avoided. The abovementioned dynamic experimental results validate the advanced seismic protection provided by the 20 mm-thick PUFJ to the infill and the RC frame. An undesirable interaction was avoided, and the base shear drift behavior was remarkably ductile, while the potential of the structure was expected to be higher (tests at higher accelerations were avoided to prevent the structure from collapse and continue with PHASES 2 and 3). Similarly, the results in PHASE 2 validate the advanced retrofit of the predamaged infills by the FRPU jackets, while the potential of the structure was expected, again, to be higher.

The advanced FE analyses allowed for a comparative investigation of the analytical response of the bare RC structure (model 1) against analytical models 2 and 3. The infilled structure with seismic protection developed a three times higher maximum acceleration (or base shear) than the bare RC structure at PHASE 1. The corresponding increase in the initial stiffness was similarly remarkable while maintaining the high displacement ductility, with a 3.7% horizontal drift. At PHASE 2, the FRPU retrofitted structure still revealed a higher initial stiffness when compared with the bare structure, while it could develop, at least, around double the maximum acceleration (base shear). Model 3 reproduced the FRPU repair effects despite the local brick damage. Broad infill regions (predamaged inregions cluded) were engaged through the interaction with the highly deformable FRPU jacket and the RC frame through the PUFJ. It seems that FRPU bridged the damaged brick infill areas and, together with the PUFJ, succeeded in causing a beneficial stress redistribution within the infill.

In the case the OrthoBlock infill with PUFJ protection is considered in the intervention for a weak bare RC frame, it can ensure a remarkable increase in the elastic stiffness and base shear while providing a desirable ductile P–δ behavior, with a drift as high as 3.7% (even after multiple earthquake excitations). Simultaneously, for the case under investigation, it prevents in- or out-of-plane collapse of the protected infills. Even if the structure suffers a drift as high as 3.7%, it may receive emergency retrofitting to maintain the integrity of the infill and increase its beneficial contribution to the RC frame.

The realization of a desirable ductile behavior of infilled frames through PUFJs of only 20 mm thickness (only 0.0087 of the height of the infill, compared with alternative proposals with detailing of 65 mm seismic joints or 90 mm sliding joints) and FRPU jacketing is of high importance. PUFJ, due to its high deformability and excellent bonding properties, constitutes a low-volume application (only 20 mm thickness of the joint) with simple detailing (only an interface layer). The application is straightforward in the retrofitting of existing walls (by injection) or in the construction of new walls with the use of prefabricated PUFJs. The polymer cures within hours (see [20]), and therefore it can serve as an emergency measure for retrofitting through injection or in new wall constructions (even with the construction of the prefabricated joints in situ ready to be used). No shear key detailing is required to exclude out-of-plane failure of the infill. The simplicity of the detailing, the low volume and the fast curing, compared with alternative seismic joint solutions such as the abovementioned ones presented in [6,22,23], reveal this solution's feasibility of practical application, economic advantages and fast full protection activation. Similarly, FRPU, due to the high-deformability polyurethane involved, can be applied without any prior treatment of the infill substrate, cures within hours and maintains the bond with the substrate despite extensive infill cracking during loading. Such advantageous alternatives may contribute to building collapse prevention as well as preventing in- and out-of-plane infill collapses. Both are common causes of detrimental human injuries or human loss during severe earthquakes. The developed FE models provided a unique insight on the variable interactions and effects of the innovative interventions and may serve as a solid basis for further analytical investigations such as the utilization of combined retrofits.

Author Contributions: Conceptualization, T.R.; investigation, T.R., V.V., T.F. and E.A.; writing—review and editing, T.R., V.V., T.F. and E.A.; supervision, T.R.; project administration, T.R.; funding acquisition, T.R., T.F. and E.A. All authors have read and agreed to the published version of the manuscript.

Funding: This research was co-financed by Greece and the European Union (European Social Fund-ESF) through the Operational Programme «Human Resources Development, Education and Lifelong Learning 2014–2020» in the context of the project "Advanced Material Retrofits of Infilled Framed Reinforced Concrete Structures with Predamages Against Collapse" (MIS 5050146).

Institutional Review Board Statement: Not applicable.

Informed Consent Statement: Not applicable.

Data Availability Statement: Not applicable.

Conflicts of Interest: The authors declare no conflict of interest.

References

1. Mehrabi, A.B.; Benson-Shing, P.; Schuller, M.P.; Noland, J.L. Experimental evaluation of masonry-infilled RC frames. *J. Struct. Eng.* **1996**, *122*, 228–237. [CrossRef]
2. Al-Chaar, G.; Issa, M.; Sweeney, S. Behavior of masonry-infilled nonductile reinforced concrete frames. *J. Struct. Eng.* **2002**, *128*, 1055–1063. [CrossRef]
3. Almusallam, T.H.; Al-Salloum, Y.A. Behavior of FRP Strengthened Infill Walls under In-Plane Seismic Loading. *J. Compos. Constr.* **2007**, *11*, 308–318. [CrossRef]
4. Akhoundi, F.; Lourenço, P.B.; Vasconcelos, G. Numerically based proposals for the stiffness and strength of masonry infills with openings in reinforced concrete frames. *Earthq. Eng. Struct. Dyn.* **2016**, *45*, 869–891. [CrossRef]
5. Hak, S.; Morandi, P.; Magenes, G. Prediction of inter-storey drifts for regular RC structures with masonry infills based on bare frame modelling. *Bull. Earthq. Eng.* **2017**, *16*, 397–425. [CrossRef]
6. Morandi, P.; Hak, S.; Magenes, G. Performance-based interpretation of in-plane cyclic tests on RC frames with strong masonry infills. *Eng. Struct.* **2018**, *156*, 503–521. [CrossRef]
7. Huang, H.; Burton, H.V.; Sattar, S. Development and Utilization of a Database of Infilled Frame Experiments for Numerical Modeling. *J. Struct. Eng.* **2020**, *146*, 04020079. [CrossRef]
8. Dawe, J.L.; Seah, C.K. Out-of-plane resistance of concrete masonry infilled panels. *Can. J. Civ. Eng.* **1989**, *16*, 854–864. [CrossRef]
9. Angel, R.; Abrams, D.; Shapiro, D.; Uzarski, J.; Webster, M. Behavior of Reinforced Concrete Frames with Masonry Infills. Civil Engineering Studies: Structural Research Series. 1994. Available online: http://hdl.handle.net/2142/14210 (accessed on 15 May 2021).
10. Walsh, K.Q.; Dizhur, D.Y.; Giongo, I.; Derakhshan, H.; Ingham, J.M. Effect of boundary conditions and other factors on URM wall out-of-plane behaviour: Design demands, predicted capacity, and in situ proof test results. *SESOC J.* **2017**, *30*, 57.
11. Stavridis, A.; Koutromanos, I.; Shing, P.G. Shake-table tests of a three-story reinforced concrete frame with masonry infll walls. *Earthq. Eng. Struct. Dyn.* **2011**, *41*, 1089–1108. [CrossRef]
12. Uva, G.; Porco, F.; Fiore, A. Appraisal of masonry infll walls efect in the seismic response of RC framed buildings: A case study. *Eng. Struct.* **2012**, *34*, 514–526. [CrossRef]
13. Celarec, D.; Dolsek, M. Practice-oriented probabilistic seismic performance assessment of inflled frames with consideration of shear failure of columns. *Earthq. Eng. Struct. Dyn.* **2012**, *42*, 1339–1360. [CrossRef]
14. Ruggieri, S.; Porco, F.; Uva, G.; Vamvatsikos, D. Two frugal options to assess class fragility and seismic safety for low-rise reinforced concrete school buildings in Southern Italy. *Bull. Earthq. Eng.* **2021**, *19*, 1415–1439. [CrossRef]
15. Yuksel, E.; Ilki, A.; Erol, G.; Demir, C.; Karadogan, H.F. Seismic retrofitting of infilled reinforced concrete frames with CFRP composites. In *Advances in Earthquake Engineering for Urban Risk Reduction*; Wasti, S.T., Ozcebe, G., Eds.; Springer: Dordrecht, The Netherlands, 2006; pp. 285–300. [CrossRef]
16. Altin, S.; Anil, Ö.; Kara, E.M.; Kaya, M. An experimental study on strengthening of masonry infilled RC frames using diagonal CFRP strips. *Compos. Part B Eng.* **2008**, *39*, 680–693. [CrossRef]
17. Ozden, S.; Akguzel, U.; Ozturan, T. Seismic strengthening of infilled reinforced concrete frames with composite materials. *ACI Struct. J.* **2011**, *108*, 414–422. [CrossRef]
18. Koutas, L.; Bousias, S.N.; Triantafillou, T.C. Seismic strengthening of masonry-infilled RC frames with TRM: Experimental study. *J. Compos. Constr.* **2014**, *19*. [CrossRef]
19. Gams, M.; Tomaževič, M.; Berset, T. Seismic strengthening of brick masonry by composite coatings: An experimental study. *Bull. Earthq. Eng.* **2017**, *15*, 4269–4298. [CrossRef]
20. Rousakis, T.; Ilki, A.; Kwiecień, A.; Viskovic, A.; Gams, M.; Triller, P.; Ghiassi, B.; Benedetti, A.; Rakicevic, Z.; Colla, C.; et al. Deformable Pol-yurethane Joints and Fibre Grids for Resilient Seismic Performance of Reinforced Concrete Frames with Orthoblock Brick Infills. *Polymers* **2020**, *12*, 2869. [CrossRef]
21. Fardis, M.N. Design provisions for masonry-infilled RC frames. In Proceedings of the 12th World Conference on Earthquake Engineering, Auckland, New Zealand, 30 January–4 February 2000.
22. Marinkovic, M.; Butenweg, C. Innovative decoupling system for the seismic protection of masonry infill walls in reinforced concrete frames. *Eng. Struct.* **2019**, *197*, 109435. [CrossRef]
23. Butenweg, C.; Marinković, M. Damage reduction system for masonry infill walls under seismic loading. *Bull. Earthq. Eng.* **2018**, *2*, 267–273. [CrossRef]
24. Rousakis, T. Brick walls Interventions with FRPU or PUFJ and of RC columns with FR in Brick-Infilled RC Structures with the use of Pushover Beam-Column Element Analysis and Pseudo-Dynamic 3D Finite Element Analysis. In Proceedings of the 17th International Brick and Block Masonry Conference from Historical to Sustainable Masonry (IB2MaC 2020), Krakow, Poland, 5–7 July 2020.
25. Akyildiz, A.; Kwiecień, A.; Zając, B.; Triller, P.; Bohinc, U.; Rousakis, T.; Viskovic, A. Preliminary in-plane shear test of infills protected by PUFJ interfaces. In Proceedings of the 17th International Brick and Block Masonry Conference from Historical to Sustainable Masonry (IB2MaC 2020), Krakow, Poland, 5–7 July 2020.
26. Karabinis, A.I.; Rousakis, T.C.; Manolitsi, G.E. 3D Finite-Element Analysis of Substandard RC Columns Strengthened by Fiber-Reinforced Polymer Sheets. *J. Compos. Constr.* **2008**, *12*, 531–540. [CrossRef]

27. Rousakis, T.; Karabinis, A. Fiber Reinforced Polymer Confinement of Bridge Columns Suffering from Premature Bars' Buckling–Strength empirical model. In Proceedings of the 34th International Symposium on Bridge and Structural Engineering, Venice, Italy, 22–24 September 2010.
28. Yu, T.T.; Teng, J.G.; Wong, Y.L.; Dong, S.L. Finite element modeling of confined concrete-I: Drucker–Prager type plasticity model. *Eng. Struct.* **2010**, *32*, 665–679. [CrossRef]
29. Hany, N.F.; Hantouche, E.G.; Harajli, M.H. Finite element modeling of FRP-confined concrete using modified concrete damaged plasticity. *Eng. Struct.* **2016**, *125*, 1–14. [CrossRef]
30. Youssef, O.; El Gawady, M.A.; Mills, J.E. Displacement and plastic hinge length of FRP confined circular reinforced concrete columns. *Eng. Struct.* **2015**, *101*, 465–476. [CrossRef]
31. Yuan, F.; Wu, Y.F.; Li, C.Q. Modelling plastic hinge of FRP-confined RC columns. *Eng. Struct.* **2017**, *131*, 651–668. [CrossRef]
32. Triantafyllou, G.G.; Rousakis, T.C.; Karabinis, A.I. Effect of patch repair and strengthening with EBR and NSM CFRP laminates for RC beams with low, medium and heavy corrosion. *Compos. Part B Eng.* **2018**, *133*, 101–111. [CrossRef]
33. Anagnostou, E.; Rousakis, T.; Georgiadis, N. Finite element analysis of deficient RC columns with square and rectangular section under pseudoseismic load and comparison with retrofit code predictions. In Proceedings of the ICCE-26 Conference, Paris, France, 15–21 July 2018.
34. Fanaradelli, T.D.; Rousakis, T.C. 3D Finite Element Pseudodynamic Analysis of Deficient RC Rectangular Columns Confined with Fiber Reinforced Polymers under Axial Compression. *Polymers* **2020**, *12*, 2546. [CrossRef] [PubMed]
35. Triller, P.; Kwiecień, A.; Bohinc, U.; Zajac, B.; Rousakis, T.; Viskovic, A. Preliminary in-plane shear test of damaged infill strengthened by FRPU. In Proceedings of the 10th International Conference on FRP Composites in Civil Engineering (CICE 2020/2021), Istanbul, Turkey, 8–10 December 2021.
36. *Eurocode 2: Design of Concrete Structures—Part 1-1: General Rules and Rules for Buildings*; British Standard: London, UK, 2008.
37. *Eurocode 8. Design of Structures for Earthquake Resistance—Part 1: General Rules, Seismic Actions and Rules for Buildings*; British Standard: London, UK, 2004.
38. Kwiecień, A. Stiff and flexible adhesives bonding CFRP to masonry substrates—Investigated in pull-off test and Single-Lap test. *Arch. Civ. Mech. Eng.* **2012**, *12*, 228–239. [CrossRef]
39. ANSYS® Academic Research, Release v 2021 R1. Available online: https://www.ansys.com/products/release-highlights (accessed on 15 May 2021).
40. Riedel, W. *Beton Unter Dynamischen Lasten: Meso- und Makromechanische Modelle und ihre Parameter*; Fraunhofer-Institut für Kurzzeitdynamik, Ernst-Mach-Institut EMI, Freiburg/Brsg, Eds.; Fraunhofer IRB Verlag: Stuttgart, Germany, 2004; ISBN 3-8167-6340-5.
41. Riedel, W.; Thoma, K.; Hiermaier, S.; Schmolinske, E. Penetration of Reinforced Concrete by BETA-B-500, Numerical Analysis using a New Macroscopic Concrete Model for Hydrocodes. In Proceedings of the 9th International Symposium, Interaction of the Effects of Munitions with Structures, Berlin, Germany, 3–7 May 1999; pp. 315–322.
42. Riedel, W.; Kawai, N.; Kondo, K. Numerical Assessment for Impact Strength Measurements in Concrete Materials. *Int. J. Impact Eng.* **2009**, *36*, 283–293. [CrossRef]
43. El Mezaini, N.S. Reserved Strength of Reinforced Concrete Buildings with Masonry Walls. *Comput. Civ. Infrastruct. Eng.* **2005**, *20*, 172–183. [CrossRef]
44. Facconi, L.; Minelli, F. Retrofitting RC infills by a glass fiber mesh reinforced overlay and steel dowels: Experimental and numerical study. *Constr. Build. Mater.* **2020**, *231*, 117133. [CrossRef]

Article

Assessment of Different Retrofitting Methods on Structural Performance of RC Buildings against Progressive Collapse

Barham Haidar Ali [1,*], Esra Mete Güneyisi [1] and Mohammad Bigonah [2]

[1] Department of Civil Engineering, Gaziantep University, Gaziantep 27310, Turkey; eguneyisi@gantep.edu.tr
[2] School of Civil Engineering, Iran University of Science and Technology, Narmak, Tehran P.O. Box 16765-163, Iran; seyedmohammadbigonah@gmail.com
* Correspondence: berhemheider@gmail.com

Abstract: Progressive collapse refers to the spread of primary local damages within the structure. Following such damages due to removing one or more load-bearing columns, the failure spreads in a chain and causes structural failure. This study represents a report investigating the influence of various retrofitting methods on the progressive collapse resistance of multistorey reinforced concrete (RC) structures. To this end, eight different cases were considered. The first one included a thirteen-story RC moment-resisting frame (bare frame), while the others were frames upgraded with the application of X-brace, diagonal brace, inverted V-brace, the viscous damper in the central bay, viscous damper in two inner bays, viscous damper only in certain stories and carbon fiber reinforced polymer. Moreover, three different column removal scenarios were considered as a column failure at stories one, six, and thirteen of each case study structure. The analysis results indicated that the redistribution of loads after the column's failure and the RC buildings' collapse resistance was increased depending mainly on the type of approach used for upgrading the bare frame.

Keywords: column removal; non-linear analysis; progressive collapse; reinforced concrete frame; retrofitting

1. Introduction

Progressive collapse refers to the spread of primary local damages within the structure (also known as disproportionate collapse) is a high-impact, low-probability event. After the collapse of the Ronan Point Building in London in 1968, the Murrah Federal Building in Oklahoma City in 1995, and the World Trade Center Towers in New York City in 2001, the structural engineers and government organizations became concerned about progressive collapse [1,2]. Because it is hard to describe a potentially hazardous load that causes localized damage to a building, it is common to use the decoupling technique to suppose whether the remaining structure can bridge over the removed components by removing a supporting column or wall. The remaining structures would be subjected to linear static, linear dynamic, non-linear static, and non-linear dynamic analysis [3]. Energy equilibrium [4] or equation of motion can connect dynamic and static performance [5].

Researchers have recently become interested in the effect of secondary components, such as bracing, on progressive collapse performance. The strengthening of reinforced concrete (RC) frame structures against progressive collapse have various challenges: (1) The application of typical strengthening methods for RC frame construction continues to be researched because progressive collapse is a significant deformation behavior; (2) The degree of strengthening against progressive collapse is significant, which could result in "strong beams and weak columns" and have an impact on seismic performance; (3) Anchorage is the primary issue in strengthening reinforced concrete structures against progressive collapse, and dependable and rapid construction anchorage solutions for minimizing progressive collapse strengthening are urgently needed [6].

Retrofitting is not a common practice to make an old system compliant with the new code's rules, as this option is not cost-effective. Alternatively, to ensure a set degree of collapse or to avoid the building collapsing entirely, it is advised that retrofit goals for a structure prone to progressive collapse be based on performance-based criteria. Steel braces are often employed to produce lateral stiffness and resist lateral loads in steel buildings [7–9]. Seismic retrofitting with steel bracing for existing RC frames has also received much interest due to the ease of installing steel braces [10–12]. In another study by Bigonah et al. [13], which evaluated the performance of infill types, the results show that adding infill reduces vertical displacement and improves redistribution of forces against progressive collapse. Steel braces impact on the resilience of structures to progressive collapse has recently attracted researchers' attention through two-dimensional numerical models. In addition, they investigated the progressive collapse of 10-storey braced steel frames designed following different seismic intensities and found that the frame with concentric braces is more likely to collapse than the frame with eccentric braces [14]. The effective techniques of modern strengthening methods of RC frames against progressive failure are studied, and the results show that increasing the percentage of rebar reduced vertical displacement [15]. The structural behavior of three-dimensional (3d) 20-storey braced steel frames is investigated and discovered that removing a column at a higher story increases the likelihood of the frame collapsing [16,17]. Another study experimentally investigated the progressive collapse resistance of five one-fourth scaled two-bay by three-story RC frames strengthened by four types of steel bracing [18]. They found that basically all bracing can improve progressive collapse resistance, with eccentric X braces performing the best. Steel bracing is either designed for seismic design or lateral stability. The steel braces are only situated at one or several defined spans but are continuous in elevation from the first to the thirteenth floor. Costanzo et al. [19] reviewed the design rules and requirements for XCBFS to simplify the design and improve the ductility and waste capacity of the structural system, applying bracing on the roof floor to obtain a structural response with proper distribution of plastic deformation section with height. At the same time, it can be ignored for three-story structures. D'Aniello et al. and Costanzo et al. [20,21] also investigated the effect of beam flexural stiffness on the seismic response of concentric braces. The results show that the higher the stiffness, the lower the drift ratio occurs. As a result, the deformation in the brace is limited under compression. The braces can add progressive collapse resistance if the removed column is positioned in the braced span; otherwise, their contribution is minimal [12,16,17]. This means that steel bracing designed to withstand seismic loads and provide lateral stability may be incapable of enhancing structural robustness. On the other hand, steel bracing is a viable alternative for strengthening existing buildings against progressive collapse, and the best approach to design such braces requires more research.

Similarly, the effectiveness of proposed carbon fiber reinforced polymers (CFRP) and glass fiber reinforced polymers (GFRP) strengthening strategies for improving progressive collapse behavior using a series of flat slab substructures were evaluated [22]. Moreover, the efficiency of ten RC beams employing CFRP anchors and/or U-wraps is studied [23]. Hence, CFRP is used to improve the continuity of RC beams to shift the load carried by the damaged column to an intact zone and therefore control the spread of progressive collapse [24]. They found that beams with discontinuous reinforcement improved by roughly 55 to 60%, while beams with continuous reinforcement improved by 109%.

Compared to other retrofitting methods, such as bracing and CFRP, there have been fewer researches on the influence of viscous dampers retrofitting. Viscous dampers often meant to reduce building vibration caused by wind or earthquakes, are another type of retrofit to improve a structure's resistance to progressive collapse. Kim et al. [23] examined the progressive collapse resistance of structures equipped with viscous dampers, often installed to dampen wind- or earthquake-induced vibration.

In this study, an attempt is made to evaluate the influence of various retrofitting systems on the progressive collapse resistance of a multi-storey RC building. For this purpose, eight different cases were taken into account. The first one contained a thirteen-story RC moment-resisting frame (bare frame) while the others were frames upgraded with the use of X-brace, diagonal brace, inverted V-brace, the viscous damper in the central bay, the viscous damper in two inner bays, viscous damper only in certain stories and carbon fiber reinforced polymer (CFRP). Besides, three different column removal scenarios were considered as a column failure at stories one, six, and thirteen. Non-linear dynamic analysis was conducted by using a finite element program. Parametric study results for each case were provided by considering the shear, axial, and moment of columns adjacent to the collapsed column. The moment and shear forces for the beam above the collapsed column were also investigated with the story displacements and building performance levels reported after the vertical member's loss. Finally, all the investigated parameters for each case were evaluated and discussed comparatively.

2. Methodology

2.1. Description of Structural Models

The structure studied in this research is a 13-story RC building (Figures 1 and 2). The structure is two-dimensional (2D), and it consists of five bays with a 6 m length span. The height of each story is 3.2 m. Dead and live loads are 4 and 2 kN/m^2, respectively. The section of the beams for all cases is 500 mm × 350 mm. The column sections from stories 1 to 4, 5 to 9, and 10 to 13 are 600 mm × 600 mm, 500 mm × 500 mm, and 400 mm × 400 mm, respectively. The compressive strength of the concrete is 25 MPa, and the yield strength of the steel bar is 392 MPa according to the specified factory. The structure is designed following the framework of ACI Committee 318 (2014). To evaluate the behavior of the RC buildings against progressive collapse eight different cases were taken into consideration. The first one is RC moment-resisting frame (bare frame), whereas the others are the upgraded frames, namely, X braced frame, diagonally braced frame, inverted V-braced frame, viscously damped frame in the central bay, viscously damped frame in two inner bays, viscously damped frame only in certain story and frame with CFRP. Then, three different column removal scenarios were adopted by considering the failure of the central column at stories one, six, and thirteen. Evaluation and comparison of the structural response of the eight frames against progressive collapse have been conducted. Table 1 defines the cases studied.

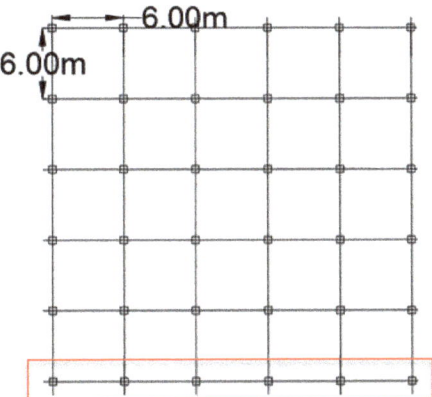

Figure 1. Plan view of the RC building.

In the case of the frames upgraded with braces, common configurations for concentric bracing systems including inverted-V (chevron)-type, X-type, and diagonal braces were considered. The force-displacement relationship of braces based on the uniaxial phenomenological model, adopted in Federal Emergency Management Agency (FEMA-(356)) [25], in which Δ_y and Δ_{cr} are the yielding and buckling displacements, and P_y and P_{cr} are the tension and compression forces, respectively. The braces are hollow steel tubes, and sections are 2UNP14 and 2UNP13 for stories 1 to 6 and 7 to 13, respectively. The analyzed structural models are subjected to the loss of the first, sixth and thirteenth-floor center column, in which the structure deforms symmetrically, and the full capacity of bracing is activated. Figure 3 shows the various bracing configurations to be analyzed.

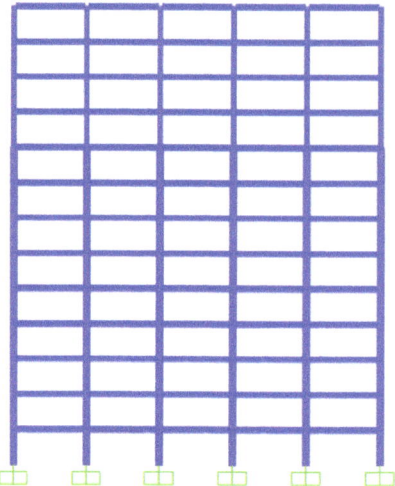

Figure 2. Elevation view of the RC building.

Table 1. Definitions for the different frames.

Case No	Frame Model
Case 1	Moment resisting frame (Bare frame)
Case 2	X braced frame
Case 3	Diagonally braced frame
Case 4	Inverted V braced frame
Case 5	Viscously damped frame in the central bay
Case 6	Viscously damped frame in two inner bays
Case 7	Viscously damped frame only in certain stories
Case 8	Frame with CFRP

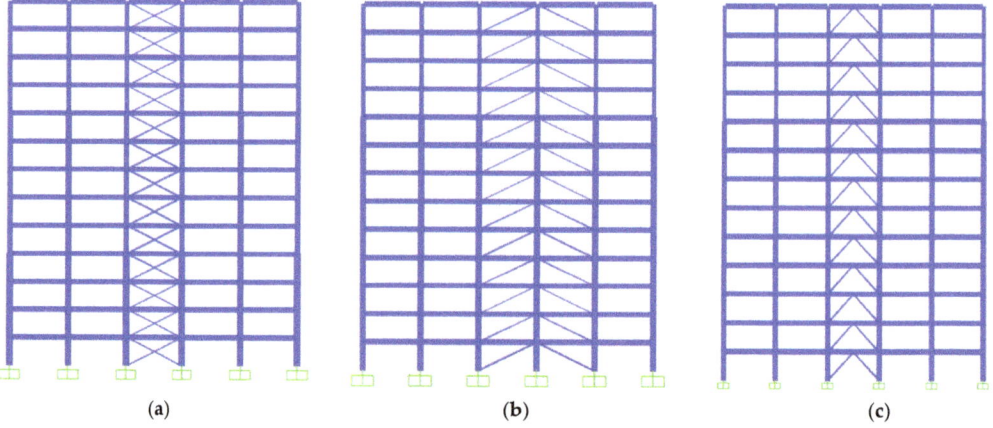

Figure 3. (**a**) X-braced frame; (**b**) Diagonally braced frame, and (**c**) Inverted V-braced frame.

The diagonal brace or chevron brace is the traditional configuration for viscous dampers. In the case of the frames with viscous dampers, the former one was used. The cases for viscous dampers are viscously damped frame in the central bay, viscously damped frame in two inner bays, and viscously damped frame only in a certain story, as shown in Figure 4. Damper's capacity has been determined according to the study of Cimellaro and Retamales [26].

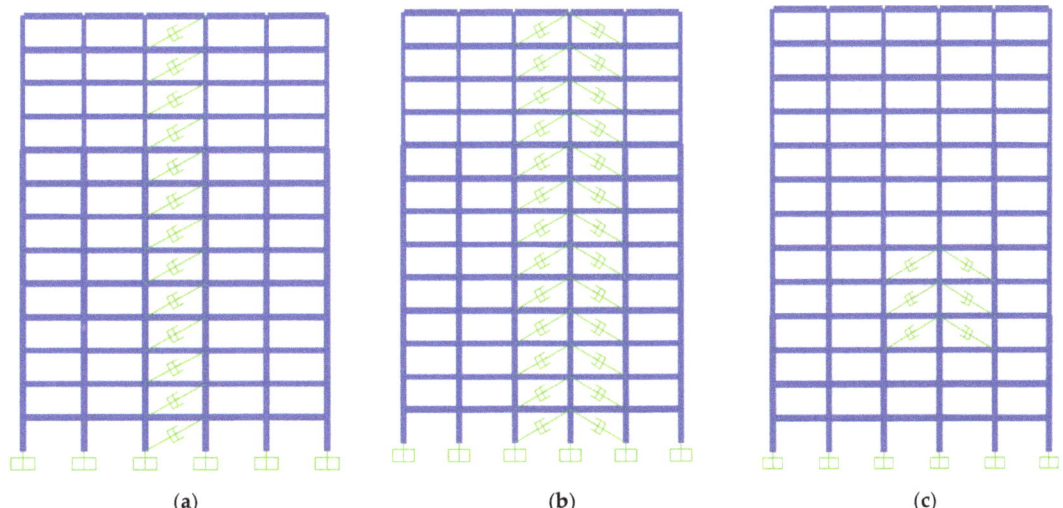

Figure 4. (**a**) Viscously damped frame in the central bay, (**b**) viscously damped frame in two inner bays, and (**c**) Viscously damped frame only in certain stories.

In the studied structure, the effect of viscous damping with 15% damping has been investigated. To determine the required damping of the structure to reach the damping percentage of the target, the stiffness of the whole structure should be determined [26]. First, the triangular pattern of the base shear force and the level of drifts of each story are calculated. Then, with the help of the sheer force of each story and the corresponding drift with the same shear force, the shear stiffness of each story can be obtained. Finally, the total damping coefficient that must be added to the structure to achieve the target damping can be calculated [27].

In the case of the frames upgraded with carbon fiber reinforcement polymer (CFRP), the CFRP is assumed to be warped around the two sides and bottom of beams. The ultimate strength is 3200 MPa, in the longitudinal direction of fibers, and the elastic modulus is 210,000 MPa. The thickness of the CFRP wrap is 1.4 mm. An investigation of the effectiveness of the proposed CFRP strengthening schemes in mitigating the progressive collapse of the structures of this study is made.

The nominal shear strength of an RC section (V_n) with CFRP is expressed in Equation (1) [28]:

$$V_n = V_C + V_S + V_f \tag{1}$$

V_C is the shear strength of the concrete, V_S is the shear strength of the steel reinforcement, and V_f is the shear contribution of the CFRP. The design shear strength, V_n, is achieved by multiplying the nominal shear strength by a strength reduction factor for shear, the factor for steel and concrete contribution from (ACI 440.2R-02) [28] is 0.85 [29], and the factor for CFRP contribution is suggested to be 0.70. Equation (2) [28] presents the design shear strength.

$$\phi V_n = 0.85(V_c + V_s) + 0.7 V_f \tag{2}$$

The expression to compute CFRP contribution is given in Equation (3) [28]. This equation is similar to steel shear reinforcement and is consistent with (ACI 440.2R-02).

$$v_f = \frac{A_f \times f_{fe}(\sin\beta + \cos\beta) \times d_f}{S_f} \leq \left(\frac{2\sqrt{f_c'} \times b_w \times d^2}{3} - V_s \right) \qquad (3)$$

2.2. Removal of the Column and Details of the Analysis

The removal process of the column should be such that the effect of a dynamic effect on the structure is seen due to the shock caused. A method to consider collapse ability in SAP2000 software was used to do this. The number of joints and elements is depicted in Figure 5. To simulate the sudden removal of columns in different scenarios in non-linear analysis, dead and live loads were first applied from 0 to 5 s, and then, they were removed, and the structure response was reviewed after this moment up to 10 s. Figure 6 shows the removal of the inner columns at different story levels. The dynamic amplification factor in the dynamic analysis is not recommended by both guidelines (GSA and UFC). To apply dynamic analysis, prior to column removal, axial force acting is calculated. Then the column is replaced by point loads equivalent to its member forces [1,2,30].

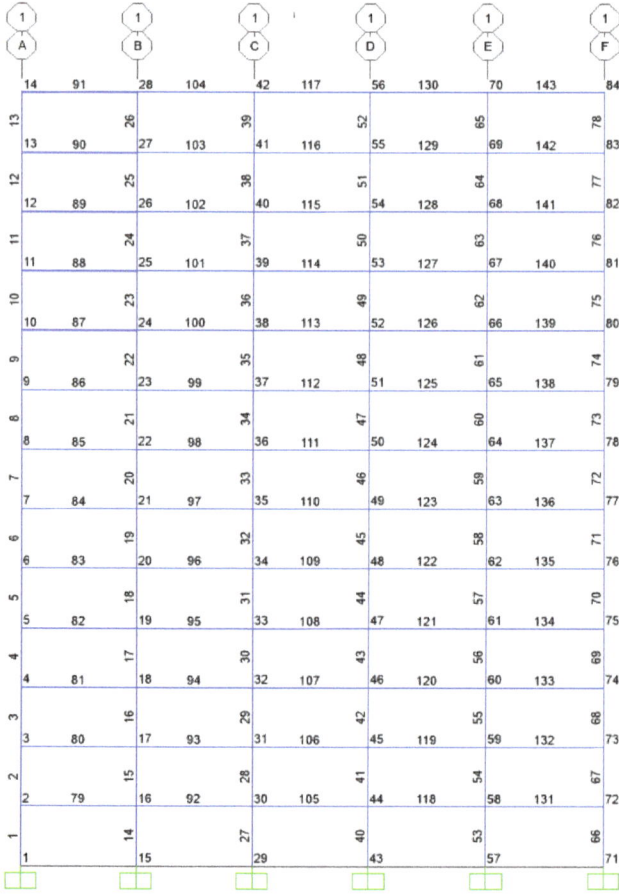

Figure 5. Numbering of elements and joints.

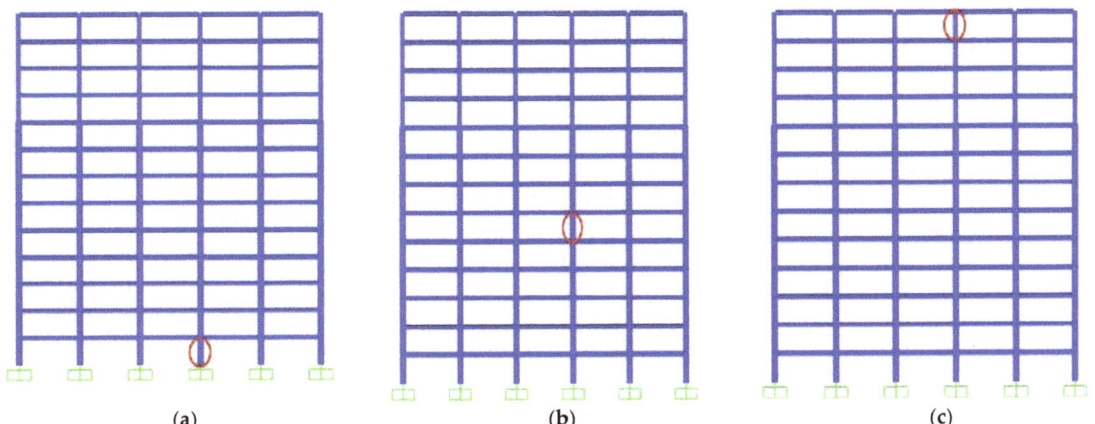

Figure 6. Column removal scenario in (**a**) story one, (**b**) story six, and (**c**) story thirteen.

3. Results and Discussion

In this section, the status removing columns are investigated, with the amount of displacement and the redistribution forces after applying the column removal scenario. Non-linear dynamic analysis (NDA) is done for inner column removal in the first, sixth and thirteenth floors, and results are presented here. Four graphs are plotted in SAP2000: vertical displacement vs. time, axial force vs. time, bending moment vs. time, and shear vs. time. Vertical displacement is taken at the point where the column was removed. The beam with maximum axial force and bending moment vs. time is taken for plotting.

All of the examples were subjected to a 2D frame analysis. As illustrated in Figures 7–14, all 2D frame analysis cases showed partial progressive collapse. The collapse zones in the 2D frame analysis were estimated directly after the collapse of beams connected to the removed column. The 2D frame analysis revealed that the collapse spreads to all levels of the structure, indicating that the structure has a high risk of progressive collapse and should be modified, according to the General Service Administration (GSA). In addition, because the main support was removed, the static system of beams became longer, resulting in collapse due to insufficient reinforcing.

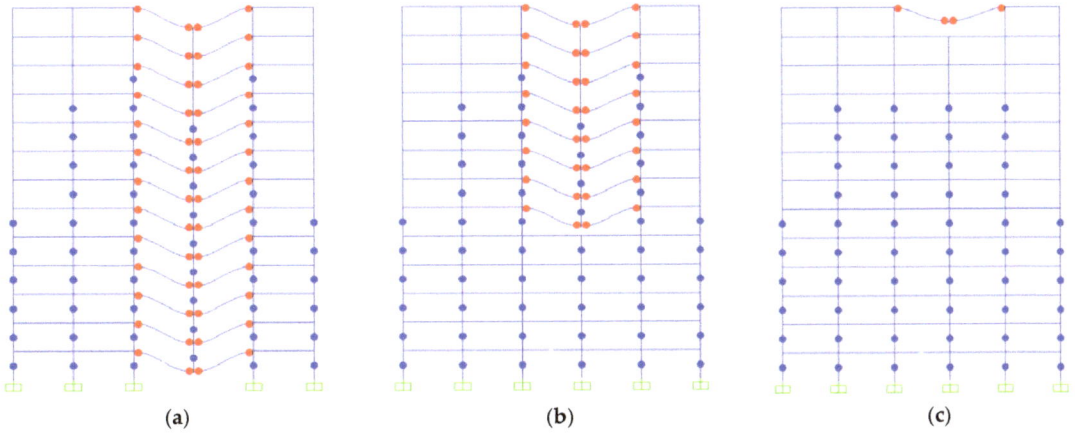

Figure 7. Hinge formation in the moment-resisting frame (Bare frame) exposed to column removal in (**a**) story one, (**b**) story six, and (**c**) story thirteen.

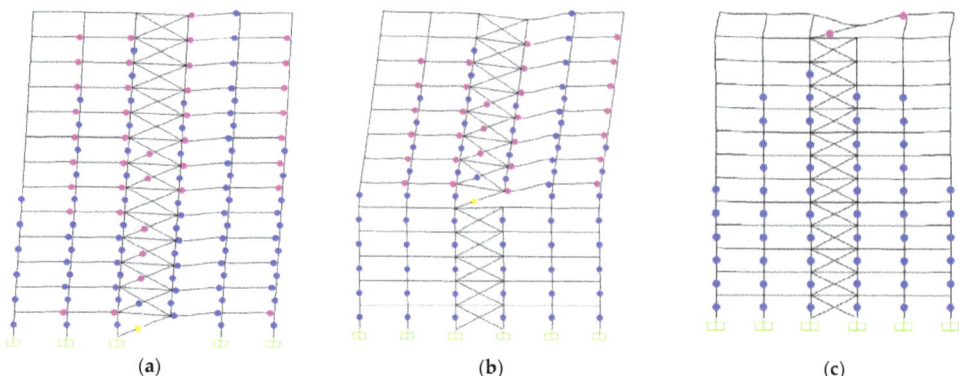

Figure 8. Hinge formation in the X braced frame exposed to column removal in (**a**) story one, (**b**) story six, and (**c**) story thirteen.

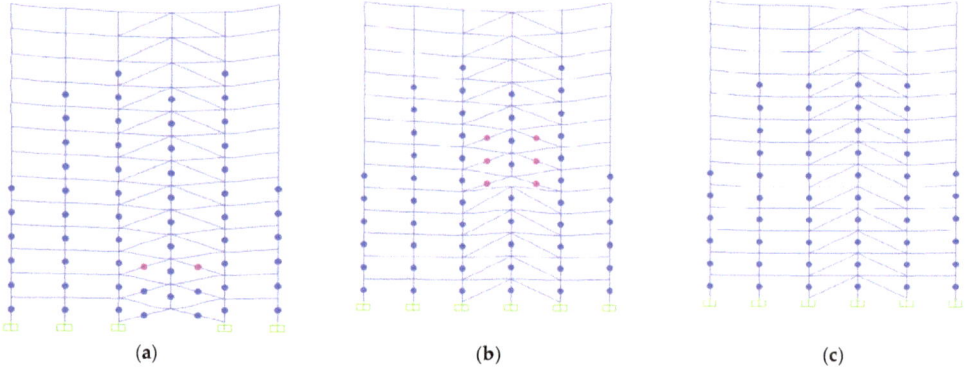

Figure 9. Hinge formation in the diagonally braced frame exposed to column removal in (**a**) story one, (**b**) story six, and (**c**) story thirteen.

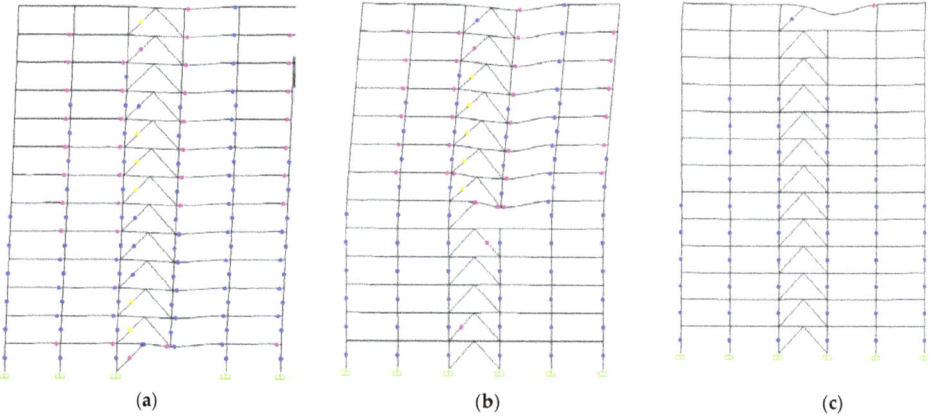

Figure 10. Hinge formation in the inverted V-braced frame exposed to column removal in (**a**) story one, (**b**) story six, and (**c**) story thirteen.

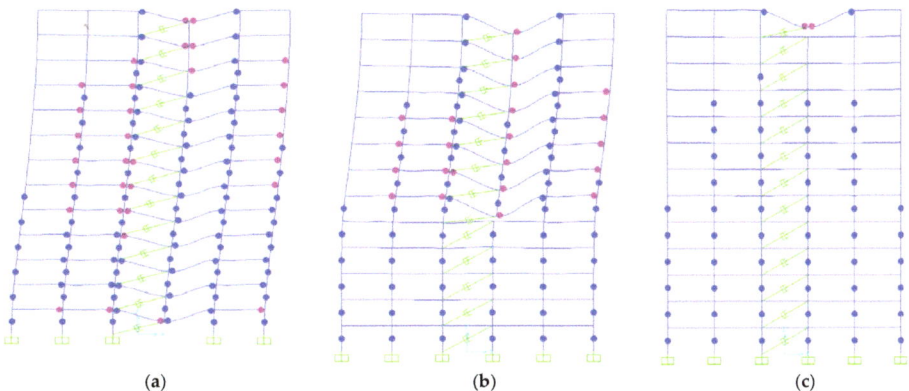

Figure 11. Hinge formation in the viscously damped frame in central bay exposed to column removal in (**a**) story one, (**b**) story six, and (**c**) story thirteen.

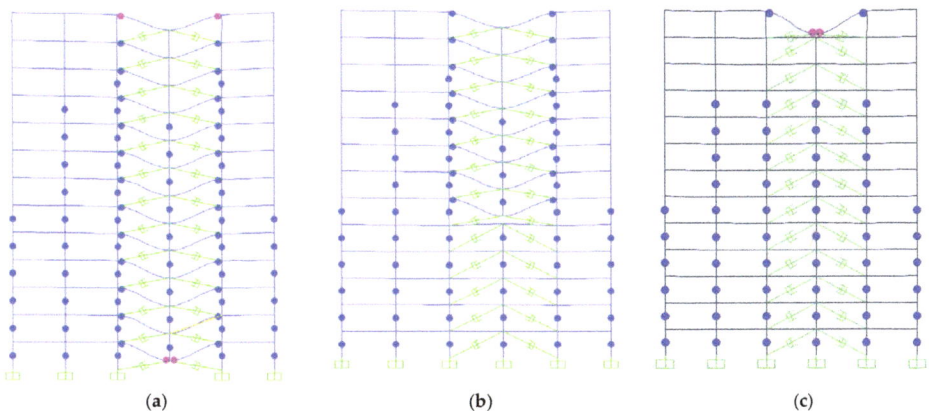

Figure 12. Hinge formation in the viscously damped frame in two inner bays exposed to column removal in (**a**) story one, (**b**) story six, and (**c**) story thirteen.

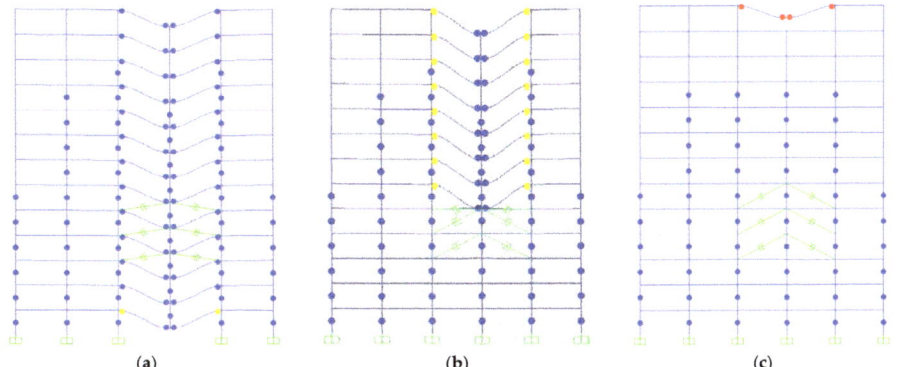

Figure 13. Hinge formation in the viscously damped frame only in certain stories exposed to column removal in (**a**) story one, (**b**) story six, and (**c**) story thirteen.

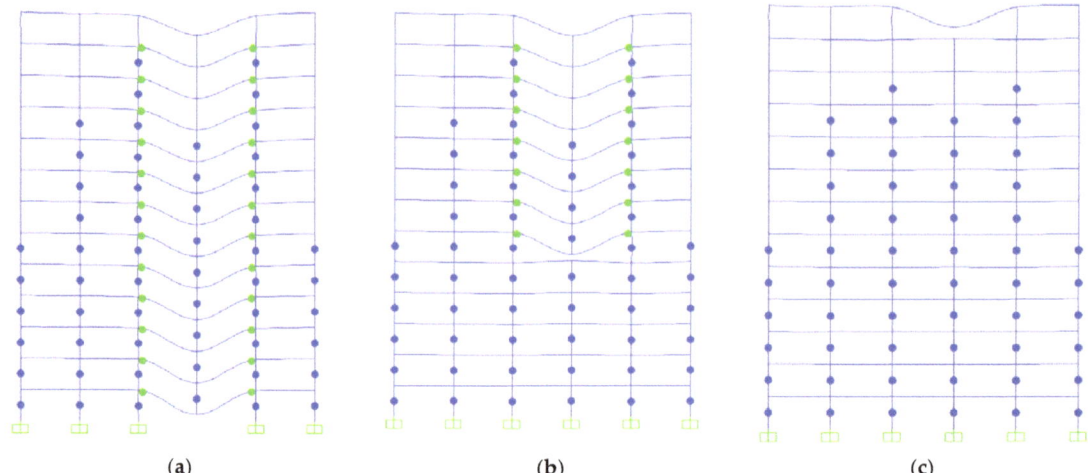

Figure 14. Hinge formation in the frame with CFRP exposed to column removal in (**a**) story one, (**b**) story six, and (**c**) story thirteen.

3.1. Performance Level of the Frames

The damage levels of members were analyzed for the non-linear analytic techniques using various performance levels such as immediate occupancy (IO), life safety (LS), and collapse prevention (CP). The performance level of the structure should not exceed collapse prevention according to GSA rules, or structural members will be classified as seriously failed [1]. In all cases, the plastic hinge formation is illustrated in 2D frames.

Figure 7 shows the removal of columns on different stories in a moment-resisting frame (bare frame). After removing the column on the first floor, all the beams in the span were failed at collapse prevention, and the span completely collapsed at all floors above. The removal of the column on floor six caused the beams to fail and reach the life safety point at all floors above, and the structure could not redistribute the forces well. Moreover, removing the column on floor thirteen puts its beam almost in a state of complete collapse. The hinges of the beams in the affected spans reach the failure point (red-colored hinge), as shown in Figure 7 in all the cases of column removal scenario. The failure of one column leads to the collapse of all the members in the affected span.

Steel bracings are used as a remedy to provide resistance against progressive collapse [31]. Although few bracing members fail by buckling in compression, a bracing system can strengthen the building to resist progressive collapse. For example, Figure 8 shows the removal of columns at different stories by adding an X brace. After removing a column and a brace, the function of the beams was out of operational performance.

The plastic hinges generated in the beams meet the acceptability level. The plastic hinge creation occurred on the diagonal braced element inside the collapse range, which is considered compression failure of the brace, as shown in Figure 8a.

In Figure 9, the use of one-way bracing on both sides improved the performance of the whole structure. The plastic hinge did not enter the non-linear area because the structure's chain function was well performed. However, even the performance of the brace did not enter the operational performance levels. As seen in Figure 9a,b, the plastic hinge formation has begun to develop in brace members. As a result, plastic hinges formed in columns and beams and were disseminated over particular building members within the immediate occupancy area. The lateral stiffness is improved as well compared with other braced frames.

In Figure 10, the use of inverted V-bracing in the structure improved the performance of the structure against progressive collapse. Although the column was removed at higher stories, the structure had a better performance, and the concentration of stiffness at the removed point improved the redistribution of forces and the chain performance of the structure. The inverted-braced arrangement increases the constraint level at the beam end, allowing catenary action. The bracing system generated a new load transfer path. The horizontal braces distributed certain gravitational stresses to the surrounding structures then carried to the foundation via the vertical braces. Following compression brace buckling, certain columns buckled before tension braces yielded, culminating in brittle failure modes. When a column adjacent to the braced bay was removed, the constructions with only single-bay braces proved extremely fragile. The progressive collapse can be avoided in this scenario by designing the frame with braces.

In Figure 11, the frame is called a viscously damped frame in the central bay; a viscous damper was used diagonally in a span. The results showed that, as shown in Figure 11a, the plastic hinges in beams on the third and fourth floors reach within range of collapse and that the application of a viscous damper prevented significant oscillation and shock to the structure. In addition, plastic rotation in the beam ends was lowered to below the immediate occupancy state, and plastic hinges were removed in numerous places.

The application of viscous dampers in two inner bays enhanced the structure's tensile performance, whereas the performance of the beams improved when compared to the viscously damped frame in the central bay example, as shown in Figure 12. As shown in Figure 12a,b, the plastic hinge creation in the area of column removal (the connection of beam and column removal) is observed. The plastic hinge formation is more common in the structure's vertical parts.

Figure 13 shows viscously damped frame only in certain stories based on the method of Cimellaro and Retamales [26], the most optimal case in 3 stories was used. The results show that it had little effect on the performance of the structure and the redistribution of forces was not well done and the plastic hinges were out of the collapse prevention.

Figure 14 shows a frame with CFRP, which uses a CFRP layer in all beams. The results showed that the redistribution of forces and tensile performance in the beams was fairly distributed, increased the structure's strength, and improved the chain performance in the beams. Moreover, relatively low cost created the best performance in the whole structure against progressive collapse. The plastic hinge formation disperses mostly on the column, which is in a state of immediate occupancy.

3.2. Frame Displacements Due to Column Loss

For internal column loss in the first, sixth, and thirteenth stories, the numerical findings from non-linear dynamic calculations up to 10 s are given in Figures 15–17. The maximum permitted ductility and/or rotation limits of beams are verified following GSA. For RC structures, GSA refers to ASCE 41 [32] approval criteria for non-linear analysis to assess the damage on a structure due to a column loss. If beam end rotations in any of the frames investigated herein surpassed the acceptance standards, then it would indicate the maximum permissible ductility of the beam, as specified in [1,32]. It is seen that the maximum displacement is varied around 45 cm for the braced frames, as indicated in Figure 15, due to the rapid inner column loss in story one. When compared to other types of frames, the abrupt columns in the first, sixth and thirteenth stories do not affect the displacement of an X braced frame.

On the other hand, the displacement of X braced, and inverted V-braced frames are often smaller than that of other types of frames. Moreover, as given in Table 2, the moment-resistant frame (bare frame) is observed to be failed for all column removal scenarios. The discrepancies in displacement values for the others are varied depending mainly on the strengthening method used.

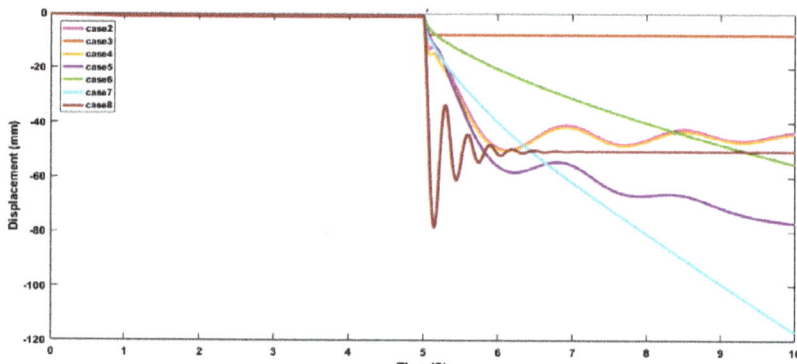

Figure 15. Vertical displacement at column removed position due to column loss in story one.

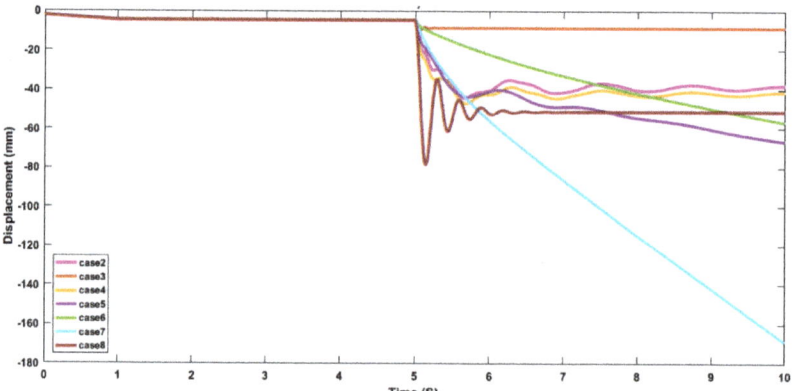

Figure 16. Vertical displacement at column removed position due to column loss in story six.

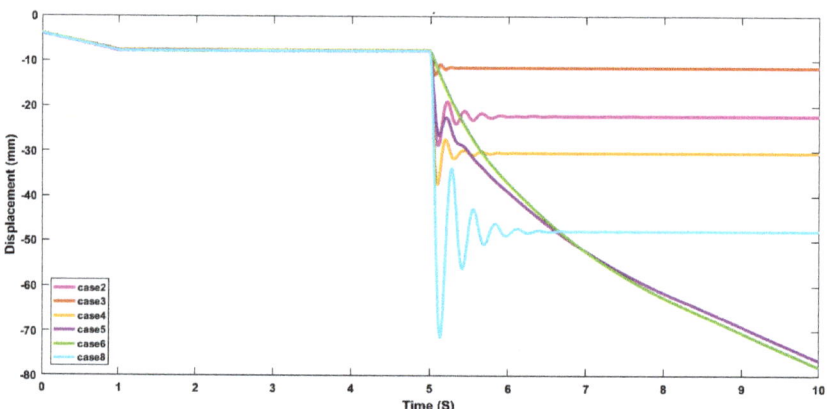

Figure 17. Vertical displacement at column removed position due to column loss in story thirteen.

Table 2. Vertical displacement after removing a column in the stories.

Type	ST1		ST6		ST13	
	Displacement (mm)		Displacement (mm)		Displacement (mm)	
	Max	Constant	Max	Constant	Max	Constant
Case 1	Fail	Fail	Fail	Fail	Fail	Fail
Case 2	−49.94	−43.38	−44.20	−37.99	−28.77	−22.18
Case 3	−8.68	−7.73	−9.72	−8.63	−13.10	−11.50
Case 4	−50.19	−44.05	−47.08	−41.33	−37.32	−30.38
Case 5	−76.75	−76.75	−66.65	−66.65	−76.48	−76.48
Case 6	−55.15	−55.15	−56.81	−56.81	−77.93	−77.93
Case 7	−116.87	−116.87	−168.79	−168.79	Fail	Fail
Case 8	−78.35	−50.45	−78.64	−51.37	−71.40	−47.77

Due to the removal of the column on the first story as shown in Table 2, the maximum vertical displacement is related to case 7 and also case 1 is damaged in all floors and the best performance is related to case 3, while in the 6th floor the most Vertical displacement is related to case 7 and the lowest is related to case 3 with a value of −8.63 mm. Therefore, in addition to removing the column on the top floor, there was a breakdown in case 3 in case 7.

3.3. Force Distribution Due to Column Loss in Beams Next to Column Removal

Bending moment and shear generated by sudden column loss were greater for story one beams than for frames with CFRP (Case 8). This is due to the higher flexural stiffness of the CFRP-framed structure, which captivates more forces. Figures 18 and 19 depict the bending moment and shear on the beam caused by column loss for the first and sixth stories. Bending moments for the bare frame and various retrofitting frames are less than for CFRP framed first and sixth stories, as demonstrated in Figure 19. This is because the joint stiffness of CFRP frames attracts more force. Apart from Case 8, forces are dispersed evenly throughout all levels in all other situations. As a result, having a CFRP framed structure reduces the bending moment demands in a simple jointed beam, increasing the CFRP frame's progressive collapse resistance owing to column loss.

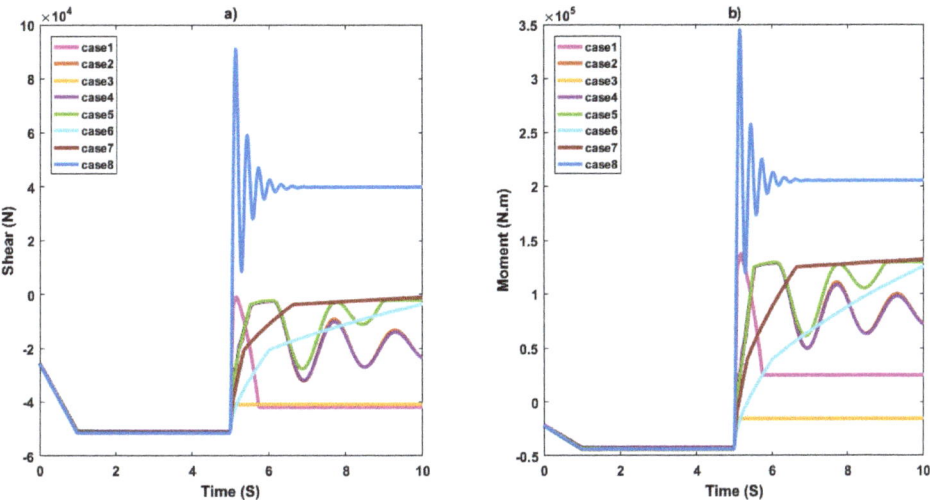

Figure 18. Beam behavior after column removal in story one: (**a**) shear and (**b**) moment.

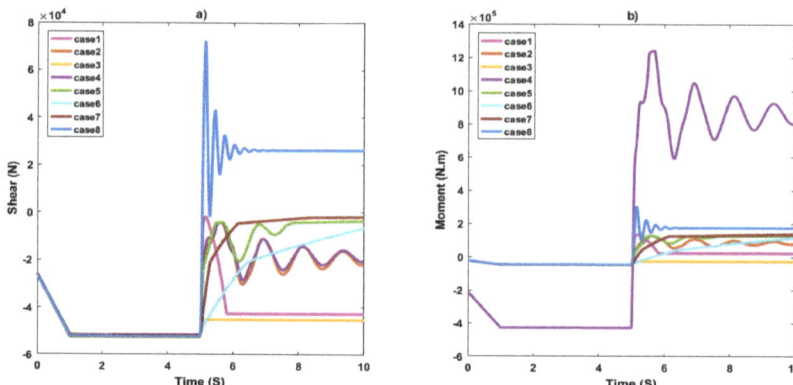

Figure 19. Beam behavior after column removal in story one: (**a**) shear and (**b**) moment.

In the first and sixth stories, the beam shear force for the bare frame (Case 1) is less than the other cases (see Figures 18 and 19). This is because the joint stiffness of CFRP frames attracts more force. In other circumstances, pressures are evenly dispersed overall plot levels. As a result, having a CFRP framed structure minimizes the axial force demands in a simple jointed beam, enhancing the CFRP frame's progressive collapse resistance due to column loss. The bending moment value for the viscously damped frame in two inner bays on floor thirteen is more than the bending moment for CFRP framed, as shown in Figure 20. Because of the stiffness, Case 8 performs well against progressive collapse, although there are many moments in the column adjacent to the brace, necessitating the employment of a less rigid brace. Viscous dampers were able to lower bending moment forces and provide effective energy absorption in the circumstances where they were applied.

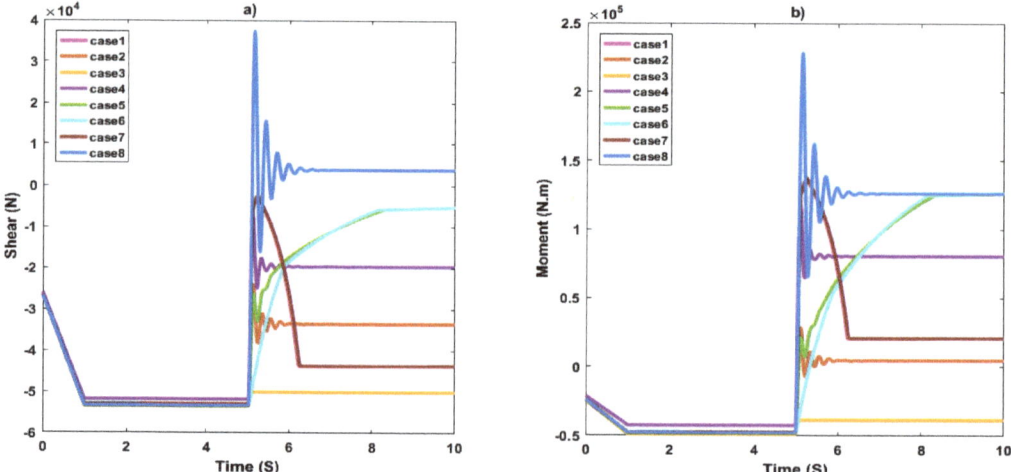

Figure 20. Beam behavior after column removal in story thirteen: (**a**) shear and (**b**) moment.

As shown in Table 3, which was compared with the removal of the column on the first floor in case 8, in beam number 118, the highest shear force is related to case 8, and the lowest is related to case 1; however, in the bending moment model is the highest force in case 8 and the lowest is related to case 3. As Figure 19 shows and values are illustrated in Table 4, after removing the column on the 6th floor, the highest shear force is created in case

8, while the lowest is related to case 3. Therefore, the bending moment force is related to case 3.

Table 3. Result force story 1 for element 118.

Type	Shear (N)		Moment (Nm)	
	Max	Constant	Max	Constant
Case 1	−954.38	−41,857	137,747	−4271.2
Case 2	−2409.6	−23,488	129,084	−2396.7
Case 3	−25,662	−40,876	−11,803	−4171
Case 4	−2531.8	−23,531	128,930	−2401.2
Case 5	−2031.4	−2044.9	130,093	−208.66
Case 6	−3693	−3693	125,379	−376.84
Case 7	−1123.5	−1123.5	132,357	−114.65
Case 8	91,115	39,827	345,480	4063.98

Table 4. Result force story 6 for element 123.

Type	Shear (N)		Moment (Nm)	
	Max	Constant	Max	Constant
Case 1	−1860.2	−42,805	137,632	23,002.7
Case 2	−4208.8	−21,984	126,347	78,079.4
Case 3	−26,358	−45,391	−23,574	−26,477
Case 4	−4254.7	−20,611	1,242,119	802,693
Case 5	−3646.9	−3646.9	127,990	127,990
Case 6	−6524.5	−6524.5	115,720	115,720
Case 7	−1857.7	−1857.7	137,318	137,318
Case 8	71,997.3	26,287.7	304,069	175,567

As shown in Figure 20 and tabulated in Table 5, case 8 has the highest shear force, and its lowest shear force is related to case 7 with a value of −2622.87 N. In the case of bending moment, when it is proven after the oscillation, the highest bending moment is related to case 4; however, the lowest is related to case 7 and its value is 20,838.58 N.

Table 5. Result force story 13 for element 130.

Type	Shear (N)		Moment (Nm)	
	Max	Constant	Max	Constant
Case 1	−2625.25	−43,752.4	137,480.4	20,838.6
Case 2	−24,073.4	−33,571.7	28,612.19	4591.762
Case 3	−26,772.2	−19,694.6	−24,506.6	−38,748.3
Case 4	−7316.24	−19,694.6	118,550.3	80,526.13
Case 5	−5218.5	−5218.5	126,341.4	126,341.4
Case 6	−5199.03	−5199.03	126,519.3	126,519.3
Case 7	−2622.87	−43,752.4	137,660	20,838.58
Case 8	37,417.97	3799.409	228,482.2	126,166.3

As shown in Figure 21 and depicted in Table 6, after removing the first-floor column, the results show that the highest axial force in the case of Case 4 is with the −1,009,521 N value, whereas the lowest in case of Case 6 is that of 18,863.2 N in the case of flexural anchor, is the most applied force associated with Case 5 and the lowest in case of Case 7.

As shown in Figure 22 and Table 7, after removing the column in the 6th floor, the maximum axial force created in the column in case 3 is −594,235 N, while in the shear force created in the column, the maximum is related to case 2, and in the case of bending moment, most of it is related to case 2, although it has good performance in some members and has enhances improved the chain performance as well as the forces in the members.

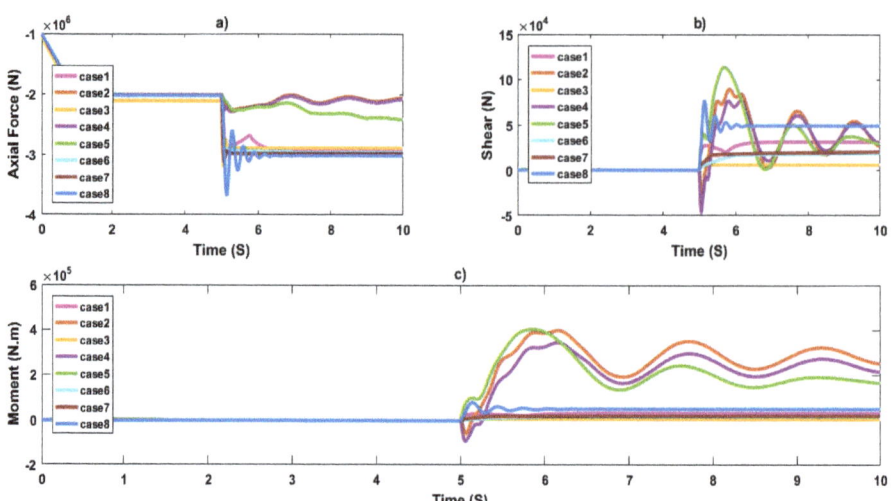

Figure 21. Column behavior after column removal in story one: (**a**) axial force, (**b**) shear, and (**c**) moment.

Table 6. Result force story 1 for element 53.

Type	Shear (N)		Axial force (N)		Moment (Nm)	
	Max	Constant	Max	Constant	Max	Constant
Case 1	31,745.6	31,742.8	−1,005,789	−2,413,732	32,907.1	32,878.7
Case 2	90,041.5	26,483	−1,006,285	−2,059,768	398,733	252,295.7
Case 3	6833.02	6028.07	−1,055,734	−2,890,048	7058.88	6225.3
Case 4	77,242.5	29,757.9	−1,009,521	−2,080,616	345,715	215,709.8
Case 5	114,224	29,249.3	−1,005,789	−2,413,732	404,512	166,427.8
Case 6	18,863.2	18,847.7	−1,005,789	−2,964,215	19,528.7	19,506.1
Case 7	20,330.7	20,318.1	−1,005,789	−2,984,732	21,128.2	21,106.5
Case 8	77,252.2	49,470.8	−1,008,649	−3,013,599	80,140.9	51,206.8

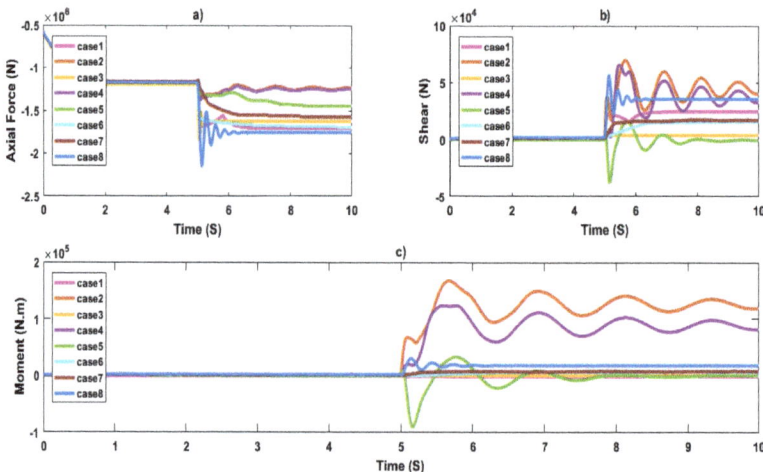

Figure 22. Column behavior after column removal in story six: (**a**) axial force, (**b**) shear, and (**c**) moment.

Table 7. Result force story 6 for element 58.

Type	Shear (N)		Axial force (N)		Moment (Nm)	
	Max	Constant	Max	Constant	Max	Constant
Case 1	24,731.7	24,718.8	−581,912	−1,702,429	0.9	−428
Case 2	69,741.2	40,499.1	−582,285	−1,228,997	168,017.6	118,980.8
Case 3	4886	3924	−594,235	−1,619,619	1640.2	629.3
Case 4	65,370.1	32,649.9	−584,647	−1,241,878	124,436.8	81,744.3
Case 5	16,127.4	−522.1	−581,912	−1,444,982	33,400.3	329.2
Case 6	16,423.7	15,919.1	−581,912	−1,687,495	7708.1	6959.7
Case 7	17,723.2	17,067.8	−581,912	−1,567,263	8725.8	7948.4
Case 8	56,651.4	35,584.6	−583,447	−1,752,195	30,784.9	18,377.1

In Figure 23 and Table 8, the results show that the maximum bending force generated in Case 2 is 92,318.4 N, while the lowest bending force is in Case 3. The lowest is related to case 3. In the case of shear force, the highest force is related to case 8, and the lowest is 5244.2 N in case 3.

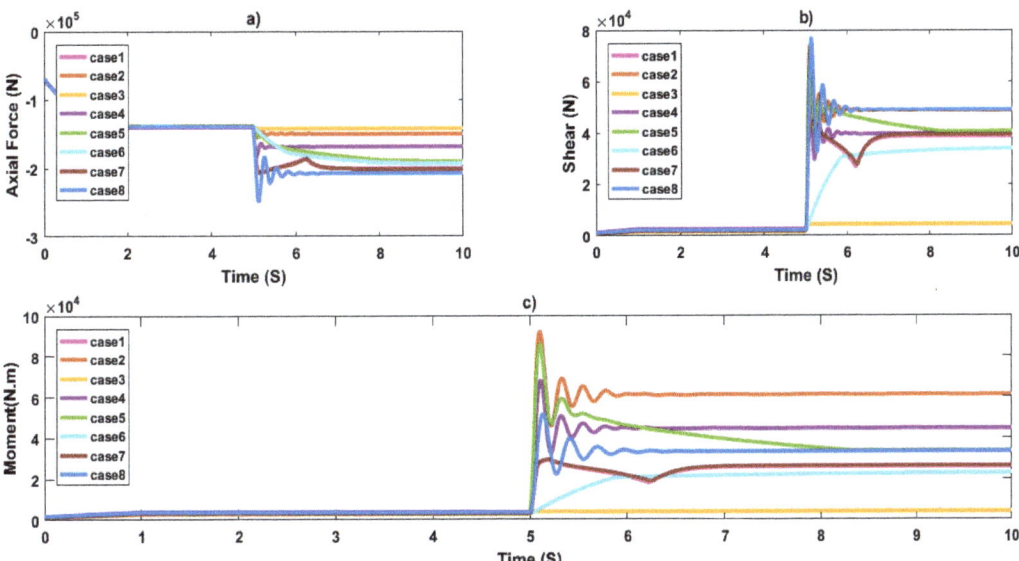

Figure 23. Column behavior after column removal in story six: (**a**) axial force, (**b**) shear, and (**c**) moment.

Table 8. Result force story 13 for element 65.

Type	Shear (N)		Axial force (N)		Moment (Nm)	
	Max	Constant	Max	Constant	Max	Constant
Case 1	43,963	38,738.9	−69,268.5	−200,110	29,336.2	25,964.6
Case 2	74,027.3	48,891	−69,311.3	−149,529	92,318.4	61,104.8
Case 3	5244.2	4527.3	−68,916.7	−141,667	4554.1	4132.5
Case 4	60,214	39,615.7	−69,922.7	−168,470	68,195.2	44,635.9
Case 5	68,762.7	40,588.4	−69,268.5	−190,002	85,899.2	33,461.9
Case 6	33,930.3	33,930.3	−69,268.5	−193,062	22,858	22,858
Case 7	44,140.8	39,111	−69,268.5	−200,602	29,511.1	26,201.8
Case 8	77,010.4	49,050.8	−69,591.2	−207,391	51,915.9	33,394.9

4. Conclusions

In this study, the assessment of different retrofitting approaches on the structural performance of RC buildings against progressive collapse was studied by comparing the eight different cases. The following conclusions are drawn from the results of the analysis:

- Overall, in structures reinforced with various braces, the use of a diagonal bracing system has performed better than other braces and has been able to maintain the level of immediate occupancy performance, and the distribution of plastic hinges in structures has been improved. The most important limitation of this method is that the brace's frame has a better performance against progressive failure. However, 2D models do not redistribute forces properly and affect other frames.
- In X-braced frames, the plastic hinge creation occurred on the braced element inside the collapse range, which is considered compression failure of the brace.
- In diagonal bracing, the plastic hinge formation has begun to develop in brace members. As a result, plastic hinges formed in columns and beams and were disseminated over particular building members within the immediate occupancy area.
- The inverted V-braced arrangement increases the constraint level at the beam end, allowing catenary action. In addition, the horizontal braces distributed certain gravitational stresses to the surrounding structures, which were then carried to the foundation via the vertical members.
- In structures reinforced with viscous dampers, the results show that dampers in certain stories have lower performance than other methods.
- In the viscously damped frame in the central bay, the plastic hinges in beams on the third and fourth stories reach within range of collapse, and that the application of a viscous damper prevented oscillation and shock to the structure.
- Using the viscously damped frame in two inner bays enhanced the structure's tensile performance, and improved the performance of the beams compared with the viscously damped frame in the central bay.
- The use of CFRP to retrofit the results shows that it improves the structure's overall performance, increases its chain performance, and improves the redistribution of forces in the structures. The advantages of this method are that it can be easily implemented in the whole structure and does not cause architectural problems. On the other hand, the disadvantage of this method is that it could create much force after removing the column; this issue could be problematic with increasing the possibility of damage in the structure.
- In general, various retrofitting schemes can be applied to strengthen the structure and increase the resistance of the structures under progressive collapse. In this regard, their combined use could be utilized furtherly to redistribute forces more quickly and achieve the best performance in the structure in terms of chain performance under progressive collapse.

Author Contributions: Conceptualization, B.H.A.; methodology, B.H.A.; software, B.H.A.; validation, M.B and B.H.A.; formal analysis, B.H.A.; investigation, B.H.A. and E.M.G.; resources, B.H.A. and M.B; writing—original draft preparation, B.H.A. and M.B; writing—review and editing, E.M.G.; visualization, B.H.A.; supervision, E.M.G.; project administration, E.M.G.; funding acquisition, B.H.A. All authors have read and agreed to the published version of the manuscript.

Funding: This research received no external funding.

Institutional Review Board Statement: Not applicable.

Informed Consent Statement: Not applicable.

Data Availability Statement: Not applicable.

Conflicts of Interest: The authors declare no conflict of interest.

References

1. GSA. *General Services Administration Alternate Path Analysis and Design Guidelines for Progressive Collapse Resistance*; US General Services Administration: Washington, DC, USA, 2013.
2. UFC. *UFC 4-023-03: Design of Buildings to Resist Progressive Collapse*; Department of Defense: Washington, DC, USA, 2009.
3. Marjanishvili, S. Progressive Analysis Procedure for Progressive Collapse. *J. Perform. Constr. Facil.* **2004**, *18*, 79–85. [CrossRef]
4. Izzuddin, B.; Vlassis, A.; Elghazouli, A.; Nethercot, D. Progressive collapse of multi-storey buildings due to sudden column loss—Part I: Simplified assessment framework. *Eng. Struct.* **2008**, *30*, 1308–1318. [CrossRef]
5. Yu, J.; Yin, C.; Guo, Y. Nonlinear SDOF model for progressive collapse responses of structures with consideration of viscous damping. *J. Eng. Mech.* **2017**, *143*, 04017108. [CrossRef]
6. Pan, J.; Wang, X.; Dong, H. Strengthening of Precast RC Frame to Mitigate Progressive Collapse by Externally Anchored Carbon Fiber Ropes. *Polymers* **2021**, *13*, 1306. [CrossRef]
7. American Concrete Institute. *ACI 318, Building Code Requirements for Structural Concrete (ACI 318-05) and Commentary (ACI 318R-05), ACI Committee 318, American Concrete Institute, Farmington Hills, MI, 2005 ACI 530, Building Code Requirements for Masonry Structures (ACI 530-05/ASCE 5-05/TMS 402-05)*; American Concrete Institute: Farmington Hills, MI, USA, 2005.
8. AISC. *ANSI/AISC 360–05: Specification for Structural Steel Buildings*; AISC: Chicago, IL, USA, 2005.
9. ANSI. *AISC 341-05: Seismic Provisions for Structural Steel Buildings*; American Institute of Steel Construction Inc.: Chicago, IL, USA, 2005.
10. Badoux, M.; Jirsa, J.O. Steel Bracing of RC Frames for Seismic Retrofitting. *J. Struct. Eng.* **1990**, *116*, 55–74. [CrossRef]
11. Nateghi-A, F. Seismic strengthening of eightstorey RC apartment building using steel braces. *Eng. Struct.* **1995**, *17*, 455–461. [CrossRef]
12. Maheri, M.R.; Sahebi, A. Use of steel bracing in reinforced concrete frames. *Eng. Struct.* **1997**, *19*, 1018–1024. [CrossRef]
13. Bigonah, M.; Soltani, H.; Zabihi-Samani, M.; Shyanfar, M.A. Performance evaluation on effects of all types of infill against the progressive collapse of reinforced concrete frames. *Asian J. Civ. Eng.* **2020**, *21*, 395–409. [CrossRef]
14. Khandelwal, K.; El-Tawil, S.; Sadek, F. Progressive collapse analysis of seismically designed steel braced frames. *J. Constr. Steel Res.* **2009**, *65*, 699–708. [CrossRef]
15. Shayanfar, M.A.; Bigonah, M.; Sobhani, D.; Zabihi-Samani, M. The Effectiveness Investigation of New Retrofitting Techniques for RC Frame against Progressive Collapse. *Civ. Eng. J.* **2018**, *4*, 2132–2142. [CrossRef]
16. Fu, F. Progressive collapse analysis of high-rise building with 3-D finite element modeling method. *J. Constr. Steel Res.* **2009**, *65*, 1269–1278. [CrossRef]
17. Fu, F. Response of a multi-storey steel composite building with concentric bracing under consecutive column removal scenarios. *J. Constr. Steel Res.* **2012**, *70*, 115–126. [CrossRef]
18. Feng, P.; Qiang, H.; Ou, X.; Qin, W.; Yang, J. Progressive Collapse Resistance of GFRP-Strengthened RC Beam–Slab Subassemblages in a Corner Column–Removal Scenario. *J. Compos. Constr.* **2019**, *23*, 04018076. [CrossRef]
19. Costanzo, S.; D'Aniello, M.; Landolfo, R. Proposal of design rules for ductile X-CBFS in the framework of EUROCODE 8. *Earthq. Eng. Struct. Dyn.* **2019**, *48*, 124–151. [CrossRef]
20. D'Aniello, M.; Costanzo, S.; Landolfo, R. The influence of beam stiffness on seismic response of chevron concentric bracings. *J. Constr. Steel Res.* **2015**, *112*, 305–324. [CrossRef]
21. Costanzo, S.; D'Aniello, M.; Landolfo, R. Seismic design rules for ductile Eurocode-compliant two-storey X concentrically braced frames. *Steel Compos. Struct.* **2020**, *36*, 273–291.
22. Qian, K.; Li, B.; Zhang, Z. Influence of Multicolumn Removal on the Behavior of RC Floors. *J. Struct. Eng.* **2016**, *142*, 04016006. [CrossRef]
23. Kim, J.; Lee, S.; Choi, H. Progressive Collapse Resisting Capacity of Moment Frames with Viscous Dampers. *J. Comput. Struct. Eng. Inst. Korea* **2010**, *23*, 517–524. [CrossRef]
24. Orton, S.L.; Chiarito, V.P.; Minor, J.K.; Coleman, T.G. Experimental testing of CFRP-strengthened reinforced concrete slab elements loaded by close-in blast. *J. Struct. Eng.* **2014**, *140*, 04013060. [CrossRef]
25. Council, B.S.S. *Prestandard and Commentary for the Seismic Rehabilitation of Buildings*; Report FEMA-356; Federal Emergency Management Agency: Washington, DC, USA, 2000.
26. Cimellaro, G.P.; Retamales, R.M. Optimal softening and damping design for buildings. *Struct. Control Health Monit. Off. J. Int. Assoc. Struct. Control Monit. Eur. Assoc. Control Struct.* **2007**, *14*, 831–857. [CrossRef]
27. Hu, X.; Zhang, R.; Ren, X.; Pan, C.; Zhang, X.; Li, H. Simplified design method for structure with viscous damper based on the specified damping distribution pattern. *J. Earthq. Eng.* **2020**, 1–21. [CrossRef]
28. Soudki, K.; Alkhrdaji, T. Guide for the design and construction of externally bonded FRP systems for strengthening concrete structures (ACI 440.2 R-02). In *Structures Congress 2005: Metropolis and Beyond*; American Concrete Institute: Farmington Hills, MI, USA, 2005; pp. 1–8.
29. Khalifa, A.; Belarbi, A.; Nanni, A. Shear performance of RC members strengthened with externally bonded FRP wraps. In Proceedings of the (CD-ROM) of Twelfth World Conference on Earthquake, Auckland, New Zealand, 30 January–4 February 2000.
30. Kim, J.; Kim, T. Assessment of progressive collapse-resisting capacity of steel moment frames. *J. Constr. Steel Res.* **2009**, *65*, 169–179. [CrossRef]

31. Qian, K.; Weng, Y.-H.; Li, B. Improving Behavior of Reinforced Concrete Frames to Resist Progressive Collapse through Steel Bracings. *J. Struct. Eng.* **2019**, *145*, 04018248. [CrossRef]
32. ASCE. *Seismic Evaluation and Retrofit of Existing Buildings*; American Society of Civil Engineers: Reston, VA, USA, 2014.

MDPI
St. Alban-Anlage 66
4052 Basel
Switzerland
Tel. +41 61 683 77 34
Fax +41 61 302 89 18
www.mdpi.com

Applied Sciences Editorial Office
E-mail: applsci@mdpi.com
www.mdpi.com/journal/applsci

www.ingramcontent.com/pod-product-compliance
Lightning Source LLC
LaVergne TN
LVHW070444100526
838202LV00014B/1664